Stratification of a tropical forest as seen in dispersal types

Tasks for vegetation science 17

Series Editors

HELMUT LIETH

University of Osnabrück, F.R.G.

HAROLD A. MOONEY

Stanford University, Stanford, Calif., U.S.A.

1 Box, E.O. *Macroclimate and Plant Forms.* An introduction to predictive modelling in phytogeography. ISBN 90 6193 941 0
2 Navin Sen, D. & Singh Rajpurohit, K. *Contributions to the Ecology of Halophytes.* ISBN 90 6193 942 9
3 Ross, J. *The Radiation Regime and Architecture of Plant Stands.* ISBN 90 6193 607 1
4 Margaris, N.S. & Mooney, H.A. (eds) *Components of Productivity of Mediterranean – Climate, Regions.* ISBN 90 6193 944 5
5 Müller, M.J. *Selected Climatic Data for a Global Set of Standard Stations for Vegetation Science.* ISBN 90 6193 945 3
6 Roth, I. *Stratification in Tropical Forests as Seen in Leaf Structure.* ISBN 90 6193 946 1
7 Steubing, L. & Jäger, H.J. *Monitoring of Air Pollutants by Plants: Methods and Problems.* ISBN 90 6193 947 X
8 Teas, H.J. *Biology and Ecology of Mangroves.* ISBN 90 6193 948 8
9 Teas, H.J. *Physiology and Management of Mangroves.* ISBN 90 6193 949 6
10 Feoli, E., Lagonegro, M. & Orlóci, L. *Information Analysis of Vegetation Data.* ISBN 90 6193 950 X
11 Sesták, Z. (ed) *Photosynthesis during Leaf Development.* ISBN 90 6193 951 8
12 Medina, E., Mooney, H.A. & Vázquez-Yánes, C. (eds) *Physiological Ecology of Plants of the Wet Tropics.* ISBN 90 6193 952 6
13 Margaris, N.S., Arianoustou-Faraggitaki, M. & Oechel, W.C. (eds) *Being Alive on Land.* ISBN 90 6193 953 4
14 Hall, D.O., Myers, N. & Margaris, N.S. (eds) *Economics of Ecosystems Management.* ISBN 90 6193 505 9
15 Estrada, A. & Fleming, T.H. (eds) *Frugivores and Seed Dispersal.* ISBN 90 6193 543 1
16 Dell, B., Hopkins, A.J.M. & Lamont, B.B. (eds) *Resilience in Mediterranean-type Ecosystems.* ISBN 90 6193 579 2
17 Roth, I. *Stratification of a Tropical Forest as Seen in Dispersal Types.* ISBN 90 6193 613 6

Stratification of a tropical forest as seen in dispersal types

by

INGRID ROTH

1987 **DR W. JUNK PUBLISHERS**
a member of the KLUWER ACADEMIC PUBLISHERS GROUP
DORDRECHT / BOSTON / LANCASTER

Distributors

for the United States and Canada: Kluwer Academic Publishers, P.O. Box 358, Accord Station, Hingham, MA 02018-0358, USA
for the UK and Ireland: Kluwer Academic Publishers, MTP Press Limited, Falcon House, Queen Square, Lancaster LA1 1RN, UK
for all other countries: Kluwer Academic Publishers Group, Distribution Center, P.O. Box 322, 3300 AH Dordrecht, The Netherlands

Library of Congress Cataloging in Publication Data

Roth, Ingrid.
Stratification of a tropical forest as seen in dispersal types.

(Tasks for vegetation science ; 17)
Bibliography: p.
Includes indexes.
1. Rain forest ecology. 2. Tropical plants--Ecology. 3. Seeds--Tropics--Dispersal. 4. Fruit--Tropics--Dispersal. I. Title. II. Series.
QK938.F6R675 1986 581.5'2642 86-10489
ISBN 90-6193-613-6

ISBN 90-6193-613-6 (this volume)
ISBN 90-6193-897-X (series)

Copyright

PRINTED IN THE NETHERLANDS

Die Schöpfung des Lebendigen ist
auf eine Entwicklung hin geplant,
sonst hätte der Tod keinen Sinn.

Dedicated to my beloved mother

PREFACE

Our knowledge of the structure and dispersal of tropical fruits and seeds is very limited up to the present day, though richness of species and variety of forms is overwhelming in the tropical forests. Morphology of tropical fruits and seeds has always attracted the curiosity of botanists from many countries and information may be obtained from books and publications concerned with taxonomy. Ulbrich's "Biologie der Früchte und Samen" (1928) gives examples of tropical fruits and seeds and their dispersal methods. The two volumes by Van der Roosmalen (1977) dealing with the description of tropical plants were of utmost value to me, as the area considered, Surinam, is close to Venezuelan Guiana and because both regions have many species in common. However, the inner structure of tropical fruits and seeds is less known. Some new examples of the structure and development of tropical fruits and seeds have been presented in my book "Fruits of Angiosperms" (1977). Corner's "Seeds of Dicotyledons" (1976) offers many detailed structural peculiarities of seeds in drawings. Unfortunately, the least information is available regarding detailed descriptions of fruit and seed dispersal of tropical plant species. In his bible "Plant dispersal throughout the world" Ridley (1930) gives a general survey over the most important facts. Guppy (1917) deals with "Plants, seeds and currents" in the Westindies and Azores. Van der Pijl with his extensive knowledge of tropical plants offers a great selection of detailed information on the subject "Principles of dispersal in higher plants" (1972, and earlier papers). The author who has earned most merits in the field of seed and fruit predation, chemical defenses of plants, and animal-plant interactions is Janzen. He and his collaborators have thrown new light on this subject. Nonetheless, a large unknown field still remains ahead of us, especially regarding a detailed knowledge of fruit and seed dispersal of tropical plant species.

The great opportunity for my own studies was the fruit and seed collection of Dr. Bernard Rollet to whom I am very much obliged for placing the material at my disposal. He made the collection during his stay in Venezuelan Guiana when he directed the tree inventory project in this area. My thanks go also to Dr. P.R. Mohana Rao who started the morphological studies together with me in Venezuela, but unfortunately was compelled to return to India soon after his arrival for health reasons. Finally, I have to thank my former student and collaborator, Licenciado Ricardo Castro-Penso, for the preparation of the slides used for the work reported in this volume.

CONTENTS

CHAPTER 1

SOURCE OF THE MATERIAL AND SAMPLE COLLECTION

The material investigated mainly comes from the humid tropical forest in Venezuelan Guiana where an inventory was initiated by the Ministry of "Agricultura y Cria" (MAC) of Venezuela together with the Food and Agriculture Organization (FAO) and the UNO. In this inventory all arboreous species of this area were statistically covered. The total area studied for the inventory amounted to about 67 km in lenght and approximately 25 m in width, covering in total about 155,5 ha. A clearing of the above mentioned length and width was cut through the forest to collect trees, shrubs, and partly also lianas. In total, 67.777 individual trees (and shrubs) with a diameter of not less than 10 cm at breast height were included.

The tropical humid forest from which leaf samples were collected corresponds to a dense evergreen rain forest which Beard would call "seasonal" to indicate the incidence of distinct dry seasons. Geographically it is situated on the shield of Venezuelan Guiana at the so-called massif of Imataca (south of the Orinoco river and east of the junction with the Caroni river). This extreme northern part of the Guianan shield has an undulated ground relief which fluctuates between 150 m and 550 m in altitude. The shield of Guiana is composed of rocks of the Precambrian Era, essentially of granites and gneisses. The soil corresponds to a red latosol. The annual precipitation is estimated at about 2000 mm or more, with a dry season of varying length between January and April. Special floristic studies were carried out around the camps "Rio Grande", "El Paraiso", and "El Dorado".

In the interior of the forest, some special edaphic formations are included such as periodically inundated soils or permanently inundated river banks, rocky sites, etc., which may well be recognized by their floristic composition. In its western part, the forest is in contact with a belt of semideciduous and deciduous forests, bordering in their turn, savannas. Although the tropical humid forest is called "evergreen", Rollet (1964) was able to count about 60 species which drop their leaves, mainly in April. However, the defoliation is somewhat ephemeral and irregular, even considering one and the same species so that the forest as a whole preserves its evergreen aspect. Flowering and fructification are not very clearly limited to certain periods of the year, but according to Rollet a certain proportion of the species flowers and fruits with more intensity four times a year, in February-June-August-December and in February-May-August-December, respectively. For more details of the humid tropical forest in the State of Bolivar, Venezuela, and its floristic composition, see Rollet (1964, 1969a, b), from which publications the above data are taken.

Most of the species were identified by Dr. Julian Steyermark, Instituto Botanico, Caracas. Those samples which could not be identified in Caracas were sent to specialists in the USA, England, France, the Netherlands, Belgium, Sweden, Colombia, and Brazil. Some of the samples, however, are still in the process of being identified, although the inventory was finished approximately in 1967. Most difficulties in the identification arose among the Burseraceae (the genus *Protium*), the Sapotaceae (*Pouteria*), Myrtaceae, and Rosaceae (*Licania*). In spite of all the efforts of taxonomists, we still expect some misinterpretations and unidentified species. For these reasons, the vernacular names are still of great value. In the book "Structural patterns of tropical barks" I tried to explain these names and to trace their origin (Roth 1981).

The principal objective of the dendrological studies was the collection of wood samples for technological purposes. These samples were processed in the Instituto Forestal in Mérida, Venezuela, and in the Instituto Technológico Forestal of the University of Mérida. For further information the reader is referred to B. Rollet "Inventario forestal en la Guayana Venezolana", MAC, Informe No. 3, parte 2, Caracas 1967. Trees and shrubs were classified into three categories:

1. Very tall trees of 30 m height or more (category A),
2. Trees between 10 m and 29 m in height (category a), and
3. Small trees and shrubs less than 10 m high (category aa).

Most samples were taken from trees with an average diameter between 10 cm and 1 m at breast height. However, some trees do not even reach 20 cm DBH in the adult stage, such as some Rubiaceae, Capparidaceae, and Melastomaceae. These trees were also included, however, in an attempt to cover the maximum number of genera and species.

The collection of fruit and seed samples was carried out by the forestry engineer Dr. Bernard Rollet, who directed the inventory program of the FAO. In total, about 56 different plant families with 370 species are included in these studies. As far as it was possible each family with its species is described in the special part concerning fruit and seed morphology, inner structure and dispersal. The great advantage of this collection is the exact knowledge of the number of individuals of each species, genus and family. In this way it is possible to make exact calculations of the number of individuals and their dispersal types, of the distribution of certain dispersal types in the different forest strata, of the number of individuals in homogeneous groups, and of the number of species per hectare, to mention only a few of such possibilities.

CHAPTER 2

INTRODUCTION

"Dispersal is but one of the factors determining the presence of plants"

Van der Pijl

In view of the great forest destruction in the whole world the tropical forests are given more and more consideration. Destruction of tropical forests is mainly due to the felling of trees for the conquest of agriculturally profitable land (e.g. in Brazil) or to useless burning of forests by pyromaniacs (e.g. in Venezuela). Settlement on cleared land, construction of roads, and cementing of large areas as well as smog in the form of SO_2, NO_x and ozone escaping from factories or waste-gas of cars in industrialized regions play an increasingly important part in the destruction of the tropical flora. One of many examples is the flora of lichens in the city of Caracas, Venezuela, which was superabundant in the fifties, but has completely disappeared now. The waste-gas escaping from the unbounded phalanx of cars and its following conversion into ozone and triethyl lead must be the reason of this debacle, as the emission of the little industrial smoke is practically negligible. Another very interesting phenomenon disappeared in Caracas as a consequence of the high degree of air pollution. The rosettes of the epiphyte *Tillandsia recurvata*, Bromeliaceae, normally live on branches of trees and in Caracas abundantly occupied the overhead wire system in the air resembling musical symbols on lines as in a musical manuscript. It is commonly accepted that the seeds are dispersed by wind, but the way the rosettes are distributed on the wires makes bird dispersal probable. Possibly the seeds stick to the feathers of birds which later take a rest on the wires. However, this airy pattern of rosettes on wires was present everywhere in the whole of Caracas during the sixties, but at the present day it has disappeared completely due to air pollution. Epiphytes are particularly sensitive to polluted air and consequently are the first plants to dissappear. They may be considered bioindicators. Certain species of Loranthaceae (*Phoradendron*) have long ago disappeared in the small town of Maracay, Venezuela, for the same reason.

Realizing that large areas of tropical forest are destroyed every day and that more and more species are extinguished for ever, we may think of saving genetic material in seed banks, possibly in freezers, as tropical seeds are usually not dormant. We still have only a slight idea of the species abundance in tropical forests and no one can appreciate the size of the genetic information stored therein, nor are we able to estimate the extent of secondary substances lodged in tropical plants in the form of chemical compounds which may be used in medicine, pharmacy and industry and of their curing effects becoming apparent by their application in popular medicine.

It is, therefore, an exciting idea to have a more or less complete material of a well-known area in the tropical forest at hand to study fruit and seed structure as well as dispersal methods related not only to the different plant families, but also to the height categories of the trees. In the studies presented here the question arose whether forest stratification together with microclimatic conditions possibly influence outer morphology and inner structure of fruits and seeds and whether the dispersal mechanisms differ from one another in different forest strata. In my book "Stratification of tropical forests as seen in leaf structure" I drew attention to the importance of the microclimate in the forest for the interpretation of the varying leaf structures at different forest levels. It is well known that a humid shady environment characterizes the lower forest strata, while at the higher forest levels drought and insolation play an increasingly important part. This change of the microclimate characteristically expresses itself in the leaf structure not only during the development of the individual tree, but also at the different levels of one and the same adult tree and of different species pertaining to different height categories. In the shady and humid environment of the undergrowth the structure of the hygromorphic shade leaf develops, while a transformation into the xeromorphic sun leaf takes place in the canopy. The investigations of fruit and seed structure and of dispersal will show that forest layering also exercises its influence upon dispersal methods in the different strata. The knowledge of the exact number of individuals of each species present in the area studied enables us further to calculate percentages for an entire forest area covering about 155,5 ha in total.

CHAPTER 3

GENERAL PART

Fruit and seed types

To start with the description of the structure and dispersal methods of the studied fruits and seeds we have to give first a general survey of the existing fruit and seed forms and types as well as of their possible dispersal strategies. In my book "Fruits of Angiosperms" (1977) I had the opportunity to present an exact definition of the fruit so that it seems sufficient here to draw attention to this bibliography. A classification of fruits on an evolutionary basis is likewise presented in the same text. In spite of our knowledge of fruit evolution which permits the performance of a phylogenetic classification (presented on p. 5 l. c.) this can hardly be utilized for practical purposes. Consequently, we have to deal more or less with the old classification of dehiscent and indehiscent fruits. Within the indehiscent fruits we distinguish the nut with a hard pericarp and a single seed, the single-seeded drupe with a fleshy-stony pericarp, and the many-seeded berry with a fleshy pericarp. The aggregate fruit represents a special fruit type, as well as the compound infrutescence. During the phylogenetic development of the fruit a reduction of seeds may have taken place. We observe this process in the dehiscent as well as in the indehiscent fruits. A many-seeded capsule may become single-seeded in this way and even loose the ability to dehisce as a consequence of monospermy. It may transform in this way into a monospermic nut. The polyspermic indehiscent fruit with a hard or dry pericarp may undergo reduction of seeds during evolution and additionally develop a fleshy outer pericarp part so that it becomes a drupe. Even polyspermic berries may reduce their seed number to only one during evolutionary processes. However, nature never draws strict insuperable limits between forms, and transitional types may be found everywhere. Consequently, the distinction between a monospermic berry and a drupe is sometimes difficult. Furthermore, the hard endocarp postulated for a drupe may not always be present in the desired form. I, therefore, suggest to use the term "drupe" for a fleshy single-seeded fruit in general. The reader will find many examples of seed reduction in fleshy fruits within the below cited families. One could possibly make a distinction between a syncarpic drupe with several stones ("pyrene"), and a drupe originating from a single carpel by seed reduction ("putamen"). Winged fruits called "samaras" and other wind-dispersed fruits with hairs of different kinds or with a persistent enlarged calyx or perigone are probably derived forms of the nut-type. This opinion coincides with our concept of the evolution of wind dispersal. In many families we find examples of secondary wing formation or even beginning wing development, e.g. within the papilionacean species *Pterocarpus*. Wind dispersal is often related to the height of the tree or the special environment of the species. Likewise other fruit types with stiff hairs, hooks, glands or similar devices have to be considered derived forms. Apocarpic fruits are certainly primitive and the follicle has undergone the same processes of reduction and transformation as the paracarpous and syncarpous fruits: i.e. reduction of the number of seeds, loss of dehiscence, transformation of the nutlet into a samara or modification of the pericarp wall, possibly of the fleshy fruit wall into a dry one. Most derived are infrutescences such as of *Pandanus* (Fig. 1), the pineapple or the fig.

When the fruit dehisces, the seed itself may take over dispersal. Winged seeds or seeds with hairs are destined for wind dispersal. In other cases, the seeds may develop fleshy parts, possibly in the form of a sarcotesta or aril to provide animal dispersal. The sarcotesta seed is regarded as a primitive type originally dispersed by means of endosaurochory (van der Pijl). But this dispersal method was later replaced by endoornithochory. By reduction the sarcotesta gradually transformed into a more specialized aril originating at the seed base. Caruncle ("micropylar aril") and strophiole, originating from the raphe or funicle, are derived forms of the aril, according to Takhtajan. The caruncle of myrmecochorous plants as well as the elaisome, the latter more frequently found in seeds than in fruits, bring about a very special dispersel mechanism by ants. Hooks and glands on seeds are destined to adhere to the fur or feathers of animals and to serve for epizoochory.

In this short survey we are not able to discuss all possible fruit and seed forms and their morphology. The reader finds a description of the most important fruit types and their dispersal mechanisms in Ulbrich (1928). Takhtajan (1959) introduces evolutionary aspects into his discussion of fruits and seeds, and van der Pijl's book (1972) is mainly concerned with evolutionary problems of dispersal. The inner structure of fruits is thoroughly treated in my book "Fruits of Angiosperms" (1977). In summary we may say that the specific fruit and seed type is determined by the morphology of the gynoecium, on the one hand, and by the method of dissemination, on the other.

Factors influencing the presence of plants

Van der Pijl (1972) emphasizes in his "Principles of dispersal in higher plants" that "Dispersal is but one of the factors determining the presence of plants". For this reason we shall first discuss in short all the factors which are possibly responsible for the distribution of plants. One of the factors studied in this book, dispersal of fruits and seeds, will be discussed later in detail. It is, however, not the most important factor in the area in question.

Fruits and seeds develop from the ovary and ovules of flowers. The number of ripe fruits and seeds, present on a tree, therefore, depends on the formation of flowers on the tree. There is a difference whether flowers appear singly on the branches or whether they occur clustered in the form of inflorescences (racemes, cymes, heads etc.). The usual experience in the humid tropical forest is that flowers develop in more or less compact influorescences (see also Pennington & Sarukhan 1968). However, the number of ripe fruits developed depends only to a certain degree on the number of flowers present on the tree. Usually only a few ovaries reach fruit maturity in a large inflorescence, while most developing fruits are shed while still small (e.g. mango). Hartshorn (1972) counted 651 inflorescences on a tree of *Pentaclethra macroloba* corresponding to about 1 million of flowers. But only 231 seeds from the same tree could be harvested, as in each inflorescence not more than one legume develops. This abundance of flowers possibly points towards actual difficulties of pollination in the forest. The more the size of the fruit increases until maturity, the more the number of fruits becomes reduced on the tree. In certain families, such as the Lecythidaceae, Guttiferae, Bombacaceae, for example, larger flowers and fewer fruits are developed. Formation of large flowers in few-flowered inflorescences and development of large fruits is probably a characteristic of relatively primitive plant families. Another prerequisite of fruit formation is pollination and the availability of the adequate pollinators. Even the time and duration of flowering may exercise its influence over fruit setting: in certain Lecythidaceae pollination lasts only a single day.

The position of the fruits on a tree is not indifferent for the dispersal. Some fruits are clustered in compact infrutescences, others appear in "stories" on the tree *(Carica papaya)*. Caulicarpy, a phenomenon not seldom observed in tropical trees *(Theobroma cacao, Couroupita, Swartzia)* is realized when the flowers develop on the tree stem or on old branches. The opposite takes place when the fruits draw away from the stem becoming exposed and pendulous as it happens with the mango (Fig. 1) or with fruits destined for bat dispersal which develop elongated peduncles. Particularly important for animal dispersal is the seasonality of fruiting, i.e. the months in which fruits are available for certain, more or less specialized animals. Even for anemochorous dispersal units the fruiting time may be of consequence. Dispersal may be most effective at the time when winds are most vigorous. Phenology of tropical trees tells us that many trees do not flower every year, but in biennial intervals or even in intervals of several to many years. Of the bamboo we know that synchronous flowering takes place about after 15–60 (120) years depending on an "internal physiological calendar" (Janzen 1976). Suessenguth (1938) made the accumulative occurrence of sun-spots responsible for gregarious flowering of bamboos. Finally, the entire life span of a tree influences fruit and seed production, as well as the age in which the tree starts to fruit.

When the seed is dispersed its viability and life span are the next to affect the survival of a species. It is well known that seeds of tropical trees, especially of the humid tropical forest, generally have no dormancy at all or a very short one. However, the longer the latency, the more increases the chance of the seed to survive adverse environmental conditions and to be able to germinate. Viability is generally very good in many seeds of Leguminosae.

Frequently a light gap is necessary for the germination of seeds in the darkness of the forest floor. When germination has taken place, the survival of the seedling and the young plant is in danger. Many seedlings and young plants have to wait in the darkness of the undergrowth – so-to-say remaining on the "waiting list" – until an old tree dies so that they can continue growing. However, most of them never reach this second stage of growth. Tolerance of drought and poor soils as well as formation of a strong radical system helps the young, and even the grown-up plant to survive in an adverse environment. Growth rate of the young plant is a further factor influencing survival. Young plants that are able to grow fast have a better chance to reach the upper forest strata where illumination is more favourable early enough. Fast-growing young plants have thus an advantage over slow-growing ones.

Species which have a good adaptability to environmental conditions such as soil and climate, have a better chance of survival. Competition of other individuals of the same species or of other species endangers the development of seedings and young plants. Coincidence of many factors plays thus its part. Some species have the ability to keep alive better than others, e.g. by secretion of certain substances through the radical system which inhibit the growth of other species. In a more or less optimal environment such as the humid tropical forest many enemies of plants likewise enjoy a luxuriant development. Leaves, buds, fruits and seeds easily become a prey of insects and other harmful animals. Fungi and bacteria may attack fruits and seeds lying on the forest floor. In certain cases, plants are able to develop special defensive methods to prevent predation, such as bitter-tasting chemicals, poisonous substances, or simply hard shells. Other means of escape from predation are keeping long distances between individuals of one and the same species, special fruiting times, and other methods which will be discussed later in detail (see also Janzen, many publications cited in the bibliography). Last but not least we have to consider historic facts such as, for example, the development of the flora in a certain area and time, as well as geologic events which may have exercised some influence on the composition of the flora. In this connection the refuge theory may play its part in the region in question.

Many of the above mentioned factors may have affected the actual distribution of species and individuals in the primary forest more significantly than the more or less successful dispersal methods at the present day.

Seasonality of flowering and fruiting

As may be expected, flowering in tropical humid forests is more or less spread out over the whole year. There are always some species flowering. While certain species flower continuously *(Moringa)*, others set flowers several times or only once a year *(Gustavia augusta)*. Some species have biennial rhythms (most species of the Lecythidaceae) or rhythms of several years. Examples of annually fruiting trees in the tropics are rare, e.g. *Anacardium excelsum, Apeiba tibourbou, Bursera simarouba, Cecropia peltata, Cordia alliodora, Enterolobium cyclocarpum, Genipa caruto, Guazuma ulmifolia, Luehea candida, Parkinsonia aculeata, Pithecellobium saman, Sterculia apetala.* Most trees "skip" a year because of interactions with dispersal agents, pollinators or seed predators, according to Janzen (1978). Bamboos need about 15–60 (120) years to come to flower. The wide-spread lack of synchronization among species and individuals is the reason why some species are always found flowering in the humid tropical forest. However, also gregarious flowering occurs. According to Medway (1972) gregarious flowering occurred in response to drought in 1963 and 1968 in a tropical rain forest of Malaya. Corner (1954) likewise suspects a certain climatic stimulus which regulates this phenomenon. Suessenguth has pointed to the coincidence of cumulative sun flecks with gregarious flowering of bamboos. In many cases, flowering may chiefly be stimulated in dry weather. Accumulative flowering may be observed at the beginning or end of the dry season. A great number of flowering (ornamental) trees start to flower at the end of the dry season in the surroundings of Caracas, according to my own observation. For British Guiana Schomburgk could show a flowering peak at the beginning and end of the short dry season (February and April), and at the beginning of the long dry season (September). For a Malayan rain forrest McClure (1966) reports that some species timed their fruiting with the January-February dry season, while defoliation and flowering followed extended periods of lessened rain fall. Some species did not flower or fruit in 5 years, others fruited regularly. Medway stresses the importance of external conditioning factors such as the stimulating effect of drought for the intraspecific synchronization of flowering. The peak in flowering and fruiting of tree species in the lowlands of Central America is the result of selection for sexual reproduction of the most opportune time in the year, rather than the result of immutable physiological processes, according to Janzen. The advantage of the dry season is the lack of interference with vegetative processes – maximizing the vegetative competetive ability and the use of dispersal agents (Janzen 1967). Inner physiological factors, on the other hand, may be responsible for mast fruiting. *Hymenaea courbaril* shows mast fruiting after several years. Fruiting here depends on the amount of stored reserves. Seed predation is the major force for deviation from annual fruiting, and also for setting the timing and variance of seeding distribution within the fruiting year (Janzen 1978). The frequency of flowering and fruiting very obviously depends on climatic factors in many species. *Pithecellobium saman* flowers and fruits once a year in Central America, but twice in Malaysia where it is introduced. However, this seems to depend on the predators which are absent in Malaysia. The annual frequency of flowering and fruiting, on the other hand, is certainly dependent on the climate in the case of the mango. While mangos set fruit twice a year in a very warm and humid climate, such as of the Caribean islands or of the Venezuelan lowlands, they fruit only once a year in a cooler and drier climate, e.g. at higher altitudes of the hilly areas around Caracas at about 1000 meters, according to my own observations. For the Far East, Whitmore (1975) states that regular periodic flowering with two annual peaks (first peak in March-July, second peak at the end of the year) is probably the rule for most tree species of primary lowland forests.

Koelmeyer (1959) observed a certain relation between the time of leaf change and flowering. Of 45 species studied in the dipterocarp forest of Malaya, Medway found only one, *Ficus sumatrana*, with floral activity continuing in a regular succession of cycles, apparently under endogeneous control. In all other species, flowering was intermittent. A certain alternation of flowering in the two halves of the year with a relatively high incidence in February-July and a low one in August-January has been reported by Medway. Not infrequently, two peaks of flowering during the year were observed in tropical humid forests, possibly depending on two distinct dry seasons, a smaller and a larger one, according to this author.

Layering of the forest becomes also apparent in the flowering of the plants from different strata. As Richards (1952) points out, there is some evidence that flowering in the upper stories is more stimulated by seasonal changes than in the lower stories. Davis and Richards (1933) observed in British Guiana that A and B story trees show two main flowering seasons conditioned by seasonal changes, while among the C story trees, the shrubs and herbs, flowering takes place equally all the year round. This fact is attributed to the less variable humidity and temperature in the lower stories.

When flowering is more bound to the dry season, fruiting, and especially development of fruits, may occur during the wetter season. Furthermore, the type of fruit developed may be dependent on the season. Dry fruits more frequently ripen during the dry season, while fruits dispersed during the rainy season are often fleshy. Species which bear fruits throughout the year may produce different types of fruits depending on the season: i.e. smaller fruits in the dry season and heavier fruits during the wet season. All dry diaspores and particularly anemochorous fruits and seeds have a better dispersal chance during the dry season. As seasonality of fruiting is important for the dispersers, seasonality of fleshy fruits is frequently related to the presence and life cycle of animals.

Vertical stratification of the forest, on the other hand, may exercise a certain influence on seasonality of flowering and fruiting. According to Davis (cited in Richards 1952), seed production is usually seasonal in the canopy, but in the undergrowth flowers and fruits are often borne almost continually in small quantities or at irregular intervals. As we have seen above, the climate is more balanced and more uniform in the understory, whereas the high trees are more exposed to climatic changes, such as wind, drought, high temperature, insolation, daily and seasonal changes.

While our knowledge of the seasonality of autochorous and anemochorous species is relatively bare, particularly concerning fruiting seasons, many examples of synchronization of fruit setting and animal behaviour may be cited. Contemporaneous flowering and fruiting of many canopy trees in the rain forest of the Far East results in seasonal peaks (possible two peaks) of food supply for herbivorous, frugivorous and insectivorous animals, according to Whitmore (1975). A peak of nesting of insectivorous birds in March/April may be the consequence of the attraction of large masses of pollinating insects when dipterocarps flower. Medway (1972) has pointed to the correlations existing between pregnancy of rats and the supply of fallen fruits in a forest of Malaya. Rats show elevated pregnancy in July-December. The peak in their reproductive cycle coincides with the seasonal peak of fruiting. The rats are to a large extent dependent on fallen fruit. Likewise, the annual breeding of birds can be related to the observed seasonality of the plant community in the forest studied. A bimodal pattern in the periodicity of flowering and fruiting was also observed by Snow (1965) in Lauraceae with a main peak in April-June and a minor peak in November which fact he deduces from the climatic seasonality. A main dry season in January-May not only occurs in Trinidad, but the same seasonality is characteristic of a great part of Central America (and Venezuela). A bimodal periodicity becomes also obvious in other seasonal phenomena, such as bird breeding. Smythe (1970) who studied the Panamá canal zone observed seasonal changes in the behaviour of cavimorph rodents. The reason why some fruiting species are more seasonal than others is that animals act as dispersal agents for most tree species whose diaspores do not possess some wind dispersal mechanisms. Smythe distinguishes two types of fruits: larger ones with a dimension greater than 1,5 cm and smaller ones less than 1,5 cm long. In May, June, July he observed an increase in the number of large fruits falling, while smaller fruits fall more evenly throughout the year. Wind-dispersed diaspores, on the other hand, fall chiefly from December to April, i.e. during the dry season which is also the time of the strongest winds. The water content of the fruits falling during the wet season (September, October,

November) is somewhat higher as compared with that of fruits falling during the dry season (January). Most species ripen their fruits in May and June. The smaller fruits are mainly dispersed by animals which swallow them and pass the seeds undamaged. Rodents usually destroy the embryo of the seeds they eat, but agoutis *(Dasyprocta punctata)* do not – they "scatterhoard". The larger fruits (e.g. of *Anacardium*) do not fall in such quantities as the small fruits and often are encountered singly by agoutis. The fruit is carried away by them, not infrequently to an open spot in the undergrowth. *Spondias mombin* ripens its fruits at a different time than other large-fruited species, i.e. at a time when there is a shortage of food for agoutis. The population of agoutis is highest, when *Spondias* fruits ripen. As the seeds of *Spondias mombin* are not edible, they are dispersed by agoutis while undergoing reduced competition for dispersal with other large-fruited species. Although Smythe admits that physical environmental factors may be the most important ones in determining fruiting seasons, competition for dispersal agents may act as an additional selective pressure. In summary, small fruits the seeds of which pass unharmed through the guts of animals tend to be non-seasonal and thus avoid competition for dispersal. Fruits with large seeds, on the other hand, that are too large to be swallowed and are dispersed by hoarding rodents benefit from fruiting as synchronously as possible. The larger seeds of larger fruits are destroyed by animals when eaten. But the agouti being unable to devour the entire synchronous crop scatterhoards much of it and since it later fails to find many of the buried seeds it aids in the dispersal of its food species. While the fruit of *S. mombin* is of an attractive orange colour and has a sweet tasty pulp, that of *S. radlkoferi* is green and tasteless. Nevertheless, fruits of *S. radlkoferi* find enough visitors, because they mature at a time when competition for food is greater. The fruits are eaten by monkeys. This delayed fruit maturation is possibly an adaptation to the occurrence of food supply in the forest (Croat 1974). In general, the cause of periodically enhanced breeding behaviours of animals is a periodical increase in food supply which may follow from a response of plants to seasonal changes of the climate. Murphy (1973) believes that there are not only two fruiting peaks above ground, but even two peaks in the activity of the entire soil microflora at the time of seasonal fruit body production by higher fungi. An entire cycle of synchronous phenomena is thus closed, concerning not only the plants and animals above ground, but also the flora (and fauna?) below.

Dispersal methods

In the following chapter the different dispersal methods are discussed, starting with the "mechanical or abiotic" methods such as dispersal by water or by wind, and mechanisms developed by the dispersal unit of the plant itself (autochory). These are followed by dispersal methods carried out through living dispersal agents such as different kinds of animals. Van der Pijl (1972) has given an exhaustive survey on the subject in his "Principles of dispersal in higher plants". The role of man in the dispersal of tropical plants is very doubtful. The dispersal effect exerted by man lies firstly in new open habitats, as van der Pijl emphasizes. Of the closed rain forest regions in South America, and especially in Venezuela, we know almost nothing concerning the influence of man in plant dispersal. We know that Indians have selected certain plants and cultivated them for thousands of years. However, in the virgin forests of Venezuelan Guiana they may have played a very subordinate role or possibly never penetrated the densest areas. On the other hand, many species of this region received vernacular Indian names, and Indians may have inhabited certain forest margins (see also André 1964). But as a rule, they may have preferred the vicinity of large water ways such as the Orinoco.

Hydrochory

In contrast to van der Pijl (1972), Ridley (1930) considers hydrochory perhaps more important than anemochory and certainly one of the most ancient dispersal mechanisms which do not require much specialization of the dispersal units. However, in certain advanced types, e.g. hydrochastic capsules or fruits and seeds with particular floatation tissues, hydrochory has to be regarded as derived. Water dispersal comprises rain wash as well as the transport by rivers and streams or by currents of lakes and of the sea. Floods may play an important part in certain types of tropical forests such as periodically inundated regions, and even ice can transport seeds and fruits as well as pollen grains. Dispersal units may be carried floating on the water surface or drifting under water. For floating seeds and fruits buoyancy is of most importance and it is attained by reduction of the specific weight through air-filled intercellular spaces and lacunas or by air-filled dead cells (see also Roth 1977). Air-filled free spaces between the seeds of a pod, for example, or other cavities filled with air give many fruits and seeds a good buoyancy. Water-dispersed fruits usually contain few or only a single seed and buoyancy is not infrequently due to a seed cavity partly filled with air. Free spaces between seed coat and embryo, or spaces left between the cotyledons filled with air are often responsible for buoyancy. A corky seed coat composed of dead air-filled cells is an exception (e.g. in *Carapa guianensis).* In these cases, the floating tissue or the air-filled spaces are located either in the seed or in the fruit itself (e.g. in the pericarp, locule, seed coat, etc.). But even floating organs outside the fruit and seed may be responsible for the buoyancy, such as sepals, bracts, pedicels *(Anacardium).* Ridley as well as Guppy (1917) point at the variation of diaspore buoyancy in one and the same species. But relatively few diaspores support prolonged floatation without injury. Protection against wetting is achieved when the surface of diaspores provides waxy or cuticular layers. Interspaces between papillas or hairs may hold the air so tenaciously that wetting of the surface is prevented. Ulbrich (1928) gives a number of examples with special floatation devices. Many water plants and plants from swampy regions have floating fruits or seeds. However, floating fruits and seeds have to be protected not only against water logging but also against decay in water. Impregnation with tannins may prevent rotting of many fruits and seeds of tropical humid forests. Besides, it is advantageous for drifting seeds that they do not germinate before reaching their final destination. The germination capacity of *Entada scandens,* for example, a "drifting fruit", may be preserved for years. However, these examples are exceptions.

Rain wash. Rain wash is an important method of transportation in tropical rain forests. Rainstorms which are often very heavy carry away not only small and light seeds and fruits, but also heavy ones. Ridley emphasized the effectiveness of rain wash in the dissemination of herbaceous plants. However, we know very little of this type of hydrochory in the tropics.

Some plants with capsular fruits are adapted to dispersal by rain, such as desert and steppe plants with hygrochastic capsules. In these cases, an autochorous mechanism is associated with dispersal by rain. A very derived type of hydrochory indeed.

Floating fruits and seeds. Floating fruits and seeds are to some extent characteristic of tropical plants growing at the riverside. Some of them are mentioned by Ridley and Van der Pijl. Most fruits and seeds adapted to wind dispersal by reduction of the specific weight and an increasing surface area are also apt to be transported by water.

Submerged transport of diaspores in the water. Van der Pijl (1972) draws attention to aquatic and shore plants with diaspores that are not able to float, but are transported submerged in the water. In this connection, however, mainly species of temperate regions are listed.

Dispersal by the sea. Guppy (1917) studied this subject profoundly. From our point of view, the equatorial current from the Amazon and Orinoco to Trinidad and from the estuaries and shores of Guiana and Brazil is of some interest. Within the Jamaican beach drift Guppy found three species most abundant: *Carapa guianensis, Saccoglottis amazonica,* and *Manicaria saccifera.* Other diaspores found in tropical beach drift are *Symphonia globulifera, Andira inermis,* and *Hymenaea courbaril. Entada scandens, Mucuna urens, Dioclea reflexa, Spondias lutea,* and *Cassia grandis* endure prolonged floatation without injury. *Carapa guia-*

nensis is occasionally associated with mangroves. However, its seeds are killed in sea water. *Cassia fistula* - introduced from the old world - is carried over long distances. Trinidad and Tobago have drift from the Orinoco, from Guiana, and the Amazon. However, many seeds of the riverside have the tendency to germinate when afloat and have no effective value for over-sea dispersal. Not even 1/5 of the seeds and fruits brought down by Westindian rivers from island districts to the sea would be capable of reproducing the plant after a traverse of a 100 miles of ocean, according to Guppy. Ridley (1930) emphasizes that dissemination of plants by sea-currents from coast to coast and from island to island is one of the oldest forms of dispersal. The distances that seeds can travel unharmed by sea without loosing their viability are the longest of any method of transport, according to Ridley. The diaspores must be able to float for a long time without absorbing water and must resist the action of salt. One of the most famous examples in this respect is *Cocos nucifera*. Guppy and Ridley list a number of tropical terrestrial plants dispersed by water, fresh water as well as sea water. Within these species some are mentioned which are also included in the present studies. Terrestrial flowering plants of tropical regions with seeds dispersed by sea currents are among others, two species of *Calophyllum* (Guttiferae), *Carapa guianensis* (Meliaceae), and other species of *Carapa, Sapindus saponaria, Anacardium occidentale, Andira inermis* (Leguminosae) which has been mentioned already, *Terminalia catappa* (Combretaceae), *Guettarda speciosa* (Rubiaceae), *Cordia subcordata* (Boraginaceae), *Vitex trifolia* (Verbenaceae), *Hernandia, Ricinus communis* and *Hevea brasiliensis* (Euphorbiaceae) and others, according to Ridley. *Caesalpinia bonduc* and *C. bonducella* seeds can float for years, thus providing an almost pantropical distribution (Van der Pijl 1972). Species studied in this book which are also dispersed by water (sea) will be mentioned in the related chapters. In most of these cases other dispersal mechanisms besides water transport are observed, e.g. in *Hevea brasiliensis*. In quite a few instances buoyancy of dispersal units and water transport may be more or less incidental (e.g. in *Anacardium occidentale)*. However, it is certainly advantageous for any plant species to use more than one dispersal method.

Anemochory

In contrast to other dispersal methods (e.g. rain wash) wind-dispersal is a well studied type of dissemination. Anemochory is often found in open areas where it is easy to study. It is natural that wind-dispersal is more effective in those areas where wind blows, such as coastal regions, small islands, steppe and desert, or mountains. The strategies of wind-dispersal are manyfold. Very small "dust seeds" are one method, while reduction of the specific weight and increase of the surface area may lead to very different forms and to distinct flight strategies of dispersal units. Hairs and wings are most effective in increasing the surface area and fruits as well as seeds may act as anemochorous diaspores. Not seldom, wind-dispersal is connected with an autochorous dispersal mechanism. For example, dust-seeds, winged seeds, or seeds surrounded by hairs are released by dehiscing capsules and transported further by wind (orchids, Bignoniaceae, *Ceiba pentandra, Swietenia,* and many others). Wind-dispersed capsular seeds may be carried further by water (e.g. rain wash) thus becoming triplochorous. Anemochory has developed as a very effective dispersal method, for example, in the pappus-furnished achenes of Compositae (e.g. *Taraxacum officinale*). Dispersal distances are sometimes very great, very obviously so when islands are colonized by new anemochorous species from the continent (island of Krakatau).

Ulbrich (1928) considers anemochory the most important and most effective dispersal method. He describes all the distinct devices and modifications of wind-dispersed diaspores as well as their flight path. There are bladder-flyers with thin-walled air-filled bladders, fruits and seeds with wings of different kind and shape (with one to several wings), diaspores furnished with hairs which may be arranged in different ways (e.g. hairs all around, in the form of a wreath, or a tuft of hair: shuttlecock type. See also Roth (1977). For the flying diaspores the point of gravity as well as the shape of the diaspore are of utmost importance. Not only the fruit itself (e.g. the pericarp) or the seed develops flying devices, but also structures or organs outside the ovary may supply hairs or wings (e.g. sepals, bracts, glumes, pedicels). Well known is the flying calyx of the Dipterocarpaceae or the pappus of the Compositae. Concerning the path of the flight Ulbrich distinguishes gliding flyers, whirling flyers, screw flyers, rotating flyers, rotating roller flyers, shuttlecock flyers, flying discs, rotating discs, and sailing flyers.

Dingler (1889) was the first to present a classification of wind-dispersed diaspores concerning their path of flight and patterns of motion. Aerodynamics of anemochorous seeds and fruits have been studied recently by Burrows, Sheldon, and Norberg. Burrows (1975 b) discusses the movement of different types of wind-dispersed diaspores. He furthermore gives calculations of the primary trajectories of dust seeds, spores, and pollen in unsteady winds (1975 a) and of plumed seeds in steady winds (1973). The dispersal effectiveness of the achene-pappus units of selected Compositae in steady winds with convection is discussed by Sheldon & Burrows (1973). Efficiency of dispersal is determined more by the fine details of the pappus geometry which directly affects its aerodynamic properties than by the size ratio of pappus to achene. Under steady horizontal winds increased height of fruit release increases dispersal distance. While increased wind velocities increase the trajectory distance, dispersal of many Compositae is hampered by the pappus response to humidity. Increased dispersal distances are given by steady convection currents, whereas turbulence may either curtail or prolong transportation. Occurrence on islands of species with plumed fruits or seeds gives evidence for long-distance dispersal. The slower the rate of fall, the greater the effect of upward currents of air and the greater the chance of longdistance dispersal. However, seedling establishment may influence species distribution more than dispersal. The flight of winged fruits and seeds has often been compared with the aerodynamics of aeroplanes in unpowered flight. Norberg (1973) studied the autorotation, self-stability, and structure of single-winged fruits and seeds in comparison with the animal flight. As the wing is to reduce the falling speed of the diaspore, increasing in this way the distance it may be transported, selection pressure is directed towards improvement of the dispersal potential. In other words, mainly tall plants (trees, lianas) have winged diaspores so that a loss of altitude is allowed on rapid initial fall before the diaspores enter autorotation and a slow descent. The autodynamics of samaras (Norberg regards winged fruits as well as seeds as samaras) are in many respects similar to those of helicopters and autogyros. It is not only the gravity but also the centrifugal force (excentric point of gravity) of the blade (or wing) that interacts with the aerodynamic force to give pitch stability. The direction of the resultant force is the ratio between its two component forces: lift and drag. The sinking speed of a samara is similar to or slower than that of most birds in gliding flight (about 0,9 m/sec in *Acer*). The samara stands out as a simple and very efficient structure in braking the sinking speed of the diaspore.

Ridley (1930) gives a great many examples of anemochorous fruits and seeds, particularly from tropical regions. It is well known that anemochorous dispersal units mainly occur in open regions which are exposed to wind and that the first species colonizing a new area are often anemochorous. Anemochory is, therefore, more frequently found in deciduous forests where fruit dispersal cumulatively occurs during the "leafless" dry season. Wikander (1980) studied the diaspores of a dry deciduous tropical forest in Venezuela and found 42% anemochorous species, 30% zoochorous, 19% barochorous, and 1% autochorous species. Anemochory, consequently, prevails here. In the humid tropical forest, on the contrary, about 80% of the species are zoochorous (Foster 1973). Aristeguieta (1967) found 73% of the flora in a deciduous forest of the "Llanos Altos" in Venezuela anemochorous.

But not only open areas are characterized by anemochorous diaspores – anemochory also occurs in the forest, but here the exposed tall trees as well as tall lianas and epiphytes develop wind-dispersed seeds and fruits. It has been mentioned already that a certain layering concerning dispersal methods is obvious in the tropical forest, as anemochory mainly occurs in the upper forest story, while zoochory is much more bound to the lower stories. Hall & Swaine (1981) found 20-30% of the tall trees and lianas with winged or plumed diaspores in a tropical rain forest of Ghana, whereas only 5% of the understory trees had wind-dispersed propagules. Keay (1957) already emphasized that wind-borne seeds and fruits are characteristic of many tall trees, whereas understory trees and shrubs mostly have fleshy fruits. Lighter-seeded trees, on the whole, are more characteristic of open habitats, while species with larger seeds occupy more shaded areas. Emergent trees such as *Ceiba pentandra* may reach a height of 36-48 m and are anemochorous.

Some plant families which are mainly characterized by tree species are distinguished by anemochory, such as the Bignoniaceae, Apocynaceae, Asclepiadaceae. However, Gentry (1976) emphasized that the Bignoniaceae, for example, although wide-ranging in their distribution, show most abundance and ecological importance in dry areas. Absolute density of the Bignoniaceae increases from wet forest to dry forest. However, abundance of individuals, but not species diversity is greater in dry than in wet forest. Distinct climatic preferences are important determinants for occurence of species of Bignoniaceae. Local edaphic conditions constitute another important parameter, and substrate is likewise a factor to consider. Competition for pollinator resources is probably a major diversity determining factor, according to Gentry. Animal dispersal in Bignoniaceae is correlated with wet forest habitat, while plants with wind-dispersed diaspores are typically more prevalent in dry areas. As we shall see later, Opler, Baker, and Frankie (1975) found anemochory in tall tree species of the genus *Cordia*, while zoochory is common to small tree species and shrubs of the same genus.

Not only a drier climate and the presence of more air movement favours dissemination of anemochorous plants, but also the daily weather and the time of the day may influence wind-dispersal. Plummer & Keever (1963) found in camphorweed *(Heterotheca latifolia)* that dissemination of the achenes occurs best on clear days that are abnormally windy and dry, with winds of more than 12 miles per hour and a VPD greater than 10 mm Hg. Early afternoon conditions with turbulent air play the most important role in conditioning the plant for dissemination. Achenes with a pappus travel up to 30 m from the mother plant by a wind velocity of 1200 ft/min (14 miles per hour). A distance about 10 times longer may be covered by winged seeds. Ridley (1930) states that the longest distance which winged fruits can travel is 880 yards. However, who has ever counted the longest distances covered by all the anemochorous diaspores of tropical trees in the whole world? Furthermore, wind may be an important agent in the dispersal of "unmodified seeds" from dehiscent fruits when they are very light or flattened. This reflection leads us then to the discussion of autochory.

Autochory and explosively dispersed seeds

Autochory is understood as active seed dispersal by the plant itself, usually by a turgor mechanism or by hygroscopic movements (see also Roth 1977) so that the seeds are thrown out of a fleshy or dry capsular fruit. Besides we may distinguish geoautochory (e.g. in *Linaria cymbalaria*) and herpautochory (e.g. creeping diaspores of Gramineae, *Centaurea*). However, there are also other dispersal types which we could call autochorous, as the plant itself provides the dissemination but without using an explosive mechanism. Pyxidia with circumscissile dehiscence as well as pore capsules let the seeds fall out slowly, possibly using some air movement or other mechanical movement. Ulbrich (1928) even includes barochory, i.e. "Fallvorrichtungen durch Eigengewicht der Diasporen", within the autochorous mechanisms, although the inclination of the local situation may play its part in this dispersal type (see also Wikander 1980). As mentioned before, we cannot draw an exact line between autochory and allochory as the boundaries are fluid. Van der Pijl (1972) distinguishes between active ballists with a special dispersal mechanism of the plant and passive ballists using other means for seed dispersal, such as wind, rain, or animals.

Utmost attention has always been paid to the explosive dispersal mechanisms. In the tropical humid forest the explosive capsules of many Leguminosae and Euphorbiaceae are of much importance. However, here too we have to consider not only the explosively dehiscing capsules, but also the intermediate forms, and those with a delayed dehiscence which finally leads to indehiscent fruits. We thus find all kinds of intermediate stages between two extreme types and the intermediate stages are probably more numerous than usually accepted.

Between the most succesful "shooters" *Bauhinia purperea* (with 15 meters) and *Hura crepitans* (with 14 meters, according to former authors) are mentioned as "holders of the record". Swaine & Beer (1977) lately studied the explosive seed dispersal in *Hura crepitans*. The number of carpels per capsule oscillates between 9 and 16. The capsule measuring 50–80 mm in diameter is green, when the vascular connection between fruit and branch is sealed off. The capsule then begins to dry adopting progressively a dark brown colour. Further water loss leads to seed detachment from the carpels so that the capsule rattles when shaken. Dehiscence is very sudden and strongly related to the moisture content of the capsule which looses about 64% of its mature weight during drying. At dehiscence each woody carpel separates from the central column as well as from the neighbouring carpels splitting into two valves. Dehiscence is thus speticidal, loculicidal and columnicidal at the same time (Fig. 36). The seeds are relatively large and the authors suppose that this has some importance for the penetration of the canopy. However, it may also be advantageous for the nutrition of the seedling. The seed is offered least resistance by the air when projected "edge on". A projection angle of 34° above the horizontal is close to the optimum for maximum dispersal distance. The maximum dispersal distance measured by the authors was not less than 45 meters from the parent tree – with a model distance of about 30 meters. The projection of the seeds is so sudden and vehement that the seeds may hurt a person nearby or, at least, may frighten her or him. Beer & Swaine (1977) emphasize that dispersal by animals is subject to unpredictable influences and that wind is most unpredictable as a dispersal agent. On the contrary, explosively dispersed seed is largely determined at the instant of release from the parent and less unpredictable. The distance to which the seed travels depends on the proportion of the total energy released during dehiscence. This energy manifests itself in the initial velocity of the seed. The distance travelled is also influenced by the angle to the horizontal and the height above the ground at which the seed is projected as well as by the air resistance which is determined by the seed shape and mass. Beer & Swaine elaborated an equation valid for any seed projection providing there is no wind and no obstacle in the way (e.g. leaves, twigs). In most cases, the optimum angle is considerably less than 45°, but more than 0° (horizontal). The fact that the height of release has little influence on dispersal distance provides one of the reasons why explosive dispersal mechanisms are not confined to trees or other plants which attain a great height (lianas, epiphytes), according to the authors. On the other hand, the height of release has considerable influence on the distribution of the seeds on the ground. In case the angles of projection are limited and the velocities are also restricted, the distribution on the ground will become a narrow ring around the parent tree at some fixed distance. In all examples, the mean distance travelled is about one half the maximum, indicating that the bulk of the seeds are carried some distance and are not closely associated with the parent.

According to the above mentioned studies of Swaine & Beer, *Hura crepitans* holds the record in seed shooting with 45 meters distance from the parent tree. The authors also state that the explosive dispersal of seeds may not only be effective, but also relatively "secure" not being influenced by many unknown factors. However, it is usually a dispersal method over relatively short distances, and in many species it may be of only little effectiveness, as will be shown later. We have furthermore to admit that violently "exploding" fruits are the exception, while most capsules dehisce more slowly and less vigorously than those of *Hura crepitans*. Besides, *Hura crepitans* is not a tree characteristic of dense forests but preferably grows in more open and drier habitats. In the dense forest the "exploding" method encounters a series of severe obstacles such as stems, twigs and a multitude of leaves. The majority of dehiscing fruits releases the seeds in a much less spectacular way and in many cases autochory of this type could rather be called "barochory", as the seeds are just freed from the capsule by dehiscence but are not dispersed in this way, if not wind or some movement aids in dissemination. But even so, dehiscing fruits are called autochorous in this book for lack of information and for the inpracticability of distinction of the different transitional types.

Dispersal by fish (ichthyochory)

Ichthyochory is considered a relatively ancient way of dispersal. Fish may eat fruits and seeds equally well whether they float in the water or fall to the bottom of rivers and lakes. Well known examples of ichthyochorous plants are supplied by species of Leguminosae, particularly of the genus *Inga* and *Eperua*. Van der Pijl mentions *Ficus, Guateria* (Annonaceae), *Genipa americana* (Rubiaceae), and some palmae as fish-dispersed plants. Some fruits and seeds may even be used as "baits" by fishermen. Most of the floating fruits eaten by fish are distinguished by a floating tissue. Knöppel (1970) studied the food resources of Central Amazonian fishes in the black- and clear-waters of the Rio Negro region. Of 49 fish species more than 50% had vegetable remains – mainly fruits, seeds, and leaves – in the stomach. However, he did not identify the stomach content nor did he investigate from which plant species the vegetarian diet originated. He only speaks partly of granulous, solid fruits or seeds, of non-aquatic origing. The fruits and seeds may well be dispersed over large distances in this way. Gottsberger (1978) studied the inundated regions of Amazonia. Of 33 plant species he found 16 with seeds or fruits regularly dispersed by fish. Most of the fishes that eat fruits and seeds are among the larger species. Of 17 plant species seeds were usually destroyed by fish. The 16 genuine ichthyochorous plant species are relatively primitive, according to Gottsberger, while the remaining 17 species are phylogenetically more advanced. This observation coincides with the suggestion that ichthyochory is a relatively archaic form of dispersal. The particular structures of dispersal units exhibiting ichtyochory may be attractive, e.g. in the form of fleshy arils *(Virola)*, arilloids *(Annona)*, or tasty fruit flesh *(Annona, Genipa americana)*, while others are protective (hard shells or endocarps, and surfaces slippery through slimy substances). The reproductive rhythm of fish-dispersed várzea plants seems to be partly related to fish activities. Fruit ripening begins at a time when the water level is still high, but continues, while the level is lowering and fishes appear to become more abundant (e.g. *Annona hypoglauca)*. Gottsberger raises the question whether the fishes are resistant to toxic substances in certain seeds (e.g. of Euphorbiaceae, *Carapa, Anacardium*). Exploding pods of *Eperua rubiginosa* make a sound when falling into the water which attracts certain fishes. Fishermen imitate this sound to catch fish. Fish-dispersed fruits of *Annona hypoglauca* fall into the water immediately on ripening. Possibly also other ichthyochorous fruits and seeds fall immediately when ripe, in contrast to many fruits dispersed by birds and arboreal mammals which remain attached to be eaten on the tree. During the dry periods in the late Pleistocene and Post-Pleistocene, the forest in lowland Amazonia were widely reduced and only covered a few limited areas or refuges. The wide distribution of várzea forests today may be due to the "ease of diaspore dispersal by water". Apart from ichthyochory also saurochory (by turtles, alligators) is common in the várzeas. Even the water itself is an important disperser. Of the plant species dispersed by fishes Gottsberger mentions the following: *Annona hypoglauca, Duguetia marcgraviana, Unonopsis* aff. *matthewii*, cf. *Unonopsis* (Annonaceae), *Virola* cf. *surinamensis* (Myristicaceae), *Ficus* sp., *Cecropia membranacea* (Moraceae), cf. Elaeocarpaceae, *Neobalatia cuprea, Lu-*

cuma cf. *dissepala* (Sapotaceae), *Liciana* cf. *apetala, Licania* sp. (Chrysobalanaceae), *Protium* sp. (Burseraceae), *Simaba* cf. *guianensis* (Simaroubaceae), *Astrocaryum jauary*, Joarizeiro (Arecaceae), and possibly *Macrolobium acaciaefolium* (Caesalpiniaceae), *Caraipa laurifolia* (Clusiaceae), and *Byrsonima amazonica* (Malpighiaceae). The remaining (17) species are also eaten by fishes, but their seeds are broken up and destroyed by masticating. Consequently, about 50% of the plant species studied seem to be adapted to fish dispersal. The known homogeneity of the inundated vegetation of Amazonia might have its cause in the very effective fish dispersal throughout the whole of Amazonia and its tributaries (Gottsberger 1978).

Dispersal by reptiles (saurochory)

Of the modern reptiles only few live on leaves and fruit. The alligator apple, *Annona palustris*, has received its vernacular name from fruit-eating alligators. *Ctenosaura similis* lizards are insectivorous when young, but adults feed almost entirely on leaves and fruit. The arboreal iguana and some turtles are fruit eaters and possible dispersers. Van der Pijl (1972) suggests that dropped *Inga* fruits are dispersed by reptiles, e.g. in Guiana. Very few are the reports on modern saurochory. Glander (1979) studied feeding associations between howling monkeys and basilik lizards which eat the pedicels of *Anacardium excelsium*, but not the nut. However, the fruits are not dispersed in this way, they remain on the ground beneath the mother tree. Bats and birds are the probable dispersal agents. The pedicels of *A. excelsum* are fleshy and edible, while the fruits contain the toxic cardol oil. Toxic fruit and edible pedicel are an evolutionary response of *Anacardium excelsum* to ensure the dispersal of the fruit without injury. Epizoochory on reptiles is insignificant, as these animals have a smooth skin. Basal caulicarpy or basicaulicarpy may have developed in connection with reptiles. According to Van der Pijl (1969) fish and reptiles were the original living dispersal agents of fruits and seeds. The syndrome of characteristics of saurochory comprises: diaspores with a smell and a showy colour (possibly orange-red), large juicy seeds, fruits borne near the ground or dropped at maturity. But all in all, saurochory plays an unimportant part in today's seed dispersal. Swain (1976) supposes that herbivorous dinosaurs became extinct by the alkaloids developed in the cretaceous angiosperms being poisoned by them, because their ability to detect them in their food was limited.

Dispersal by birds (ornithochory)

Birds are the most important plant dispersers due to the abundance of birds in all parts of the world and the long distance they can fly. The fruits eaten by birds are mainly baccate and drupaceous. About 70% of the tall trees and lianas have fleshy fruits, e.g. in a rain forest of Ghana (Hall & Swain 1981). Fruits and seeds attracting birds have usually a showy bright red or crimson, yellow to orange colour or a brilliant black or blue. Contrasting colours such as black and red or black and white are also attractive. Ornithochorous diaspores are generally medium-sized to small, however, also larger fruits and seeds are eaten by birds. Large neotropical birds such as toucans eat large seeds surrounded by a thin fruit flesh. A further prerequisite to ornithochory is that the diaspores remain on the tree. The fruits or seeds have to be well "presented", possibly dangling from long stalks (e.g. mango) or being offered on dehisced valves. Even the stratum of the forest in which the seeds or fruits occur is of importance, as bird species are generally confined to certain forest layers. The seeds are frequently arillate having a fleshy and showy aril, except the mimetic seeds which are brightly coloured, but do not offer any edible reward. The plants with coloured fruits are mostly shrubs, smaller trees or climbers and this is also the stratum where most birds are active. A large number of plant species owes their distribution to birds. More regular fruit eaters are found in the tropics than in temperate regions. In the tropical rain forest no regular migrations of birds occur, as there are no very distinct seasons. However, birds can fly overwhelming long distances. Ridley

(1930) mentions the arctic tern which flies not less than from about 7,5 degrees of the North Pole to the antarctic regions. Some birds can fly as fast as 180 miles an hour. In this way seeds could be dispersed over very long distances. However, migrating birds are rarely frugivorous and their digestion is usually fast. Ridley also discusses the question for how much time a seed may be retained by a bird. He calculates between ½ hour and about 8 hours. The seeds of *Guarea glabra*, for example, are bulky and migrant birds regurgitate them promptly within 5–20 minutes (Howe & De Steven 1979). Proctor (1968) states that seeds of various plant species remained in the digestive tract of *Charadrius vociferus* for up to 144–340 hours. Seeds of many species can remain viable in the intestinal tract of some shorebirds long enough to be transported several thousend miles. Particularly seeds with a diameter greater than 1 mm and a hard coat may be retained for more than 100 hours. Ridley presents a long list of frugivorous birds. Diaspores may be dispersed by birds in very different ways. The most common way is endozoochory. Diaspores pass through the digestive tract and are ejected. In other cases, the seeds are regurgitated by the bird. Diaspores may also be carried by birds and hoarded. Seeds of semiparasites, such as of *Viscum, Phoradendron*, and other Loranthaceae, are carried in the beak and deposited on a twig. A further way of diaspore transportation is by epizoochory. In many cases several of these methods may be mixed with one another as is often observed in nature. Endozoochory may either be adapted or accidental. It is natural that obligatory dispersers are of greater value for the plant than occasional visitors. For certain seeds a passage through the intestinal tract of birds or other animals is obligatory or – at least – advantageous for germination. In this case, the seed may either be prepared chemically by digestive enzymes or mechanically by rupture of the seed coat. Mc Diarmid et al. (1977) studied the germination rate of *Stemmadenia*, Apocynaceae, with the result that it was highest in seeds from which the aril was removed either by scarifying or by passing through the digestive tract of birds, while germination of seeds with intact arils is delayed about 5–10 days. Sousa (1969) observed that 50% of all seeds of *Bursera simarouba* germinated when they were taken from the stomach of 5 different bird species, while less than 1% germinates in forestry plantations. *Cecropia* fruits are a favoured food of many tropical birds and the seeds show an enhanced viability after passing through the alimentary canal of birds (Leck 1972).

The organs, parts of organs or tissues developed for the attraction of birds are manyfold. The importance of sarcotesta and aril have been especially treated by Van der Pijl. The pericarp, septs, placentas, or special viscid layers of fruits are another means of attraction. Structures outside the ovary are discs, pedicels, sepals, petals, bracts, and even inflorescence axes. Diaspores suitable for epizoochorous dispersal have viscid layers, spines, hooks, glabrous glands and similar devices. Viscid layers are particularly well known of the adhesive fruits of Loranthaceae. Mimetic seeds attract birds by their colour not giving any reward. Red tropical "coral-seeds" deceive fruit-eating birds (Van der Pijl 1969). Van der Pijl treats this subject thoroughly in his book (1972). Mimetic seeds are often exposed by the dehiscing capsule. The gradual transformation of the juicy sarcotesta seed into a dry mimetic seed is obvious by the diminishing thickness of the testa. Mimesis is abundantly observed in Leguminosae (e.g. *Pithecellobium, Ormosia*). The large woody legume seeds are scarlet red or bicoloured red and black. Ridley calls them "imitation arils". The discrimination of mimetic seeds is weakest when birds encounter them very infrequently. Consequently, mimetic seeds evolve more often in rare species *(Ormosia)*. Polymorphism of the mimetic seeds of *Ormosia* – all red and varying red and black in a single crop – increases the effectiveness (Mc Key 1975).

Endozoic dispersal characteristically takes place over short distances. Plant species with small seeds and endozoic dispersal are disseminated by many bird species and show large dispersal areas in Middle and South America (e.g. *Bursera simarouba)*. Plant species with large seeds, on the other hand, are dispersed by few bird species having a restricted dispersal. Birds with a large beak can eat larger fruits and seeds, while those with small beaks eat smaller food. Large soft seeds which are easily destroyed by birds have frequently inconspicuous colours such as brown, or black, while bracts, peduncle or receptacle take on an attractive colour (e.g. a showy red). Those seeds which are eaten by the bird are then destroyed, but others which are dropped cannot be found any more by the bird and may escape. The inconspicuous colour is thus a means of protection for the seed (Sousa 1969). Besides by an inconspicuous colour, seeds are often protected against destruction either mechanically by a hard usually lignified coat or chemically by acids, even cyanic acid, tannins, latex, alcaloids, amino acids, and the like (see p. 25–32). Energetically, the plant expends the most calories for protection (husk) followed by expenditures for germination (reserve substances in the seed) and dispersal (aril) in *Stemmadenia*. However, on a calorie-per-gram ash-free dry weight basis, the plant puts the greatest amount of energy into the aril for dispersal, as the aril is very oily, followed by the seed's reserve substances for germination and by the husk for protection. The thick husk and the sticky white latex prevent insect damage to the immature fruit (Mc Diarmid et al. 1977). It is obvious that plants evolve to minimize the cost of their dispersal – poor nutritional quality of most fruits demonstrates this fact. Some fruits, however, contain oil and proteins. These are adapted to specialized frugivores (Mc Key 1975).

A distinction between specialized and non-specialized frugivores is therefore necessary. Specialists live more or less exclusively on fruit, favouring fruits with a high nutrient content. They usually have a "gentle digestive system" which allows toxic seeds without a thickened testa or endocarp to pass quickly through the gut unharmed. Non-specialists, on the other hand, have grizzards which would rupture the seeds favoured by specialists, thus poisoning the birds. Non-adapted birds prefer seeds which are well protected mechanically by a thick husk or are too small to be crushed, even when the fruits are not highly nutritious, as their diet includes other sources of food too. With specialized fugivores the plant can produce relatively soft seeds which quickly germinate (e.g. Lauraceae and Loranthaceae) (Mc Key 1975). Opportunistic dispersers, on the other hand, can disperse a greater quantity of seeds. Flight patterns of frugivorous birds are responsible for directional patterns of seed dispersal. These patterns influence seed mortality. Seeds dispersed away from the parent tree have a better chance to survive. Seeds carried away by obligatory specialized frugivores are less likely to be wasted, as more seeds survive under other trees than under the parent tree. Specialists thus carry seeds away from areas of high seed mortality (Janzen 1970, 1972; Howe & Primack 1975).

According to Janzen (1975), competition among trees for dispersal agents is intense. There is a mutual influence of birds upon plants and vice versa. The variety and abundance of fruits in the tropics is correlated with a diverse bird fauna showing varying degrees of specialization for frugivory and seed dispersal (Howe & Primack 1975). Plant species influenced by fruit-eating animals often show modifications in outer and inner structure (e.g. dangling fruits or seeds, mimetic seeds, arils, sweet fruit flesh, pulp with a certain smell, attractive colours). The effect of the digestive tract of birds on germination of seeds is another factor produced by animal influence. Seeds adapted to frugivores not passing through a vertebrate gut suffer retarded germination. A third effect is that produced by obligatory frugivores on the "survivorship" of seeds which suffer high animal predation (Janzen 1970). "Coevolution" has led to dispersal quality through reliability of visitation, probability of intact deposition of seeds, size of the seed a dispersal agent finds it profitable to carry, and probability that the seed is carried to a suitable site (Mc Key 1975). A high degree of correlation thus exists between plant and obligatory dispersers. The variety and abundance of fruits in the tropics is correlated with a diverse bird fauna showing varying degrees of specialization for frugivory and seed dispersal (Howe & Primack 1975). Fruits and birds have undergone "mutual evolution". This coevolution is sometimes so strong that extinction of the respective plant may lead to the dispappearance of the adapted frugivorous bird (e.g. *Casearia corymbosa* and *Tityra*: Howe 1977).

In summary, the most important characteristics of the syndrome of bird-dispersed diaspores are: Easy access to the fruit or seed (terminally placed branches, exposed position, dangling fruits or seeds), firm and permanent attachment to the branches, twigs that support the weight of birds, temporal and spatial availability of fruits (e.g. during the dry season), attractive colours (preference of red or red/black), fleshy edible parts (aril, sarcotesta, flesh of baccate or drupaceous fruits), nutritive value of the edible part (sugar, oil, protein), ease of harvesting edible parts (flesh easily to remove), protection against premature predation (green colour, acidity, tannins), protection of the seed against digestion (hard shell, bitter taste, toxic substances).

Dispersal by mammals
(mammaliochory except chiropterochory)

As with ornithochory, mammaliochory is best developed or preserved in the tropics (Van der Pijl 1972). Here too, we distinguish epizoochory and endozoochory, either accidental or adaptive. Besides, dyszoochory and scatterhoarding of rodents play an important part. Seeds of some *Lecythis* species, for example, are buried as a food reserve by agoutis *(Dasyprocta)*. Many tropical squirrels are fruit-eaters and may bury the seeds, e.g. of *Gustavia augusta, Theobroma cacao*. Of the latter they eat the inner fruit flesh (endocarp), not the "sarcotesta-pulp" (Pijl 1972), as there is no sarcotesta (Roth 1977). The endocarp pulp is sweet and also eaten by man. Epizoochory is more common in mammalia, as sticky diaspores or those with spines, hooks or burs, more easily stick to the hair of a fur than they do to feathers of birds. Fruits and seeds adapted to epizoochory are extensively treated by Ulbrich (see also Roth 1977). Some tree species with fleshy fruits are particularly adapted to the dispersal by certain kinds of animals, such as *Desplatsia, Balanites* or *Panda* to elephants (Hall & Swaine 1981). Mammalia which play an important part in seed dispersal are agoutis, squirrels and bats, the latter being treated in a separate chapter, monkeys, civet cats, bears, and elephants. Ruminants and ungulates as seed dispersers are more important in Africa than in South America. In Venezuelan humid tropical forests, marsupialia of the kind of *Didelphis marsupialis* ("Rabipelado"), a nocturnal arboreal fruit-eating animal, which is very common and almost a plague in Northern South America and the Caribbean islands, may play an important part as seed disperser. Squirrels are very active seed eaters, but also dispersers. There are several other common rodents which pass for dispersal agents. The advantage of these dispersers is that arboreal mammals do not eat fruit where they gather it, but carry it off to avoid being robbed by others. The sloth, *Bradypus tridactylus*, is a vegetarian and lives on leaves and fruit. It may be a dispersal agent, although it moves very slowly. Rodents such as of the genus *Dasyprocta*, are known as fruit eaters. The nocturnal tapir is active on the ground and also eats fruit. Pecaries such as *Tagassu pecari* and *T. tajacu*, indigenous of Venezuela, are frequently found and may be important dispersal agents. Monkeys which occur in numerous species in the tropical forest, certainly, are important seed dispersers. However, the number of bird species is incomparably much richer than that of mammalia.

The syndrome of characteristics of mammaliochory is similar to that of ornithochory. However, the fruits attracting mammalia have a smell, but do not always adopt a showy colour. They may not be firmly attached to the branch, but may be dropped to the ground at maturity. The dispersal units may also be larger than those attracting birds, as mammalia are usually larger than birds. Furthermore, the seeds have to be either protected by a hard coat, as mammalia masticate well, or by bitter taste or toxic substances. Particularly diaspores attracting nocturnal mammals, which are guided by smell, have frequently inconspicuous colours, but emanate a characteristic odour. Annonaceae, for example, have typical aromatic fruits. Improvement of germination of seeds ingested by mammals or birds is reported for many plant species. Liebermann et al. (1979) studied germination of seeds of 59 plant species removed from the dung of baboons. The fruits eaten by baboons were fleshy or partly dry (up to 43%), according to the season. Germination of ingested seeds, removed from the dung, was significantly improved in comparison to that of fresh seeds in three of four species tested. The feeding of howling monkeys on *Anacardium excelsum* pedicels has already been mentioned (Glander 1979). However, Glander assumes that the seeds are not dispersed in this way. *Cebus* monkeys eat the pulp of the large fruits of *Gustavia augusta* (Oppenheimer & Lang 1969). As an integral part of their environment, these white-faced monkeys are not only dispersers of plant seeds, but also pollinators of flowers. Furthermore, they lower the number of certain insects, and change the morphology of *Gustavia augusta* trees by feeding on terminal buds. In this way, the inhibitory effect of these buds is removed so that lateral branches may increasingly develop. *Gustavia superba* shows a peak of fruit maturation from May to July, i.e. at the end of the dry season and the beginning of the wet season. This peak coincides with the peak of species with large animal-dispersed seeds (Mori & Kallunki 1976). The large animal-dispersed seeds of *Spondias mombin* and *S. radlkoferi*, however, ripen and fall in the mid to late wet season. In this way, competition for dispersers is avoided by *Spondias*. Fruits of the genus *Spondias* are yellow, orange, purple or red. Fruits of *S. mombin* are orange and sweet. They mature from July to October with a peak in August and September. The drupe is ovoid to oblong, 2,5–3 cm long, orange at maturity, has a thin exocarp and a fleshy tasty mesocarp surrounding an inedible seed. The fruits are eaten by agoutis *(Dasyprocta punctata)*. The animals eat the edible mesocarp carrying the seed and bury it as a future food source. However, the animals may not remember where they have scatterhoarded the seeds, thus being effective dispersers (Smythe 1970). *Spondias radlkoferi* is an exception within the genus in having an inconspicuous green fruit which is not sweet, but tasteless. Nevertheless it attracts enough visitors and is as abundant in individuals as *S. mombin* by avoiding the main season of fruiting, but producing fruit later on when competition for food is greater. *S. radlkoferi* fruits are eaten by monkeys (Croat 1974). The influence of dispersal agents on fruiting seasons of plant species is well known. Smythe (1970) observed that cavimorph rodents in the humid tropical forest of Barro Colorado, Panamá, show seasonal changes in their social behaviour which are correlated with the seasonality of some fruting species. The food of the rodents almost entirely consists in fruits and seeds that fall to the ground. Very conspicuous is the relation of the increase in rainfall from May to June and the increase in weight of falling fruit during this period. Then the weight of falling fruit decreases from August onward. While the maximum number of fruit species ripen in May and June, there are other species ripening throughout the year. Fruits with smaller seeds (less than 1,5 cm long) ripen outside the peak period. The animals swallow them and pass them undamaged. By staggering their fruit seasons, the plant species reduce competition for dispersal. Small-seeded fruits thus ripen more evenly throughout the year, whereas large-seeded fruits tend to be seasonal. For fruits with large seeds that are destroyed by animals eating them it is more advantageous to fruit synchronously. The agouti being unable to devour the entire synchronous crop, scatterhoards much if it, and, since it fails to find many of the buried seeds, aids in the dispersal of its food species. This explains why some fruiting species are more seasonal than others (Smythe 1970).

Janzen (1975) has pointed out that competition among trees for dispersal agents is intense and may either result in staggered reproduction of individual trees of one and the same species over a year (e.g. *Ficus*), or more commonly, in reproductive synchronization of individuals of a tree species at a certain time of the year, for example, when fruit is scarce (Janzen 1967). Selection may also favour fruit maturation just prior to the beginning of the wet season which is suspected to be optimal for germination. A third strategy may be that the individuals of one tree species are synchronized in reproduction, while other species of the same genus (e.g. *Miconia*) have staggered reproduction, but together they span the entire year (Snow 1965, Smythe 1970).

The distance that seeds may be carried from the parent tree and the seed shadow generated by animals was thoroughly studied by Janzen. Janzen (1971 b) states that the seed shadow should be most intense near the parent for wind-dispersed seeds, for example, and for some mammal-dispersed seeds, but scatterhoarding agoutis may generate asymmetrical heterogeneity by burying seeds near trees or logs, and African duiker (*Sylvicapra*) may spit seeds into large piles while chewing their cud far from the host plant. Seeds passing through vertebrate intestines may produce a bewildering array of seed shadow morphology. An elephant with 1000 date pals seeds in its stomach is likely to generate a very patchy seed shadow. The baboons practice a kind of "inadvertent gardening" which may increase the concentration of preferred food plants within the core area of their territory (Liebermann et al. 1979). Distinction of seed shadows is largely discussed by Janzen (Janzen 1971 b, Janzen 1978 a and b, Janzen et al. 1976).

Some seeds have improved germination after passage through a vertebrate gut, either of birds or of mammals. Janzen (1971 b) suggests that these seeds have probably a retarded germination, if they do not pass through a vertebrate gut. Seeds ingested by mammalia have often a chemical or mechanical protection developed. Janzen mentions a great variety of them (see chapter on Predation).

A certain zonation concerning formation of fleshy fruits and the zones of mammalian activity may be distinguished in the forest. In a tropical rain forest of Ghana, Hall and Swaine (1981) observed that about 70% of tall trees and climbers had fleshy fruits. The proportion of species with fleshy fruits is higher among understory trees (with 80%) and lower among herbs (about 40%). Monkeys usually feed in the tree tops. Black *Colobus* monkeys *(C. satanus)* spend most of their time in the middle and upper levels of the forest, heavily feeding on seeds. In some months seeds dominate their diet by 80%, according to McKey (1978). Other arboreal mammals are more confined to the middle layer ranging up and down the tree, while ground mammals depend on dropped fruits and seeds on the ground (see also Whitmore 1975 and p. 159–162)*.

* See note added in proof, page 48.

Dispersal by bats (chiropterochory)

Dispersal of fruits and seeds by bats is mainly confined to the tropics and subtropics. As bats have a limited vision and are colourblind, fruits and seeds that are visited by them must be well exposed. Bat-fruits ripen on the tree and are "held away" from the foliage. Caulicarpy is often connected with chiropterochory (Fig. 66). Flagellicarpy and flagellispermy are other methods to expose fruits or seeds (Van der Pijl). An example of flagellicarpy is the mango (Fig. 1), while flagellispermy is realized in *Swartzia schomburgkii* (Fig. 65). Bats are nocturnal and a contrasting white or yellow may help them to find fruits and seeds. Otherwise, the diaspores may have a brownish or blue colour. However, bats have a good sense of smell prefering an odour of fermentation (musty or rancid: butyric acid). Juicy fruits are chewed and the seeds usually regurgitated near the host tree or the diaspores are transported to a suitable place, sometimes to the roosting place at a distance rarely exceeding 200 meters, as Van der Pijl emphasizes. *Pteropus* can carry fruits weighing as much as 200 gr (Whitmore 1975). The Chrysobalanaceae, Annonaceae, Sapotaceae, Anacardiaceae, and some caulicarpous Leguminosae are known to be bat-dispered families. *Andira inermis* has received its vernacular name from bats *(Andira* means "bat" in a Brazilian Indian language). Of the genus *Swartzia* with flagellispermous seeds only *S. prouacensis* is bat-dispersed. according to Van der Pijl, while other species of *Swartzia* with the same syndrome of characteristics are ornithochorous. The large dark-brown seeds have a contrasting white aril and are exposed by a long funicle, e.g. in *Swartzia schomburgkii* (Fig. 65). Van der Pijl also mentions some species of *Lecythis* which are chiropterochorous. Fox-bats *(Pteropus)* are specialized in hard-fleshed green drupes such as of Sapotaceae, *Calophyllum, Eugenia, Terminalia catappa,* according to Ridley (1930). *Annona (Rollinia) mucosa* fruits are eaten unripe by bats. Concerning the stratification of the forest, fruitbats mostly visit the upper canopy region, while insectivorous bats are more confined to the middle zones of the forest (Harrison 1962). Vazquez-

Yanes (1975) distinguishes between plant species with fruits available throughout the whole year such as *Cecropia obtusifolia* and *Ficus* spp., on the one hand, and species with fruit abundant only during a short period, such as *Spondias mombin* and *Brosimum alicastrum*, the bread-nut, on the other. All those fruits are greatly utilized by bats, which have preference for certain fruits. Bats disperse seeds over a large area. They visit secondary as well as primary vegetation playing an important role in the introduction of secondary-growth vegetation seeds to the soil of primary forest and vice versa. *Cecropia obtusifolia* is the secondary growth plant with the best representation in bat detritus and also is a conspicuous species in all the disturbed areas studied by Vazquez-Yanes. However, we have to add here that *Cecropia* is a particularly resistant plant which endures drought and insolation, being a pioneer plant, e.g. growing along roads and in very exposed areas. This species is a good example of the universal perception that not only dispersal is responsible for plant distribution, but that the resistance of a species is probably of much more importance.

Janzen et al. (1976) studied seed shadows of *Andira inermis* and their correlation with seed predation. The ripe fruits of *Andira inermis* are green and indehiscent, single-seeded and 3–5 cm long. They are borne on stems that project outside of the *Andira* canopy surface. The fruit epidermis is glabrous, the exocarp is tough, and lacks a noticable flavour and odour. The pulp is juicy and white. Fruits are mainly eaten by bats, but also parrots visit them, and on the ground, small rodents, leaf-cutting ants or small insects feed on them. *Artibeus* bats carry fruits of many species out of the parent tree canopy and consume them at a near-by feeding roost. As a roosting tree *Guazuma ulmifolia* is frequently used. Janzen et al. suggest that there should be selection for tree traits that cause vertebrate dispersal agents to leave the host (parent) tree for highly desirable feeding roost trees in the vicinity (tree and branch morphology, odour?). At the end of the ripening season of *Andira*, *Spondias purperea* and *Anacardium excelsum* fruits are maturing, which are both taken by *Artibeus* bats. Seed predation by *Cleogonus* weevils is highest below the parent tree of *Andira inermis*, intermediate under the bat feeding roosts and lowest among accidentally dropped fruits. The seed shadows in this case are generated by the dispersal agents and predation on seeds is influenced by the heterogeneity of the seed shadow. Janzen et al. could thus show that there is an interaction between tropical seed predators and tree species richness. Heithaus et al. (1975) stated a correlation between size of bat and average size of fruit eaten: small bat species feed on resources of high abundance which are often small-fruited, while large species utilize resources that are patchy in time and space and frequently have larger fruits. Reproductive behaviour of bats is seasonal and related to resource abundance. A combination of the timing of plant-reproductive behaviour and bat foraging characteristics promotes long-distance outcrossing and seed dispersal. These are critical factors in determining the low-density populations characteristic of many tropical plant species. Longer fruiting periods facilitate increasing bat specialization on fruits. Since tropical tree species are often separated by relatively long distances, there is a need for long-distance pollinators, such as bats, which visit several plant species a night. Seed dispersal is another regulator of outcrossing in plant species. The mobility of bats suggests that dispersal distances of seeds exceed 1 km or more, even when seeds are retained only for a short time. Different bat species carry seeds for varying large distances. Selection for different dispersal distances of plants is a factor in the evolution of differential attractiveness of fruits to fruit-eating bats, according to the authors. Fleming and Heithaus (1981) observed that small-seeded plant species are disseminated by birds and bats, while large-seeded species are dispersed by terrestrial or arboreal mammals. The distance at which seeds will be deposited depends on the time spent near the fruiting plant, travel speeds, distances moved and the time that seeds are retained. Germination percentages are significantly increased in bat-passed seeds, e.g. in *Cecropia peltata*. Otherwise they suffer a "retarded germination". The distance to which seeds are dispersed away from the parent tree is not only important for the probability of establishment, but also for the escape from predators, herbivores, and pathogens (Janzen 1970, Smith 1975).

Dispersal by ants (myrmecochory)

Myrmecochory is considered a relatively recent dispersal method adapted to ants by the formation of so-called "elaiosomes" which may arise on seeds, fruits, or even flowers (sepals, petals, base of the style, receptacle, flower axis), on the utricle *(Carex)* or on the base of spikelets. Ridley (1930) mentions *Cassia occidentalis* and *Carica papaya* as dispersed by ants in the tropics. Van der Pijl (1969) emphasizes that the part myrmecochorous plants play in the tropics is strangely small when compared with European plant species. This statement is really astonishing when one considers the amount of ant species existing in the tropics as well as the important role certain ant species play in the tropical forest. This casuality may eventually be connected with the adaptive position of ant fruits near the ground. More herbaceous species are related with myrmecochory than tree species, but herbs are in the minority in tropical forests. Furthermore, the abundance of myrmecochorous plants in temperate regions as compared with the tropics may be the expression of the relatively recent evolutionary development of myrmecochory. Secondly, I doubt very much that most of the myrmecochorous plant species in tropical forests are already known. Although adaptation to ants is relatively recent in the flora (preglacial in temperate regions), there may be many cases of myrmecochory still unknown to us in the tropics. However, the action of ants on plants may often be more destructive (termites). Bresinsky (1963) found that unsaturated free fatty acids are the attractants in elaiosomes (possibly with a volatile component: Van der Pijl 1972). Van der Pijl points to the fact that the herbaceous spring flora in northern forests gives rise to myrmecochorous plants. In the tropics myrmecochores grow outside the dense forest or are epiphytes ("ant gardens" of *Peperomia*, Bromeliaceae). Uphof (1942) correctly points out that ants play a far more important role in the tropics than in temperate regions. Within the myrmecophytes, plants inhabited by ants, Uphof distinguishes 1) myrmecotrophic types which provide food to ants in the form of exudations (nectaries), certain food bodies, i.e. bromatia or elaiosomes on seeds and fruits. 2) Myrmecodomic plants which give shelter to ant nests in the form of hollow cavities or in myrmecodomatia (Melastomaceae: Roth 1976) and 3) myrmecoxenic plants which are true hosts offering shelter as well as food to the ants. As a reward for giving shelter to the ants these offer protection and defense to the plants, and may even act as pollinators or as disseminators of seeds. Mainly the harvester ants, genera such as *Messer* and *Pogonomyrmex*, contribute to the dissemination of plants by collection of seeds in dry areas. They are ground-nesting species and control the germination of the seeds they collect. The tendency of ants to prefer grasses and Compositae for seed storage is obvious (Hocking in Gilbert & Raven 1975). Seeds and fruits are carried to the ant nests for different purposes: to eat off the plumule and radicle after germination, to use them for construction of their nests or to use the food source itself. *Aphaenogaster* ants are able to carry as many as 216 seeds in two hours to their nests. Myrmecophytes in the Amazon region grow in habitats which are periodically inundated offering protection to ants. Berg (1966) mentions two species of *Dendromecon*, Euphorbiaceae, which are diplochorous: The seeds with an elaiosome at the micropyle are first dispersed by an autochorous mechanism through violently dehiscing capsules and then transported further by ants. According to Berg, the autochorous-myrmecochorous dispersal type may partly explain the polymorphism of *Dendromecon rigida*, since it tends to keep local populations isolated. A similar dispersal mechanism may be characteristic of many tropical Euphorbiaceae – however, "dispersal biology ... is an underdeveloped ... sector of botany" (Berg 1966). According to Berg (1975), the presence of complex explosive mechanisms in most Euphorbiaceae indicates that the family is monophyletic and primitively capsular with a sclerified endocarp and septicidal dehiscence. In *Micranthemum* increasing desiccation of the capsule results in a sudden explosive dehiscence. The capsule splits from its base upwards into three cocci. Simultaneously each coccus splits lengthwise from the apex downwards along both dorsal and ventral suture. The explosive force is due to a sudden dilatation of the upper part of each coccus, since the apical splitting of a coccus is followed by a violent moving apart and twinsting in opposite direction of the two halves. At the same time, each coccus shoots out the seeds. To make the shot, the coccus halves remain connected at the base, forming a sort of adaxial scoop or tail. The twisting movement makes this tail come upwards with a sudden jerk, while - concurrently - the lower portion of the coccus narrows and the adaxial upper portion of the coccus opens. The tight-fitting seeds are squeezed and pushed through the opening. The detached and dehisced cocci fly off too, often splitting completely into halves at the end of the explosion. The lower part of the central column remains on the plant but without epidermis, since the epidermis partakes in the formation of the thin ventral wall of the cocci. The upper part of the central column is entirely incorporated into the cocci, and splits into 6 during dehiscence. On their dorsal side, the cocci have an outer dry crust of dead tissue derived from the outer epidermis and the chlorenchyma of the ovary wall. The tensions that result in explosion are produced by the sclerified inner epidermis and by sclerenchyma that develops from the meristematic layer surrounding each locule, differentiating into two parts: inner and outer. In the outer part, sclereids pass more or less at right angles to the long axis of the epidermal sclereids (diagonal crossing of elements). Desiccation causes the sclereids of the crossing layers to contract in different directions - setting up the tendency for each coccus half to twist in opposite directions. In the ventral and dorsal sutures sclerenchyma cells are smaller and ruptures finally occur in this zone of least resistance. The main weakness zones result from the spongy parenchyma in the septs. Other weakness zones occur besides. Numerous structures are delicately proportioned and finaly balanced to produce the explosive effect. Seed abscission probably occurs before dehiscence. The caruncle originates from the outer integument. The inner epidermis of this integument is converted into a caruncle meristem. Fatty oils are stored in the caruncle which corresponds to an elaiosome. The myrmecochorous dissemination is thus preceded by autochorous dispersal which is caused by the explosion mechanism of the capsule, and diplochory is realized in this way.

All species of the Euphorbiaceae have either one or two ovules per locule - correlated with the shooting mechanism, as the maximum number of seeds that can be successfully squeezed out from the coccus is two. A seed-shooting mechanism variously modified occurs in most Euphorbiaceae. This dispersal mechanism is relatively primitive, according to Berg, and berries and non-explosive capsules are derived. The elaiosome developed from a seed disjunctor and consequent myrmecochory is a later supplement to the originally present autochorous dispersal. Berg postulates that the Euphorbiaceae must have evolved in a seasonal dry region, because they have xerochastically opening capsules. This point will be discussed later (see p. 73, 141).

Protection against predation

Fruit and seed eating animals not only function as dispersers, they also destroy a great deal of the crop when masticating and crushing the seeds. Another part of the crop may be carried to sites where the seeds are not able to germinate. However, although the plant attracts animals as dispersers by fleshy edible fruits, infrutescences or edible parts on seeds such as sarcotesta, aril, elaisome and others, it is not in the interest of the plant that important parts of the seed such as the embryo itself or the endo - and perisperm are destroyed. For this reason the plant has "invented" a series of means to protect reproductive parts from being eaten. The simplest strategy is a superabundant crop which satiates the predators and leaves part of it for reproduction. In many cases, however, the plant protects the seed itself against predation as a consequence of longterm adaptation and coevolution together with predating and dispersing animals. The extensive studies of Janzen on this subject have stuffed this chapter.

Protection of plants against predators is mainly threefold: mechanically, chemically or by "escape in space and time", as Janzen emphasizes. Mechanical protection is probably the most ancient form, observed already in spores and long-lived cysts of bacteria, algae, bryophytes and ferns. Waxy layers, a thick cuticle, thickened cellulose walls, possibly with some impregnation such as lignin or silicia are a good protection. Some seeds of Angiosperms have evolved very complicate structures in the outer seed coat or in the endocarp to protect the embryo and endosperm from predation, water loss, and strong insolation, such as palisade layers, sclereids, mucilage cells and layers, light lines, melanine cells and many other devices (see also Roth 1977). Hairs and spines, especially when hard and stiff, are a means of protection against predation. Persistent stiff sepals protect the fruits and seeds mechanically. Netolitzky (1926) and Corner (1976) show the whole array of structures in the seed coat of most Angiosperm families including those useful for protection. Within the mechanical devices we may also include the seed size. Some seeds escape from predation being too small to be crushed so that they are easily swallowed without being damaged, but others are too large to be eaten so that they remain untouched.

Chemical protection is often very much refined, as Janzen and collaborators could show, and probably is more evolved than mechanical protection. Coevolution of plant and animal species here becomes very obvious. Tannin, either deposited in the cell content or in the cell wall, is a very common substance in plant species of humid tropical forests and often abundantly found in seeds or fruits. It is one of the important protective substances not only against rotting (Roth 1977 a), but also against predators. Resinous compounds are another kind of chemical substances frequently occurring in tropical plants and especially characteristic of certain plant families, such as the Burseraceae, Guttiferae, Flacourtiaceae, Papilionaceae, and many others (see also Roth 1980). They deter certain animals, such as squirrels and mice. Even larger animals, e.g. mammals, may dislike fruits and seeds with a large amount of resinous substances. It is an old gardening practice in Europe to scare away mice from planted onions or bulbs of ornamental plants, such as tulips, narcisses or dahlias, by putting twigs of *Thuja* in between them. Eteric oils and other oily compounds, possibly with a penetrating smell may particularly discourage certain mammels. One of these substances is the cardol oil in *Anacardium*. Latex, very common in many tropical trees, e.g. in Euphorbiaceae and Sapotaceae, is another means of protection. Mc Diarmid et al. (1977) mention the white latex in the seed of *Stemmadenia*, Sterculiaceae, as a deterring substance. Janzen and collaborators have detected a series of toxic substances which prevent animals from predating the seeds. Fish poisons abundantly used by Indians to catch fish fall into this category. The bitter taste of certain seeds and fruits likewise deters animals (e.g. in *Piranhea trifoliata*, according to Gottsberger 1978). Even a green colour "advertising" immaturity may protect the fruits from predation.

Finally, we have to consider the escape of seeds from predation "in space and time". One tool of spatial escape is the constriction of the pericarp between the seeds that some fruits undergo (e.g. pods of Leguminosae). When the fruits are opened by animals some compartments with seeds may remain undamaged and escape predation. The most important escape from predation in space is, however, realized by the distance to which the seeds are dispersed. Most seeds that fall close to the parent tree become victims of predators. The further the diaspores are removed from the parent, the greater the chance for them to escape from predation (Janzen 1972). Escape of diaspores from predation in time may either take place by gradual discharge of the seeds from the fruit or possibly by staggered fruiting or by evolution of particular fruiting seasons. Regular skipping of fruiting years is an escape strategy used by many plant species. The life history of the predator selects for the behaviour of the tree. Production of large seed crops which completely satiate the predator is another strategy. However, large seed crops in one year are usually followed by one to several years with a small crop or no seed crop at all. Most seed predators have only one species of host plant and host specifity is produced by the chemical traits of the seed that is being eaten (Janzen 1976). In this way, a complete circle is closed. By skipping of fruiting years the host specific predators are kept in check.

Mechanical protection

Mechanical protection of seeds against predation is most efficiently carried out by a hard seed coat or a woody endocarp. Examples of this kind are immense in number and the inner structure of the seed shell and the endocarp is most variable (Netolitzky 1926, Corner 1976, Roth 1977). The endocarp of drupes is usually furnished with thick-walled sclereids or fibers, the latter characteristically interwoven with one another and crossing each other in different directions, the cell walls being hardened with lignin. The hard seed coat may be supplied with a special outer layer of palisade cells or sclereids of variable shape, the cell walls are thickened and a thick cuticle and/or waxy layer may protect the outer epidermis. There may be incrustations with crystals of ca-oxalate or silica or the seed surface may be protected by stiff hairs or hard spines.

Experiments of Janzen (1977) showed that the seed coat of a number of species is a hard barrier sufficient to exclude weevil larvae, for example. Cowpea weevil larvae could not penetrate the seed coat of *Parkinsonia aculeata*, and they penetrate the seed coat of *Hymenaea courbaril* only to a depth of 0,1–0,3 mm. They penetrated, however, the seed coat of *Cassia grandis, Enterolobium cyclocarpum*, and *Hura crepitans*. The larvae died in seeds of *Pterocarpus rohrii* and *Swietenia panamensis*. These results show that the bruchids can attack some of the seeds of non-host species. But on general terms, hardness and chemistry of the seed coats may be a barrier to the entry by larvae.

An interesting study of Janzen and Higgings (1979) draws attention to the fact that absolute protection of seeds from predation may even be disadvantageous for the plant. The authors noticed a certain balance between seeds too hard to be eaten and those having a shell that may be crushed by frugivores in *Entereolobium cyclocarpum*. They found that the force required to break the seed shell increases with increased seed weight. Pecaries are important seed dispersers; in eating the fruits they crush some seeds and spit out others, while some seeds are swallowed entirely. In answering the question: "How hard are *Enterelobium cyclocarpum* seeds?", Janzen and Higgins found an equilibrium in the hardness of the seed shell, some seeds being hard enough to resist cracking – and these are used for dispersal –, while others having a shell not hard enough may be eaten by animals. The partial edibility of seeds is considered a further attractant for the dispersal agents. Otherwise, the animals would possibly eat only the fruit pulp and leave the seeds on the ground without dispersing them.

Sizes and shapes of seeds are largely discussed by Harper, Lovell and Moore (1970). According to these authors, seed size imposes a compromise with seed number. Very large seeds are usually reduced in number, while small seeds may occur in very large numbers on the tree. A good example in this respect is presented by the comparison of *Tetragastris panamensis* seeds with seeds of *Virola surinamensis* (Howe 1983). It has been emphasized already that very small seeds are swallowed entirely by animals, while very large seeds cannot be masticated by them. Seed shape, on the other hand, represents a compromise between the form most efficient for "packing", dispersal, landing, and seedling establishment. Colonizing species make a high reproduction effort producing a large number of seeds, while woody species living in a stable habitat, such as the tropical rain forest, show low values of reproductive effort. A greater part of their available energy resource is devoted to persistent vegetative organs, according to Harper, Lovell and Moore. This confers advantages in a long-term intense struggle for existence in a crowded resource-limited stable environment. European woody species usually have smaller seeds, while tropical woody species are known for their relatively large seed size. Seed and fruit weight in general very much oscillates considering the seeds of certain orchids weighing only 0.00002 gr and the 45 cm long twin fruit of *Lodoicea maldivica*, a palm, with 27.000 gr. As mega (or macro-) spermous families the Bombacaceae, Guttiferae, Lecythidaceae, Sapotaceae, and others are cited. In the Leguminosae, on the other hand, a great variation in seed size may be observed. However, within many species seed size is very stable, otherwise seeds would not have been used formerly as units of weight by man.

Chemical protection

Chemical compounds in the plants with a possible toxic effect on animals are found in vast numbers. Many of these compounds are only known by the effect they exercise on humans or animals, but their chemical structure is not yet investigated. Accordingly, they are classified as blood poisons, neurotic poisons or narcotics, muscular poisons or neuro-muscular poisons, irritants (e.g. of the skin) and the like. Many of the toxic substances correspond to alkaloids, glycosides such as cyanogenic glycosides, terpenoids such as saponins and cardiac glycosides, hydrocarbons, acids such as oxalic acid, amino acids (L-DOPA), phenolic compounds such as flavones, tannins, and quinones. In the evolution of feeding deterrents Harbone (1977) distinguishes the terpenoid, the phenolic, and the nitrogen "pathway". However, a secondary compound may be toxic for one organism, while it is not poisonous to another, because "one beast's drink is another beast's poison" (Janzen 1979). Furthermore, many toxic compounds act dosage-dependent. According to a rough estimation, about 69% of the European herbaceous genera produce toxic fruits, while trees have no poisonous fruits (Herrera 1982). Certain secondary compounds are characteristic of certain plant families or genera, e.g. mustard oils of Cruciferae and Capparaceae, steroidal alkaloids of *Solanum*, non-protein amino acids of Leguminosae, cardiac glycosides of Apocynaceae, cyclopentenoid fatty acids and cyanogenic compounds with a cyclopentenoid ring system (gynocardin) of Flacourtiaceae, amygladin of Rosaceae, hydroxy fatty acids with laxative effects of Euphorbiaceae, phytohaemagglutinins (lectins) of Leguminosea (robin, abrin), L-dihydroxyphenylalanin (L-DOPA) of *Vicia faba*, species of *Mucuna*, and other Leguminosae, dihydrobenzols of Anacardiaceae, limarin of *Manihot esculenta* (the roots are therefore only edible when cooked) and of other Euphorbiaceae, myristicin in *Myristica* – just to mention a few examples (see also Harborne 1977, Frohne & Jensen 1979). Frohne & Jensen (1979) give a good survey of the most important plant families with their most characteristic chemical compounds. Many families are characterized by resins, e.g. the Guttiferae, Rutaceae, Papilionaceae, Burseraceae, and many others, or by latex, e.g. the Sapotaceae, Euphorbiaceae, Apocynaceae. The latex of *Hura crepitans* yields the poisons hurin and creptitin, which latter resembles ricin and abrin in its chemical structure. The plant is used as a fish poison in Venezuela. The caustic sap may produce skin irritations and even cause blindness. However, the resin as well as the latex represent mixtures of several chemical compounds usually not known in detail, particularly when tropical plants are concerned. Considering the immense number of tropical plant species, we know very little of their chemical composition. However, we have some information of the physiological effects of secondary compounds upon man and animals. Toxic effects observed on man and animals are mainly caustic irritations of skin, stomach, and intestine (e.g. by eteric oils or latex), nausea, vomiting, diarrhea, cathartic effects (ricinoleic acid), cardiac intoxications, narcotic and paralysing effects (fish poisons e.g. acetylenic compounds), agglutination of erythrocytes (lectins), inhibition of protein synthesis, disturbance of muscle action (cardenolides), inhibition of enzyme synthesis, loss of hair (mimosine), loss of weight (amino acids), inhibition of cell respiration, growth inhibition, delay of reproduction, delay of sexual maturity, retardation of differentiation and development (interference with the metamorphosis of insects), shortening of life span, physiological disturbances and loss of fitness in general, and even carcinogenic longterm effects. Ingested in small amounts, many of these toxic compounds may have curative effects acting as purgatives, diuretica, antipyretics, helmintics or cardiacs. An overwhelming majority of tropical plants species are successfully used as medicinal plants in popular medicine by natives (Pittier 1926, 1939; Roth 1981 for barks). The effects of secondary compounds are dosage-dependent. Alkaloids are generally the most toxic compounds. Non-protein amino acids are more toxic than protein amino acids. Secondary compounds found in seeds are likely to be toxic to at least some animal, and likely to be responsible at least in part for the extreme host-specificity shown by seed eating insects (Janzen, Juster, & Bell 1977). Rhoades (1979) distinguishes quantitative and qualitative defense strategies of plants. Quantitative means of defense such as tannins, resins, or silica act in dosage-dependent fashion so that the degree of protection against herbivores directly depends on the concentration of the secondary compounds in the respective tissues. Consequently, the quantitative defense substances are often present in high concentrations, up to 60% dry weight in the case of tannins. Qualitative agents of defense such as cardenolides, glucosinolates, alkaloids, cyanogenic compounds and the like occur in low concentrations (<2% dry weight), but render little protection from specialized herbivores, even in high concentrations. Secondary compounds in high concentrations which contain N may exercise a dual defensive and nitrogen-storage effect, particularly when occurring in seeds. For example, lectins are antiherbivore compounds when present in high concentrations in dormant seeds and are nutrient-storage proteins at the same time (Janzen 1979). Characteristically, secondary compounds are not evenly and uniformely distributed throughout the plant body, but appear accumulated in certain organs or parts of them, e.g. in the leaves (possibly as a protection against herbivory) or in the roots, while many other secondary compounds are enriched in the bark or in fruits and seeds, and especially in the seed coat. Rhoades (1979) emphasizes that tannins, resins, alkaloids, and other secondary compounds are often located near the plant surface (e.g. in the epidermis or in subepidermal layers, and even in trichomes) a fact pointing to their protective function. Particularly the seed coat may contain a higher concentration of secondary compounds than the content of the seed (Janzen 1979). Toxic secondary compounds preferably occuring in the flesh of ripe fruits are alkaloids (e.g. in Solanaceae, Menispermaceae, Celastraceae), saponins, cucurbitacins, non-protein amino acids, steroids, terpenoids. The fleshy portions of the fruits of *Sapindus saponaria* contain saponin. Resins in the pod of *Hymenaea* deter bruchids. Resins and volatile oils are characteristic of Anacardiaceae, e.g. of the fruits of *Anacardium* in the form of anacardol which is an irritant with blistering effect and is used as an insect repellent. Volatile oils also occur in reproductive and vegetative organs of Lauraceae. Phenolics such as cinnamic acid derivatives, flavonoids, condensed polyphenols, are by far the most common secondary compounds in ripe fruits. Chlorogenic acid and its isomeres are widespread in fruits, and anthocyanins are prominent compounds of the phenolic fraction of fruits. Flavonol glycosides and condensed polyphenols are normal constituents of ripe fruits. Tannins are very widespread compounds, especially in tropical wild ripe fruits and astringency is a typical feature of immature fruits of many wild species (Herrera 1982). Where space is a limiting factor, as occuring in seeds, selection has favoured the use of toxic compounds that are simultaneously storage products, according to Janzen (1971 b). A seed toxic to most animals may have specialists that can feed on it. Janzen suggests that evolution may take place here at the level of the microflora in the seed predator's intestine.

Aronow & Kerdel-Vegas (1965) studied the action of selenocystathionine, a pharmacologically active secondary compound in the seeds of *Lecythis ollaria* which produces a toxic syndrome in man in the form of nausea, vomiting, diarrhea immediately after and loss of scalp and body hair 1–2 weeks after ingestion. Fortunately, alopecia is not permanent. Hair loss after selenium intoxication is reported among certain Indians. Many plants are able to absorb and store high concentrations of selenium without damage, e.g. certain Leguminosae. Ingestion of these plants adapted to selenium may also cause the death of cattle and sheep (Rosenfeld & Beath 1964, cited in Harborne 1977). The seeds of *Hura crepitans* are toxic to man when eaten, an experiment which was carried out by my students in Caracas in my absence. Some of the studients had to be transported to the hospital, but later recovered. The seeds contain a toxalbumin and in small amounts (2–3) have found use as a purgative in popular medicine. The seeds have been employed to poison coyotes and other animals. On the contrary, they are the favorite food of red and blue macaws in Costa Rica. The sap, the seeds, and a decoction of the bark have all emetocathartic properties. The latex has also entered into the composition of an arrow poison. Abortion may occur when pregnant cows feed on this plant (Blohm 1962 and the literature cited there). Narcotics used as fish poisons are manifold in tropical plants, one of which is *Lonchocarpus sericeus*. The seeds of *Sapindus saponaria* are poisonous and are used as a fish poison by Venezuelan Indians. They likewise serve as an insecticide when pulverized.

Agglutination of erythrocytes, induction of mitosis, and inhibition of amylases by seed protein extracts of 15 Malesian species of Leguminosae is reported by Lee (1979). Lee assumes that proteins acting as defense mechanisms against predation are particularly marked in rain forest plants. There is a higher incidence of lectin activity in legume seeds from tropical environments (e.g. in species of *Parkia* and *Pithecellobium*), according to Janzen et al. (1976). Highest concentrations of lectins are found in seeds, e.g. of *Ricinus communis*. Ricin is also a protein, but has no agglutination properties. The toxic effect of ricin is due to an inhibition of protein synthesis. The agglutinin of *Ricinus*, on the other hand, is not toxic. Lectins contribute to the mortality of bruchid beetle larvae feeding on black beans. Toxic elements play a very important part in the protection of seeds and fruits against insectes. Phytohaemagglutinins in black beans and other legume seeds have adaptative significance in protecting the seeds from attacks by insect predators, such as bruchid beetles (Janzen, Juster, & Liener 1976). As Janzen (1969) emphasizes, alkaloids and amino acids which occur in tremendous numbers in plants prevent most or all bruchid attacks. Janzen mentions in this connection *Enterelobium cyclocarpum*. *Pentaclethra macroloba* seeds rich in alkaloids cause feeding depressions and weight loss in laboratory rats. Nevertheless, *Tylomys* feeds on them (Janzen 1974). Pyrrolizidine alkaloids have a toxic effect on kidneys and on the central nervous system of vertebrates. The alkaloid andirine (N-methyl-tyrosine) is present in the seed of *Andira inermis*. Its effect is purgative, vermifuge, vomitive, and febrifuge. For centuries, the seeds have been used for these purposes by the natives in tropical South America. High concentrations of the amino acid L-DOPA (6–9%) occur in the seeds of 5 species of *Mucuna* which are derived from a common ancestral form. The seeds show a relative immunity against insect and possibly small mammal attacks. Canavanine in *Canavalia* seeds has the same action (Bell & Janzen 1971). Diterpenes in *Croton* seed oil and seeds of other Euphorbiaceae contains carcinogenic hydrocarbons. Isoflavonoids exercise an effect on mammalian reproduction. They develop estrogenic activity, "mimicking" in this way the female hormone. As has already been mentioned, tannins are very common compounds in tropical plants (see also Roth 1981). They are certainly a good protection against rotting. The astringency of tannins is repellent to higher animals and insects. Germination of fungal spores is inhibited by phenolic compounds (Howe 1953). They also change the nutritional value of the protein content in fruits and seeds (Harborne 1977). Howe and Van de Kerckhove (1981) report large amounts of soluble tannins in capsule and seed of *Virola surinamensis*. The phenolics protect fruits and seeds against insect attacks, probably inhibiting their growth, while the edible aril contains much smaller amounts of tannin. According to Rhoades (1979), approximately 80% of all woody perennial dicotyledons contain tannins, while this substance is less common in herbs. Without doubt, tannins reduce the digestibility of plant tissues. They act as antiherbivore substances by complexing with (plant)proteins and carbohydrates and equally with herbivore digestive enzymes, disrupting digestion of the herbivore. Tannins are thus not only a protection against herbivores, but also against fungal and bacterial attacks, and even exercise an effect on virus. In present-day angiosperms and gymnosperms, the highest concentration of tannin in non-living tissue of trees, such as heartwood and bark, amounts up to 40% of the dry weight. But already concentrations higher than 2% dry weight deter mammalian herbivores by their astringency and prevent insect predation. Tannins are the most ancient of successful plant chemical defensive compounds still being used by present-day apparent plants to survive, indicating thus their prime importance in ecosystems (Swaine 1979). The high dosage of tannins necessary to defend the plant against predators as well as the circumstance that tannins are the most ancient secondary compounds used as a protection of higher plants suggests the speculation that herbivores, fruit and seed eaters as well as pathogens, have become used to this plant defense in gradually rising concentrations during evolution.

Lignin has not only a supporting and strengthening function in plant tissue, but also serves as a plant defensive substance. Lignin is relatively indigestible to herbivores and microorganisms, but some fungi are able to digest it. Lignin reduces the digestibility of both cellulose and protein, in plant tissue. Possibly, lignin reduces the digestibility of plant carbohydrates and proteins by hydrogen-bonded complex formation with these substances in an analogous way to that of tannin (Rhoades 1979). Lignins occur as non-crystalline inclusions up to 40% in woody cell walls. After cellulose, lignins are the most abundant of all natural polymers. Tannins and lignins are the most important defensive compounds of plants because of their almost universal distribution, but also because their evolutionary history is longer than that of any other secondary compound (Swain 1979). Differently act non-nutritive, highly sweet proteins and terpene glycosides which occur in a variety of tropical plants. These substances mimic sugars to attract dispersers (Rhoades 1979).

Very interesting secondary compounds are hormonal mimics in plants which may cause disturbances in insect development after ingestion, interfering with insect metamorphosis. Insect hormones and certain of their pharmacobiological mimics occur in plants as well as certain steroid hormones of vertebrates (estrogens and their mimics). They affect growth and differentiation of reproductive organs. Antihormones in plants may retard or inhibit development and reproduction of insect herbivores (Sláma 1979). Certain tropical mammal-dispersed fruits are a major source of antibiotics of different kinds, according to Janzen (1978). Ripe fruits of *Andira inermis* in Costa Rica are very resistant to mold and bacterial growth. The importance of toxic secondary substances in plants is indirectly indicated by the intricate detoxification mechanisms evolved by animals (Rhoades 1979). But as mentioned above, not all toxic substances have the same effect on distinct animal species. Particularly higher animals have different detoxification systems and are able to adapt to the toxic substances of their host plants. In this way "specialists" originate.

Specialization of certain animal species in distinct plant species which contain toxic elements is explained by gradual evolutionary processes. Many animals developed detoxification systems against secondary compounds by microsomal enzymes in liver and kidney. In other animals, detoxification takes place in the gut (Freeland & Janzen 1974). Janzen (1980) studied the specifity of seed-attacking beetles in a Costa Rican deciduous forest. 75% of the beetle species were specific to a particular plant species. Seed predators are thus strikingly prey-specific, as each plant produces its own set of secondary compounds. The specifity of predator and prey becomes obvious in coevolution. Specialized herbivores with narrow host ranges are often much more resistant to the effects of defensive substances in their host plants than are generalists or herbivores that do not normally eat the plant in question. Specialists display potent detoxification and tolerance mechanisms for plant defense substances. Deterrent effects also depend on the metabolisms of the animals in question. Alkaloids may be more deterrent to mammals than to reptiles, birds or insects. Herbivores, consequently, prefer to eat familiar food. They generally feed on plants that contain small amounts of secondary compounds. But natural selection can increase the efficiency of degrading the toxic nature of particular secondary compounds (Freeland & Janzen 1974).

Clegg, Conn, & Janzen (1979) finally ask the question about the fate of the secondary compounds present in seeds when germination starts. The answer is that in wild lima beans the linamarin, a cyanogenic glycoside, is transferred intact from the cotyledons to the growing seedling, and particularly to the first leaves. Not only fruits and seeds with toxic secondary compounds, but also nutritionally unbalanced ones and those with a low calory reward are defending themselves in this way against predators. All these chemical strategies probably serve indirectly to increase the potential of plant coexistence through their diversifying effects on diet selection by individual frugivores (Herrera 1982).

The function of plant secondary compounds is thus manifold. Possibly, the same defensive compounds are used and effective against plant competitors, herbivores, and pathogens. Additionally, some of them may even function as storage substances. Others have strengthening effects on plant tissues or make fruits and seeds more attractive by their colour.

Nutritional aspects

Nutritional aspects are equally of influence in the attraction of dispersers and in the protection of seeds and fruits. It is well known that the actions of frugivores are governed by nutritional need and reward. The same toxic concentrations which make a leaf unprofitable for herbivores may be tolerated in seeds because of the higher concentrations of easily processed nutrients. Some plant species produce small numbers of highly nutritious fruits and seeds which are visited by obligatory frugivores. By acquiring reliable dispersers the plant minimizes the waste from fruit thieves. In a low investment model, on the contrary, trees produce abundant fruit of low quality whose seeds are dispersed by a large variety of opportunistic birds and mammals. Lack of competition promotes diverse assemblages and dissemination to a variety of habitat types, but also considerable waste by inefficient foragers, attraction of fruit thieves and predators that damage seeds. Howe (1979) suggests that these two models represent two extremes along which a variety of intermediate relationships might be found. A third model is offered by fearful frugivores which hide in nearby forest to process their food. Some plant families are renowned for their special storage products in the seeds, such as the Euphorbiaceae, which are rich in oil but are starch free, or the Leguminosae which contain a high amount of protein in the seeds.

Small seed size is generally associated with a high ratio of edible aril to indigestible seed in *Virola*. *Virola surinamensis* arils are among the most nutritious known with 9,2% usable carbohydrates, 14% fat and 2,5% proteins (altogether 6,5 Kcal per aril). A "dramatic" positive correlation between the proportion of seeds dispersed and the ratio of aril to seed weight indicates that frugivores tend to prefer trees offering a high reward for the amount of ballast consumed. The inedible hard seed contains 15% tannin (dry weight) and tastes bitter (Howe 1980). Two alternatives in seed size are thus confronted with one another: smaller seeds are more advantageous for dispersal, while larger seeds are appropiate for germination.

The arillate fruits of *Tetragastris panamensis* and *Virola surinamensis* are distinguished from one another in that the seeds are 1/10 as large in *Tetragastris panamensis* as those of *Virola surinamensis* and 10 times more numerous so that the total weight of fruit per tree is about the same. *Tetragastris panamensis* arils are sweeter, but less nutritious. *Tetragastris* fruits attract twice as many species of visitors most of which are opportunists by superabundant but mediocre fruit, while *Virola* offering a richer reward attracts species which disperse the seeds more efficiently (Howe 1983). Howe and De Steven (1979) found that a superabundant crop of low quality fruits or seeds attracts opportunistic birds. Seeds of *Guarea glabra* are small and the aril offers little gross nutrition. However, visitation increases with fruit production. Superabundant fruit promotes widespread use by a variety of birds species. As the number of arilloids available increases in an increasing crop, the number of seeds removed on a tree increases too. Large trees thus disperse more seeds than small ones. One advantage of the "superabundance syndrome" is dissemination of seeds to a variety of suitable sites by birds with different habits.

Large seeds dispersed by specialists are associated with pulps of a high lipid and/or protein content. Potential profitability inherent to a fruit not only depends on the quality of its pulp, but also on the seed volume as a ballast. While temperate plant species are mostly small-seeded, tropical fruits have a pulp much richer in lipid-plus-protein content than European ones. Tropical fruits, however, tend to have larger seed loads. Large-seeded species should keep pericarp thickness to a minimum so as not to displace external fruit size further from the modal class. *Virola sebifera* fruits have far smaller seeds than those of *Virola surinamensis*, but attract the same bird species. In *Virola surinamensis* certain bird species act as fruit thieves, while in *Virola sebifera* they do not. The fruits of *Tetragastris* primarily dispersed by monkeys attract more birds than *Virola sebifera* or *V. surinamensis* in spite that the arils and seeds are much larger than those of *Virola sebifera* (Herrera 1981). While *Tetragastris* produces large crops of mediocre fruit and attracts more opportunitists or generalists, *V. surinamensis* has smaller crops of larger more nutritious fruits which attract specialists. Dispersal thus results more efficient in *V. surinamensis*, as birds disperse the seeds one by one, whereas Tetragastris seeds are defacted in clumps (Howe & Kerckhowe 1981). As has been mentioned already, large seeds are frequently characteristic of tropical rain forest trees and in particular of shade plants. The large seeds of shade-tolerant trees are loaded with nutritious substances, such as carbohydrates (starch, hemicellulose, sucrose and other soluble sugars), proteins, fats and oils, besides vitamines and auxins, so that the young seedling has enough reserve substances to survive for a longer time in the shade. Furthermore, secondary compounds play an extraordinary part especially in seeds of tropical trees.

Snow (1981) in his world survey on tropical frugivorous birds and their food plants came to the result that fruits eaten by specialized frugivores are generally large, have relatively large seeds and offer a high nutritive quality. Of the plant families which have coevolved together with frugivorous birds to produce fruits of this kind, three are of outstanding importance, according to Snow, namely the Lauraceae, Burseraceae and Palmae. Specialized frugivores feed on high quality fruits rich in fats and proteins, while unspecialized or opportunistic frugivores feed on less nutritious fruits which have a watery flesh and mainly provide carbohydrates (e.g. species of *Miconia).* Fruits of the first type are large and contain few large seeds. An upper size limit is about 70 × 40 mm for oval fruits. They may be drupaceous or have arillate seeds. Fruits that exceed the size limit may be expected to be dispersed by other agents. Specialized birds void the seeds intact, either by regurgitation or by defacation. Seed defenses are either mechanical (palms) or toxic (Lauraceae). Snow concludes that many fruits important to tropical forest birds had evolved well before the present avifaunas were in existence. Groups of closely related sympatric plant species which are bird-dispersed may be expected to have rather similar fruits, as selection for dispersal by a common pool of frugivores will tend to set limits to their fruit and seed size and shape. But competition for dispersal agents may be expected to lead to the evolution of differing seasonal regimes (Snow 1966). For the botanist, the result will be a taxonomically "difficult" group of plants. It is, therefore, not surprising that the Lauraceae is a notoriously difficult family for the taxonomist, as also is the large Melastome genus *Miconia.* The present author would like to add in this connection the Burseracea and Sapotaceae as examples of "difficult families".

Escape in space and time

Besides mechanical or chemical protection the plant may use space and time to escape from predators (Janzen 1970). Generally, two classes of predators may be distinguished: short-lived predators synchronized with the phenology of the plant in which they are specialized (e.g. insects that complete their larval stages within the developing or ripe seed), and longer-lived polyphagous feeders that take seeds during the season of abundance and eat other food in other seasons or in other places (generalists). It is of importance whether the seed predators take seeds before or after dispersal. Predispersal predators reduce the crop, but leave the dispersal processes and patterns unchanged. Post-dispersal predators, on the other hand, influence the dispersal pattern by their behavioural characteristics of choice and search (Harper 1977).

Janzen (1970 and following papers), the pioneer of the spatial and temporal theory, has emphasized that the negative effect of host-specific predators and parasites on seeds and seelings declines with increasing distance from the parent tree and from other adults of the same species. The vast majority of insects that prey on seeds are host-specific and have only few species (1–3) as host plants. In almost every plant, seeds fall too close to the parent to persist. Only the seeds that are removed to a distance reproduce the species. Where too many individuals of one species grow together, they are prone to be attacked by some pest, insect or fungus, as disease can spread easily from one to the other (Janzen 1970). This fact is vividly demonstrated in our European agricultural and forestry monocultures. Many animals may be predators and dispersers at the same time. Escape from predators also depends on the distance of moving of the animals. Janzen (1971) indicates some examples of distance moving: harvester ants move 10–130 feet, red squirrels about 4 hectares, and parrots several miles (see also Ridley 1930). According to Janzen (1970), predators are either distance-responsive or density responsive. Distance-responsive predators are mainly insects and mammals. With density-responsive predators predation intensity decreases with decreasing density of seeds. In *Andira inermis*, highest weevil seed predation is observed beneath the tree, but lowest predation among seeds dropped accidentally by seed dispersers (Janzen 1976). *Sterculia apetala* seeds that fall close to the parent tree are eaten by a bug. But at a distance of 30–60 m they escape. Squirrels and monkeys carry the seeds away. Although the pods dehisce, the seeds do not move out from the edge of the canopy more than 10 meters, even with wind. When carrying the seeds, the animals loose some of them on the way and these have the best chance to survive. The predators thus exercise an effect on the density of adult trees in the habitat and make coevolution of animals and plants obvious (Janzen 1972).

Another factor working for plant protection from predators is time. Host-specific insect predators may have different hosts at different times of the year. Escape by irregular fruiting at larger intervals is one strategy of plants ("behavioural escape"). Tropical trees fruit at larger and less predictable intervals, and, consequently, produce larger crops so that the predators may become saturated (Janzen 1970). Plants that escape predation by saturation possibly produce larger amounts of smaller seeds (Janzen 1971). Seeds may thus be absent for predators for a long time. But predators cannot maintain high population densities in the prolonged absence of seeds. *Hymenaea courbaril*, for example, fruits every 3–5 years where it is attacked by weevils, but where the weevil *Rhinochenus stigma* is absent, the tree fruits every year (Janzen 1970). *Rhinochenus* is furthermore selecting for a later age of the plant at first reproduction, more resin in the pod wall and a harder pericarp. Between the ripening of the fruit and the moment of dispersal hours to months can elapse. During this time the fruit is exposed to damaging agents, such as fungi, bacteria, invertebrates. An effective strategy is, therefore, for the plant to reduce the exposure time of ripe fruits to noxious agents. In deciduous forest the seeds of *Cassia grandis* are killed by two bruchid beetles. The more rapidly the dispersal agents remove the pods, the less predation there will be. *Cassia* bears fruit every other year, and the size of the seed crop is large enough to surpass the abilities of predators that survive the "distance" in time between two crops. This is thus a model of predator saturation (Janzen 1971). Gradual discharge of seeds is another time-dependent defensive method against predators (such as squirrels). Synchronization of fruiting of different species of the same genus or group may be a further strategy to escape predators (Harper 1977). One of the best methods to escape predation is certainly the ripening of fruits when pest pressure is lowest (Herrera 1982).

In summary we may repeat: for fruits and seeds not attacked by predators a complex of traits is the reason, such as annual and longterm phenology, secondary compound and nutrient chemistry, morphological and mechanical properties of fruits and seeds, seed size, spacing and density of the plants, and the habitat occupied (Janzen 1980). The timing of fruit production within and between years probably evolved at least in part under the pressure from seed predators (Janzen 1978). And vice versa: the polymorphy in the rain forest may be regarded as a protection against pest pressure (Van der Pijl 1969). The action of predators on seeds and seedlings is thus responsible for a high number of plant species, low density of adults of each species, and long distance between conspecific adults.

Coevolution of plants and animals

Coevolution of plants and animals becomes very obvious in the tropical rain forest. Dispersal units of plants are adapted to special dispersal methods which – in their turn – are partly induced by behavioural characteristics of certain animals. Animals are moving in different strata of the forest having distinct feeding behaviours and accordingly diaspores are adapted in their outer morphology and inner structure. Outer morphology not only concerns size and shape of diaspores, but also surface structures such as spines, hairs, glands, surface sculpturing and other surface properties. Colour is an important factor for visual animals, as we have seen. Birds prefer red, orange or contrasting colours such as red/black or red/white. Bats, on the other hand, having a poor visual power search for brown, blue, yellow colours or the like. Odour is a guiding principle of mammals. Many fruits attract animals by their sweet perfume or particular odour, in special cases even by a disgusting smell (durian). The sound of falling fruit or seed may call the attention of certain ground mammals in the forest or of fish in the water (e.g. *Inga* pods). The taste and sweetness of a fruit or seed part may be of great importance to challenge the interest of animals. As animals like juicy fruits with a soft fruit flesh, this should be composed of large thin-walled parenchyma cells with large vacuoles. A small-sized seed is preferred to a large-sized seed, as large seeds have to be considered an unnecessary ballast. Referring to diaspore chemistry the animals prefer highly nutritious seeds and fruits with a high calory reward. Oily seeds and fruits are, therefore, given preference before food rich in carbohydrates. Many animals are "generalists" visiting very different plant species during the year, being attracted by large crops usually of minor quality, while others are specialists of one or a few species, which generally offer highly nutritious food but in lower quantities. Many animals are predators and dispersers at the same time, few are probably only dispersers, while others are only predators. The plant, therefore, not only adapts to the taste and preferences of animals in order to compete for dispersers, but also has to protect itself against predators. Another syndrome of characteristics supports these purposes. This end of evolution more or less works in opposite direction. Fruits and seeds become less nutritious, too small or too large to be destroyed when eaten, take on inconspicuous colours, become hard, e.g. develop a hard seed coat or a sclerenchymatous endocarp. Fructification may be avoided at times of the year when predators are superabundant. Synchronous fruiting of individuals of the same species or of different similar species of the same genus may take place to saturate predators, or dispersers which carry the dispersal units far away to save some of them from predators or from competition of the parent tree and other conspecific individuals nearby may become favoured. The most sophisticated strategy of defense, finally, lies in the development of toxic compounds which deter predators. These secondary compounds are manifold and of very different chemical origin and composition. Particularly unripe fruits and seeds become protected against predation by a surplus of astringent tannins, presence of acids, lack of sugars, and even green colour. Adaptation of plants thus works in two different directions and the correct balance between the two possibilities brings about most success.

Regarding coevolution of plants and animals one has furthermore to consider the long life of rain forest tree species which oscillates between 100 and 500 years, exceeding 1000 or frequently many more years so that only a few generations develop at a period of time when animals form a generation every year. Animals thus have many more possibilities for adaptation and coevolution than plants within a given time.

Germination and establishment

With the dispersal of seeds a new cycle begins and although dissemination is of importance for the survival of the species, germination and more so establishment are much more critical, especially in a dense primary tropical rain forest. Supposing the seed reached an adequate area concerning soil and climate, there may not be enough light reaching the ground to stimulate germination. It is commonly known that light gaps are of great importance for seed germination in a dense forest. As latency in tropical rain forests is generally very short or none, this is the first crucial point in the life of the new plant. A great majority of seeds may, therefore, be destined for death. Those embryos which have the opportunity to germinate may only reach the stage of a seedling for lack of reserve substances or deficiency of light and grow no further. It is a common statement that the seeds of rain forest trees are often very large (and heavy) having stored a large amount of water and reserve substances, such as starch, sugar, proteins, slimy substances, and oils either stored in the cotyledons (Roth 1977 a) or in the endosperm. The deep shade of the forest floor imposes the need for heavy fruit with ample food reserves for the young seedling and sapling, raising problems for effective dispersal. Light wind-borne seeds, on the other hand, are produced by few rain forest species, according to Ashton (1969). However, the actual percentage of very large-seeded species is not so overwhelming when we consider percentage values of all the individuals of a certain community (see p. 136 ff.). Certainly, seeds that are stuffed with storage food can survive for a longer time and grow to a certain extent, e.g. those of *Gyranthera caribensis* (Roth 1977 a). But more critical is the further development from the seedling to the young plant. While a larger percentage of the seedlings may still reach the stage of a small young plant, only few will have the chance to develop further. The germinating power of seeds is another aspect of importance in the establishment of seedlings. It is good, for example, in many leguminous seeds. Those germination prerequisites as the passage of seeds through the intestine of animals have already been mentioned in the chapter "Dispersal methods". The effect of this passage may only be of mechanical nature or more frequently of chemical character, particularly in tropical rain forests. Seeds with a soft permeable seed coat, such as of *Pentaclethra macroloba*, germinate rapidly (Hartshorn 1972). According to Hall and Swaine (1980) only few primary species found in Ghanaian forest soil have mechanical dormancy resulting from a thick testa or endocarp. Impermeable seed coats restrict the gaseous exchange. Hard seed coverings may prevent absorption of water and limit germination in this way, as it occurs in many Leguminosae. Seeds of this type remain viable for long periods and individual seeds become permeable at different times. A hard seed coat thus favours dormancy (Crocker & Barton 1953). In many cases, germination power, i.e. readiness for germination, may only be stimulated by the passage of the seed through an animal, but in others it may be obligatory. In *Dipteryx odorata* germination from the pod takes six weeks, but from extracted seed only a few days (Williams 1951). In this case, the fruit wall itself seems to delay germination and the removal of the pod favours germination. Not infrequently, the simple removal of a hard pericarp may be the reason of the favourable influence of fruit eating animals on the germination readiness of seeds. Seeds of *Pithecellobium saman* eaten by cattle pass through the gut unharmed. Picked out of cattle dung they immediately germinate in moist soil, whereas seeds taken from the pods do not germinate for at least three months or may even wait longer (Janzen 1976). An example of chemical influence of the mammalian intestine on the germination power of seeds is given here by Janzen. In other cases, the influence of soil microorganisms may promote the rotting or breaking of the seed coat.

When the seedling is established, further growth may be stopped by lack of room or absence of sufficient light. The accumulation of large numbers of seedlings on a small spot around the mother plant is a very familiar aspect to anybody who studies tropical forests: e.g. seedlings around *Gyranthera caribensis* in the cloud forest or of *Chrysalidocarpus* sp., also observed as an ornamental tree in gardens. The seedlings are too close together to be able to continue their growth, remaining – often for years – almost at the same stage, so-to-say in a waiting position or "on the waiting list". The death of the mother plant or the sudden formation of a light gap may give the chance to develop further. In many instances, seeds and seedlings may be present in sufficient quantities for regeneration, but the favourable conditions for further development are missing. According to Hartshorn (1972) mortality of seedlings and saplings is often caused by falling leaves, branches and other debris.

Seed size

Many of the leguminous trees of the tropical rain forest possess large seeds, according to Harper (1977). *Mora* seeds reach a weight of up to 0,4 kg. Many palm trees bear heavy fruits and seeds. *Cocos nucifera* fruits with a weight of up to several kilos and a length of 20–30 cm are full of very nutritive substances. The twin fruits of *Lodoicea maldivica,* a palm tree, may weigh as much as 40 pounds, according to Hodge 1949. *Caryocar nuciferum* forms another well-known large seed. Small seeds, on the other hand, tend to have much greater longevity than large ones and the very large seeds have very short lives. They just take the chance of the instant, germinating at once and taking advantage of their food reserves with which they may survive for a longer time in the stage of the seedling. However, there are also many species with small seeds weighing less than a milligram (*Musanga cecropioides*) in the tropical rain forest and those with a weight of 20–30 g are more or less an exception. Salisbury (1942) recognizes a direct relation between seed size, on the one hand, and dispersability as well as germination power, on the other. Small seeds are generally produced in larger numbers and their chance to occupy a suitable habitat is more probable, since dispersal of small seeds is more efficient. Large seeds, usually few in number, contain sufficient food material for germination, compensating in this way for the less favourable dispersal conditions. They thus have a better opportunity to withstand competition. However, seed weight is only a rough approximation to the amount of food supply available for the seedling. Besides nutritive reserve substances larger quantities of water may be stored in large seeds. With increasing density of the plant community and diminishing illumination at the ground level, a marked tendency to augment seed weight may be observed. Salisbury also mentions the possibility that seed weight increases in autopolyloids. Furthermore, some species show a high degree of variability in seed weight. Favourable or unfavourable environmental conditions, on the other hand, rather affect seed number than seed size. Not only seed size, but also shape and surface structure of seeds are important factors for germination. Seeds with a rough surface are more sensitive to water loss. A mucilage layer protects the seeds of many species from excessive water loss (see also Grubert 1974). Hairs help in the anchorage of seeds.

Latency and viability

Whitmore (1975) states for the rain forest of the Far East that many species have no dormant period, germinating within a few days after being dropped. Longman and Jenik (1974) report that under forest conditions hard-shelled seeds in indehiscent pods, such as of Leguminosae, may remain viable for several months (for *Hymenaea courbaril*: Marrero 1942). Seeds of this type usually remain dormant until mechanical abrasion or decay of the shell by rotting or action of microorganisms has rendered them permeable to water. Seeds of the type of *Mora excelsa*, on the other hand, germinate immediately when fallen to the ground. Seed viability may drop from 80% on collection to 15% after two weeks storage. However, in cold storage viability of seeds may last up to 6–12 months instead of only 1–2 months under natural conditions (e.g. in *Swietenia macrophylla, Cedrela odorata*). Moreno (1977) distinguishes two types of tropical tree seeds: those with very short viability and a lack of dormancy and others with a longer viability, although shorter than that of most seeds from temperate forests, and with presence of latency. Both types seem to have adaptive value. In the first type, the seeds germinate quickly, diminishing the possibility of being picked up by predators or of rotting. In the second type the seeds undergo these risks but have a better chance of becoming established during the longer viability period. Most seeds studied by Moreno lost their viability in the first 100 days. A concise survey over this subject is given by Harper (1977). See also Medina, Mooney, & Vázquez-Yanes (1984).

Germination

Species that are physiologically adapted to very low light need substantial food reserves in the seed in order to establish. Mycorhizal fungi may be of importance for the development of some species, so that ectomycorrhiza is more dependent on the environment. As Salisbury (1942) emphasizes, seed (and fruit) weight augments with the degree of shade to which the plant is exposed. According to Richards, germination seldom seems to be a critical stage in the establishment of regeneration, as a large surplus of seeds is produced by most species. Seeds of rain forest trees generally show a high percentage of germination which may be of the simultaneous or the successive type. However, the rate of mortality between seedlings is very high and mainly due to lack of light and to root competition. Seeds of rain forest trees usually need high humidity of the air and elevated temperatures. Germination of coffee seeds takes a longer time at 17 degrees C than at 30 degrees C (Longman & Jenik 1974). The embryo of *Inga feuillei* is very resistant, e.g. to drought and to elevated temperatures (50–60 degrees). It is also protected against light and predation. When dried out, the embryo takes on a horny consistency. Most seeds germinate at the soil surface and need light as a stimulus for germination. Only seeds with extensive food reserves are able to germinate when deeply buried (Harper 1977). From the seedling to the adult plant there is a long way and most of the seedlings do not reach this goal. Besides the demand for room and light, root competition, pests, and predators may be critical factors for the survival of seedlings and young plants in the tropical rain forest.

Vivipary

Vivipary understood as seed germination on the mother tree is a very peculiar phenomenon. The best-known examples are supplied by the mangroves *Bruguiera* and *Rhizophora*. Another example is *Sechium edule*. Vivipary is also known of *Pithecellobium racemosum* where about 80% of the fruits show germinating seeds already on the parent tree (Leite & Rankin 1981). Incidental vivipary occurs in the large fruits of *Carica papaya, Persea americana* and in some species of *Inga*. Mango seeds may germinate within the stone. When the seeds of *Inga feuillei* remain within the pod, they germinate there. Gindel (1960) observed germination of seeds within the fruit itself in tomato, apple, *Cucurbita mexicana*, avocado, *Annona*, grapefruit and lemon. Within the last two mentioned fruits green cotyledons developed from the germinating seeds, when the fleshy pericarp was still entirely closed. The roots of the germs were in all cases tightly attached to and had partly penetrated into the body of the fruit. The so-called inhibiting substances which prevent the seed from germination modify, according to Gindel, during fruit ripening transforming into new compounds which favour seed germination. The conclusion that fruit in nature is an inhibitor to seed germination or should be destroyed, e.g. by insects or fungi, before germination takes place, seems not to be warranted in these cases, according to Gindel's point of view. The roots of the seedling make use of the food in the pulpy pericarp, many roots branching within the fruit. However, the cases Gindel describes may not refer to real vivipary, but mainly represent examples of fruits already separated from the parent plant. Particularly in tropical and subtropical fruits, a continuous transition may exist between vivipary, germination of seeds within the still closed but from the mother plant removed fruits, and immediate germination on the ground without latency. Everywhere in nature we are confronted with these "fluid boundaries" and all kinds of transitional stages are possible.

With this short survey over the most important facts of fruit and seed structure, dispersal methods, protection against predation, and germination we may enter the main part, the study of a distinct forest area in Venezuelan Guiana. It is evident that the introductory part is not exhausting and that some topics were dealt with very superficially. However, for the further discussion of the resulting problems the performed data are sufficient. The reader interested in more detailed information is referred to the quoted literature.

In the following, the results of my studies in the humid tropical forest of Venezuelan Guiana are described. The first part comprises a description of all the families, genera and species studied, each larger family provided with a short general introduction followed by the detailed description of the species studied concerning fruit and seed morphology and structure as well as means of dispersal. A summary of the most important facts concludes the description of each larger family.

In the second part, the number of species and individuals as well as their height categories are considered (Table 1). Then the different dispersal types, as there are, anemochory, autochory, autozoochory, and zoochory are related to the number of species and individuals studied (Table 2). This is followed by special remarks on certain characteristics of families, genera and species and the number of individuals. The distinct dispersal mechanisms are related to their special fruiting times in Table 3, a with d. Fruiting times are related to the three different forest stories (A-a-aa) in Table 4, a with c. Furthermore, the families, their fruit types and fruiting times are considered. Fruiting times are also related to geographical distribution and habitats. Animal life in the forest and forest layering, as well as seasonality of animal life is given a short consideration.

Finally, special reflections on fruit and seed dispersal in the tropics, forest layering and dispersal strategies, dispersal types in dry and humid regions, homogeneous groups or series of species with their special characters related to the number of individuals, the refuge theory and some evolutionary considerations are dealt with.

Table 1. Indicating the families studied with the number of species and individuals and the height categories to which they belong. Of some species it is doubtful to which category they belong. These could not be placed in a certain height category.

Family	Total species	Species			Total individuals	Individuals		
		A	a	aa		A	a	aa
1. Acanthaceae	1	1	–	–	2	2	–	–
2. Anacardiacecae	5	4	1	–	175	94	81	–
3. Annonaceae	7	2	4	1	991	43	572	376
4. Apocynaceae	9	8	1	–	3.047	3.043	4	–
5. Araliaceae	3		2		42		41	
6. Bignoniaceae	6	5	1	–	899	195	704	–
7. Bixaceae	1	–	–	1	34	–	–	34

Table 1. Cont.

Family	Total species	Species			Total individuals	Individuals		
		A	a	aa		A	a	aa
8. Bombacaceae	3	3	–	–	821	821	–	–
9. Boraginaceae	7	7	–	–	1.114	1.114	–	–
10. Burseraceae	12	6	6	–	5.449	3.824	1.625	–
11. Capparidaceae	3	1	1	1	11	3	7	1
12. Caryocaraceae	1	1	–	–	10	10	–	–
13. Celastraceae	3	3	–	–	294	294	–	–
14. Cochlospermaceae	1	–	1	–				
15. Combretaceae	3	3	–	–	429	429	–	–
16. Dichapetalaceae	1	1	–	–	1527	1527	–	–
17. Ebenaceae	3	3	–	–	185	185	–	–
18. Elaeocarpaceae	6	4	2	–	457	223	234	–
19. Euphoribiaceae	24	12	9	1	6.873	4.718	1.804	349
20. Flacourtiaceae	8	3	4	1	210	51	130	29
21. Guttiferae	14	7	6	–	2.513	167	2.345	–
22. Hernandiaceae	1	1	–	–	9	9	–	–
23. Hippocrataeaceae	1	–	–	1	43	–	–	43
24. Humuriaceae	1	1	–	–	37	37	–	–
25. Lacistemaceae	1	–	1	–				
26. Lauraceae	14	6	5	–	968	248	715	–
27. Lecythidaceae	11	10	–	1	8.824	8.671	–	153
28. Leguminosae Mimosaceae	33	19	11	3	5.616	1.643	3.129	844
29. Caesalpiniaceae	25	20	5	–	1.831	1.137	694	–
30. Papilionaeceae	20	20	–	–	2.145	2.145	–	–
31. Linaceae	1	1	–	–	17	17	–	–
32. Malpighiaceae	3	2	1	–	103	61	42	
33. Melastomaceae	5	3	1	1	204	144	6	54
34. Meliaceae	10	3	6	–	2.511	945	1.537	–
35. Moraceae	12	5	3	–	518	130	167	–
36. Myristicaceae	3	3	–	–	114	114	–	–
37. Myrtaceae	14	4	10	–	1.554	604	950	–
38. Nyctaginaceae	2	1	1	–	455	105	350	–
39. Ochnaceae	2	1	1	–	96	36	60	
40. Olacaceae	1	–	1	–	29	–	29	–
41. Opiliaceae	1	1	–	–	7	7	–	–

Table 1. Cont.

Family	Total species	Species			Total individuals	Individuals		
		A	a	aa		A	a	aa
42. Palmae	7	3	4	–	327	227	100	–
43. Polygonaceae	4	1	3	–	384		384	–
44. Quiinaceae	1	–	1	–	16	–	16	
45. Rhamnaceae	1				4			
46. Rhizophoraceae	1	–	1	–	418	–	418	–
47. Rosaceae	15	12	2	–	4.797	4.740	29	–
48. Rubiaceae	8	4	1	3	492	318	77	97
49. Rutaceae	9	5	1	2	342	81	25	236
50. Sabiaceae	1	–	1	–	20	–	20	–
51. Sapindaceae	6	3	2	1	1.125	284	795	46
52. Sapotaceae	21	15	5	–	6.550	4.885	1.598	–
53. Simaroubaceae	2	2	–	–	329	329	–	–
54. Sterculiaceae	4	3	–	1	1.003	1.003	–	–
55. Tiliaceae	4	3	1	–	196	161	35	–
56. Verbenaceae	3	3	–	–	74	74	–	–
57. Violaceae	2	–	1	1	91	–	91	
58. Vochysiaceae	4	4	–	–	645	645	–	–
58 Families (total)	375	233	107	19	66.977	45.543	18.814	2.262

Table 2. Indicating the dispersal types of the different families with the number of species and individuals

No.	Family	Anemochorous		Autochorous		Auto-zoochorous		Zoochorous	
		Spec.	Individuals	Spec.	Individuals	Spec.	Individuals	Spec.	Individuals
1.	Acanthaceae			1	2				
2.	Anacardiaceae	1	30					4	145
3.	Annonaceae			2	382			5	609
4.	Apocynaceae	6	3.022					3	25
5.	Araliaceae							3	42
6.	Bignoniaceae	6	899			1	34		
7.	Bixaceae								
8.	Bombacaceae	2	54	1	767				
9.	Boraginaceae	1	16					6	1.098
10.	Burseraceae					12	5.449		
11.	Capparidaceae							3	11

Table 2. Cont.

No.	Family	Anemochorous		Autochorous		Auto-zoochorous		Zoochorous	
		Spec.	Individuals	Spec.	Individuals	Spec.	Individuals	Spec.	Individuals
12.	Caryocaraceae							1	10
13.	Celastraceae	1	39	1	237			1	18
14.	Cochlospermaceae	1							
15.	Combretaceae	2	429					1	
16.	Dichapetalaceae							1	1.527
17.	Ebenaceae							3	185
18.	Elaeocarpaceae			3	75	3	382		
19.	Euphorbiaceae			20	6.391	1	14	3	468
20.	Flacourtiaceae			1		2	46	5	164
21.	Guttiferae			3	2.233	1		10	280
22.	Hernandiaceae							1	9
23.	Hippocrataecaceae								
24.	Humiriaceae							1	37
25.	Lacistemaceae					1			
26.	Lauraceae							14	968
27.	Lecythidaceae	2	669			8	8.002	1	153
28.	(Leguminosae) Mimosaceae	1	185	4	118	5	3.328	20	1.985
29.	Caesalpiniaceae	4	631	8	455	5	236	8	509
30.	Papilionaceae	12	559	3	1.450	2	106	3	30
31.	Linaceae							1	17
32.	Malpighiaceae	1	42					2	61
33.	Melastomaceae							5	204
34.	Meliaceae	2	73	1	29	7	2.409		
35.	Moraceae							12	518
36.	Myristicaceae					3	114		
37.	Myrtaceae							14	1.554
38.	Nyctaginaceae							2	455
39.	Ochnaceae							2	96
40.	Olacaceae							1	29
41.	Opiliaceae							1	7
42.	Palmae							7	327
43.	Polygonaceae	1						3	384
44.	Quiinaceae							1	16

Table 2. Cont.

No.	Family	Anemochorous		Autochorous		Auto-zoochorous		Zoochorous	
		Spec.	Individuals	Spec.	Individuals	Spec.	Individuals	Spec.	Individuals
45.	Rhamnaceae			1	4				
46.	Rhizophoraceae					1	418		
47.	Rosaceae							15	4.740
48.	Rubiaceae	1		2	315			5	177
49.	Rutaceae			9	342				
50.	Sabiaceae							1	20
51.	Sapindaceae	1	795	2	46			3	284
52.	Sapotaceae							21	6.550
53.	Simaroubaceae							2	329
54.	Sterculiaceae			2	908	2	95		
55.	Tiliaceae	1				3	196		
56.	Verbenaceae							2	74
57.	Violaceae			2	91				
58.	Vochysiaceae	4	645						
58	Families (total)	50	8.088	66	13.845	57	20.829	197	24.115

Table 3. Indicates the fruiting species in the different months (1-12). Each species is represented by the number of individuals. When a species fruits in several months, the number of individuals is repeated and marked by a "dot". Species with no number of individuals are indicated with a +

Anemochory

1	2	3	4	5	6	7	8	9	10	11	12
14	650	61		550	+		50			+	
	49			18	+					206	
				704							
	+			390							
	•39			•39							
	185			499							
	•170			•170							
•421	•421			•421							
	•487			+				•487			
	•22			•22	•22		•22				
		•135						•135			
	•18										•18
	•55										•55
	1			+							
435	2.097	196		2.813	22		72	622		206	73

Autochory

1	2	3	4	5	6	7	8	9	10	11	12
767				+	8	33			142		237
		•240									•240
					66	12			190		
							•1	•1			•1
	•3.679				•3.679						•3.679
				•1.706			•1.706				•1.706
	•95			•95	•95						
4					315				+	+	
							•5		•5		
•1.893							•1.893				•1.893
		264			76						
					•50					•50	•50
		19		908	+						
								•73			•73
							•26		•26		
				•1				•1			
		•1.262		•1.262							
2.664	3.774	1.785		3.972	4.289	45	3.631	75	363	50	7.879

Auto-zoochory

1	2	3	4	5	6	7	8	9	10	11	12
247	19			142	43		35			+	+
	5			95							
	•418			•418							
	•36										•36
	+			885							
	•2	•2									
				•1.376	•1.376						
•17							•17				
	•217				•217				•217		
	2.534								16		
•91		•91									
	•100			•100	•100						
	•6			•6				•6			
	•3.446			•3.446							
	•1.025						•1.025	•1.025			
		•2.374			•2.374						
							•803		•803		
	•44						•44				
	•14					•14			•14		
	•168			•168							
	293	167					47				
	•2.669			•2.669	•2.669					•2.669	
	•26						•26				
	•408			•408							
355	11.430	2.634		9.713	6.779	14	1.997	1.031	1.050	2.669	36

Zoochory

1	2	3	4	5	6	7	8	9	10	11	12
722	12	144		43	1	1	36		3	13	72
				•28							•28
		•376									•376
		•240									•240
98	409	404		29	40		7		28	3	17
	•36										•36
	•4			•4							
					+	+					
	624	2		+	207		+			282	30
					•263	•263					•263
	138	+		7	+		3	44			139
	•7			•7							
	•10				•10						
					•1.527		•1.527		•1.527		
	176	231		+	8		153				1
		•423				•423			•423		
							•43			•43	
				•5		•5					
				•29	•29						
+	85				77		135			+	
	9			101	20	25	614				54
					•50					•50	•50
	37			6	315		276				2
	•134	•134			•134						
							•34				•34
	•43									•43	
	•240						•240				
	•287	•287									
	•14			•14							
	•42										•42
							•2			•2	
	•226			•226							
		•3									•3
					•160			•160			
	•140	•140									
	•7					•7					
	•506								•506		•506
	•95					•95					
	•105			•105							
•40							•40		•40		
					•28			•28			
	•679			•679							
•3.258	•3.258								•3.258		
						•24	•24		•24		
	24										
	133										
	•53			•53			•53			•53	
•3	•3			•3							
	•458										•458
	28										
					•1.658						•1.658
•387	•387					•387					
				•47			•47				•47
	•164			•164							
							274				
		•103				•103					
				•795			•795	•795			
				•51	•51						
4.508	8.573	2.487		2.396	4.578	1.333	4.303	1.027	5.809	489	4.056

Table 4. Indicates the fruiting species with their no. of individuals in the different strata A-a-aa. First sheet = highest stratum (A). 2nd sheet = Stratum a. 3rd sheet = lowest stratum (aa)

a: Height category A

1	2	3	4	5	6	7	8	9	10	11	12
	409	404		7	+	1			16	+	17
	•42										•42
	+	19									
	185	+								13	54
	+										
•91		•91									
	12	61		+	43					206	
								•73			•73
							•26		•26		
				•1				•1			
	•217				•217				•217		
•421	•421			•421							
		•1.262		•1.262							
•17							•17				
		•3									•3
	138			101					3	282	
	85			885						3	
	•18										•18
	•36										•36
	•506								•506		•506
	•95					•95					
	•105			•105							
•40							•40		•40		
	•679			•679							
•3.258	•3.258								•3.258		
						•24	•24		•24		
	24										
	133										
	•53			•53			•53			•53	
	•458										•458
	28										
					•1.658						•1.658
•387	•387					•387					
				•47			•47				•47
767	650	167		550	8	33	36	44	190		237
				•28							•28
247	49			18	20	12	7				
	•36										•36
	624			390		25	50				
					+	+					
					•263	•263					•263
	•2.669			•2.669	•2.669					•2.669	
	293			499			3				
	•10				•10						
							+				
	•39			•39							
					•1.527		•1.527		•1.527		
	176				8		47				1
							•1	•1			•1
	•3.679				•3.679						•3.679
		•423				•423			•423		
							•43			•43	
	•95			•95	•95						
							•5		•5		
				•5	•5	•5					
	•44						•44				
+											+
	9										
	37										
	•134	•134			•134						

Table 4. Cont.

a

1	2	3	4	5	6	7	8	9	10	11	12
							•34				•34
	•43									•43	
	•170			•170							
	•100			•100	•100						
	•6			•6				•6			
	•3.446			•3.446							
	•1.025						•1.025	•1.025			
		•2.374			•2.374						
							•803		•803		
98		231		95			274		28		
		•103				•103					
	19			908			276				
							135				
				•795			•795	•795			
1				142							
	+			+	+						
				•51	•51						
	•487							•487			
	•22			•22	•22		•22				
		•135						•135			
5.327	21.151	5.407		13.589	12.883	1.371	5.334	2.567	7.066	3.312	7.191

b: Height category a

1	2	3	4	5	6	7	8	9	10	11	12
14		264		+	1		35		142		72
		•240									•240
	•4			•4							
722		144		29	40		614				
	•26						•26				
	•408			•408							
										+	+
	5	2		704	66						139
	•7			•7							
	•168			•168							
	•14					•14			•14		
				•1.706			•1.706				•1.706
•1.893							•1.893				•1.893
				6	76						30
					•50					•50	•50
					207						2
	•287	•287									
				+	77						
	•14			•14							
	2.534				315						
							•2			•2	
	•226			•226							
	•55										•55
	•2	•2									
				•1.376	•1.376						
	•140	•140									
	•7					•7					
	•418			•418							
				+							
•3	•3			•3							
	•164			•164							
2.632	4.482	1.079		5.233	2.208	21	4.276		156	52	4.187

Table 4. Cont.

c: Height category aa

1	2	3	4	5	6	7	8	9	10	11	12
	•240	•376		•29 43	•29		153 •240		+		•376
	240	376		72	29		393				376

Note added in proof stage

Charles-Dominique et al. (1981) listed all the mammals present in a forest of French Guiana including 6 species of marsupials (*Didelphis marsupialis, Philander opossum, Caluromys philander,* 2 species of *Marmosa,* and *Monodelphis brevicaudata*), 3 species of primates (*Saguinus, Alouatta, Saimiri*), 7 species of rodents (*Sciurus, Coendu prehensilis, Agouti paca, Dasyprocta agouti, Proechimys, Echimys, Oryzomys*), one game, jaguar and ocelot, one species of mustelidae and 2 procyonidae (*Potos flavus* and *Nasua nasua*). The distribution of the frugivorous nocturnal tree mammals at different forest heights and in different biotopes, as well as the supports utilized were studied. The itineraries in the forest of 4 marsupials and of *Potos flavus,* the "cuchicuchi, were reconstructed. The favoured seeds and fruits of seed and pulp eaters were listed. Furthermore, 3 types of fruiting cycles could be related to 3 different foraging strategies of the sympatric frugivorous mammals. 1. Species with a low productivity extending their fruiting period over the entire year. The small seeds are eaten by small vertebrates. 2. Species with synchronous, but irregular cycles of fruit production with large seeds attracting seedeating rodents which destroy part of the fruit crop. The massive, iregular and unpredictable fruiting cycles reduce seed predation. 3. Species with synchronous and regular cycles of fruit production with large seeds ingested by frugivorous vertebrates. Tree species belonging to this category have successive and partly overlapping periodes of fruit production. Food is thus available for pulp-eaters most of the year. This pattern is probably the end result of a longlasting mutual interaction between plants and seed-scattering animals.

CHAPTER 4

RESULTS

Description of the families studied concerning fruit and seed outer and inner structure, dispersal methods, fruiting time, height of the tree, number of individuals, and special habitat

The families studied are alphabetically arranged. Each scientific species name is followed by the vernacular name it received in Venezuelan Guiana, e.g. *Bravaisia integerrima* = Naranjillo (vernacular name). The vernacular names are of most importance, because certain species could not be scientifically identified. Furthermore, the species are well known by their vernacular names in the country, and anybody who wants to collect material in Venezuelan Guiana possibly needs the help of a so-called "bacchiano" who knows plants only by their vernacular names.

In the following descriptions the below indicated abbreviations are used:

Disp. = Type of dispersal
Fruit = Fruit type
Fr. time = Fruiting time
Hab. = Habitat
Height = Height category
Ind. = Number of individuals
RF = Rain forest

The height categories are:
A = 30 m or above
a = 10-29 m
aa = below 10 m

Acantheceae

Bravaisia integerrima = Naranjillo

Hab.: Lowland and semideciduous forest	Height: A
	Ind.: 2
	Fr. time:
Fruit: Capsule	
Disp.: Autochorous	

Bravaisia integerrima: The dry dehiscent capsule contains few seeds. The oblong fruit reaches 1-2 cm in length and usually contains 2-4 flattened seeds. It opens with 2 valves.·The brilliant surface is of brownish colour. Bracts, calyx, and sometimes the style are persistent. The seeds with a diameter of 0,2-0,4 cm are brown and brilliant. The tree is resistant, has a profound radical system, and adapts to different soils and climates. It also occurs in deciduous forests, and is appreciated as an ornamental tree. The species has a dispersal mechanism advantageous for high trees.

Anacardiaceae

The Anacardiaceae are represented by high trees. The fruit is often a drupe with a resinous mesocarp in which secretory cavities are dispersed.

1. *Anacardium giganteum* = Merey montañero

Hab.: RF	Height: A
Fruit: Nut	Ind.:
Disp.: Zoochorous	Fr. time:

2. *Astronium lecointei* = Pata de zamuro

Hab.: RF	Height: a
Fruit: Drupe	Ind.: 81
Disp.: Zoochorous	Fr. time:

3. *Loxopterygium sagotii* = Picatón

Hab.: RF	Height: A
Fruit: Samara	Ind.: 30
Disp.: Anemochorous	Fr. time:

4. *Spondias mombin* = Jobo

Hab.: RF	Height: A
Fruit: Drupe	Ind.: 36
Disp.: Zoochorous	Fr. time: 8

5. *Tapirira guianensis* = Patillo

Hab.: RF	Height: A
Fruit: Drupe	Ind.: 28
Disp.: Zoochorous	Fr. time: 5/12

1. *Anacardium giganteum:* The true fruit is flattened egg-shaped, up to 4 cm long, and represents a nut. Similar as in *Anacardium occidentale* (Roth 1974, 1977), the pedicel is transformed into a pseudoberry which is eaten by animals, while the true fruit is poisonous. The pseudo-berry is fiery-red and tastes less adstringent than that of *A. occidentale.* Howling monkeys and lizards eat the pedicels, but never the fruit. But the fruits are not well dispersed in this way. Birds and bats are more likely dispersal agents. The combination, toxic fruit (cardol oil) and edible pedicel, probably represents an evolutionary correlation (Glander 1979). In Surinam the tree also occurs in the Savanna forest as well as along creeks.

2. *Astronium lecointei:* The fleshy indehiscent fruit is drupaceous and contains a single seed. Dispersal takes place by animals.

3. *Loxopterygium sagotii:* The sickle-shaped samara (3 × 0,9 × 0,05 cm) develops a large membranaceous wing which shows dichotomous venation. The small calyx is persistent. The single seed is attached with a long funicle. Dispersal of the samara takes place by wind. In Surianm the species is found in the savanna and in the mountainous savanna forest too.

4. *Spondias mombin* (Figs. 42, 43): The plum-like ovoid 5-locular drupe (ca. 4 × 2,5 cm) contains 1-3-4 seeds. It has a yellow to orange surface and a thin mesocarp coating the woody endocarp. The fleshy mesocarp has a sour-sweet taste and is edible. The inedible seed is about 1,5 cm long. The endocarp, about 0,2 cm in thickness, consists in fibers which are arranged according to the principle of "crossing-over" fibers. The fiber bundles

cross each other in different directions and run either perpendicular to one another or form a plaited network to give the endocarp more strength against pressure and tearing forces. The innermost fibers run more or less parallel to the locule surface. Fruit development takes place within 5 months. Agoutis (*Dasyprocta punctata*) eat the mesocarp and carry the seed, burying it for a future food source (scatterhoarding method). However, the animals do not remember where they have buried the seeds. *Spondias* fruits fall when there is a shortage of food for agoutis. The population of agoutis is highest, when *Spondias* fruits ripen. In this way, agoutis are good dispersers of *Spondias mombin* (Smythe 1970). Besides, bats and birds eat the fruits and disperse the seeds. Although native of South America where it also occurs in the dry forest (Frankie, Baker & Opler 1974), the tree grows in dry deciduous forests in Ghana (fire zone) and is widely planted for its fruit in villages (Hall & Swaine 1981). Having a good distribution rate even in secondary forest, the species has only few individuals (36) in the forest studied. In Surinam, the tree is frequent in the ridge forest and in the swampy forest. In Amazonian forests 0–5 individuals occur on one hectare (Black et al. 1950).

5. *Tapirira guianensis* (Fig. 6): This ovoid drupe (0,8 × 0,5 up to 1,2 × 1 cm) contains a single stone with a single seed. The endocarp is covered by a thin edible fruit flesh. The fruit surface is blue green to black. The ovoid seed is about 1 cm long and has a papery coat. The embryo consists in 2 large cotyledons. The calyx is persistent.

The fruit surface is furnished with small protuberances arising from the elongated palisade-like outer fruit epidermis. Beneath lies a small-celled collenchymatous tissue in which small druses are embedded. The major part of the mesocarp consists in thin-walled large-celled parenchyma. Dispersed within this fundamental tissue are large secretory cavities with a resinous content surrounded by a wreath of secretory cells. Isolated sclereids of irregular shape occur scattered in the vicinity of the endocarp. The endocarp itself is composed of 2 fiber layers which cross each other, the outer layer being cut transversely in a transverse section, while the inner layers are cut longitudinally. Owing to the secretory cavities, the fruit has an aromatic smell. Dispersal takes place by animals (birds). In Surinam the species occurs in the Savanna forest and at the riverside.

The Anacardiaceae studied are of very little importance in the forest studied. Three of the species develop drupes and are dispersed by animals, e.g. birds and possibly also bats. They are represented by high trees and show a very low number of individuals, perhaps owing to their dispersal mechanism which may be disadvantageous for high trees in a dense forest. Although *Spondias mombin* is a very resistant species which supports adverse climatic factors, it shows a very low rate of distribution in the forest studied. It may be repeated here that zoochory of high trees in a dense forest seems to be disadvantageous and from this the low number of individuals may be explained. *Spondias mombin*, for example, has a much higher rate of distribution in more open regions and in secondary forest. The lower tree, *Astronium lecointei*, with fruits dispersed by animals shows a somewhat higher occurrence.

The fruit types of all 5 species are considered derived by reduction of the locules and the number of seeds. This applies to the drupe as well as to the nut and the samara.

Annonaceae

This family is considered relatively primitive. The numerous carpels are usually free. Single fruits are occasionally united in a dry or fleshy aggregate fruit. Fleshy aggregate fruits of *Annona* spp. such as soursap, cherimoya, sweetsop, custard apple, and ilama, are well known as delicious edible fruits. In *Rollinia* and *Annona*, the carpels are laterally united to form a fleshy mass in which also the receptacle takes part. Transitional types between aggregate and apocarpious fruits are not uncommon. A dry aggregate fruit is developed by *Anaxagorea* sp. (Yarayara amarilla). In several-seeded fruits the follicles are constricted between the seeds (e.g. in *Xylopia*). The follicles usually remain closed, the fleshy or brittle fruit wall decomposing at maturity. Fruits of Annonaceae are almost exclusively dispersed by birds, often aided by fruit bats and mammals. The aromatic smell of the fruits may assist in their attractiveness. The aromatic substances are located in secretory cells.

1. *Anaxagorea* sp. = Yarayara amarilla
   Hab.: RF, Dry forest — Height: a
   Fruit: Aggregate follicles — Ind.: 142
   Fr. time: 10
   Disp.: Autochorous
2. *Duguetia megalophylla* = Yarayara negra chiquita
   Hab.: RF — Height: aa
   Fruit: Aggregate drupes — Ind.: 376
   Fr. time: 3/12
   Disp.: Zoochorous
3. *Guatteria poeppigiana* = Yarayara morada
   Hab.: RF — Height: a
   Fruit: Aggregate drupes — Ind.: 72
   Fr. time: 12
   Disp.: Zoochorous
4. *Rollinia multiflora* = Anoncillo
   Hab.: RF — Height: a
   Fruit: Juicy aggregate follicles — Ind.: 118
   Fr. time:
   Disp.: Zoochorous
5. *Unonopsis glaucopetala* = Yarayara negra (grande)
   Hab.: RF — Height: a
   Fruit: Aggregate fruit — Ind.: 240
   Disp.: Autochorous — Fr. time: 3/12
6. *Xylopia frutescens* = Fruta de burro negro
   Hab.: RF — Height: A
   Fruit: Aggregate follicles — Ind.: 7
   Fr. time: 8
   Disp.: Zoochorous?
7. *Xylopia nitida* var. *nerviosa* = Fruta de burro montañero
   Hab.: RF — Height: A
   Fruit: Aggregate follicles — Ind.: 36
   Fr. time: 2/12
   Disp.: Zoochorous?

1. *Anacagorea* sp. (Fig. 7): The aggregate fruit is composed of 60–80 follicles of different size which dehisce at the ventral suture. They are flattened by one another so that each fruitlet shows several edges. The hard-shelled mericarps contain 1–(2) seeds. Dispersal takes place after dehiscence by the individual seeds. The entire length of the apocarpic fruit is 5,4 cm, the width 5,3 cm.

2. *Duguetia megalophylla:* The aggregate fruit is composed of drupaceous mericarps, each with one seed. The aromatic and fragrant fleshy fruit attracts mammalia.

3. *Guatteria poeppigiana:* The aggregate fruit is composed of drupaceous mericarps each with one seed. The fleshy fruitlets are eaten by animals (birds, monkeys).

4. *Rollinia multiflora:* The globular aggregate fruit is composed of many follicles. The fruitlets have a hairy rusty-brown surface. Each follicle is one-seeded, and has a juicy fruit flesh. Dispersal takes place by animals (birds).

5. *Unonopsis glaucopetala:* The aggregate fruit consists in about 10 mericarps. Each mericarp is dry and dehiscent and contains several small seeds. This description does, however, not coincide with that of Roosmalen where *Unonopsis* mericarps are described as one-seeded. Dispersal takes place in an autochorous way, as the mericarps are dehiscent, according to Rollet. In Surinam, the species frequently occurs, especially along creeks.

6. *Xylopia frutescens:* Up to 6 free light green to red follicles are united in a small receptacle, each follicle being pear-chaped or rhombic, woody, and with (1)–2–(3) seeds. The fruitlets, about 1 cm long, remain closed. Dispersal takes probably place through the seeds (by animals?) when the fruit wall is rotting. In Surinam the species frequently occurs in the ridge forest, in the savanna forest, and in the secondary forest.

7. *Xylopia nitida* (Fig. 7): This aggregate fruit, about 4 cm in diameter, consists in up to 25 free green to yellowish follicles which are slightly

curved and constricted between the seeds (Fig. 7). Each fruitlet contains about 2–6 (7) seeds which become free by rotting of the pericarp. In Surinam the species is infrequent.

The Annonaceae studied are of little importance in the rain-forest concerning their number of individuals. The species richest in individuals is *Duguetia megalophylla* (376 individuals) with drupelets dispersed by animals. It is the smallest tree of all species studied and belongs to height category aa. This species is followed in number of individuals by *Unonopsis glaucopetala* (240 individuals) with an autochorous dispersal mechanism through dehiscence of the fruitlets. Next in number of individuals is *Anaxagorea* (142 individuals), likewise with autochorous dispersal by dehiscence of the mericarps. *Rollinia multiflora* is represented by 118 individuals and shows dispersal of the aggregate fruits by animals. *Guatteria poeppigiana* with 72 individuals has drupelets dispersed by animals. All these species except *Duguetia megalophylla* belog to height category a. Least individuals show both species of *Xylopia*. They belong to the largest height category (A) and their aromatic fruits are probably dispersed by animals. The low number of individuals may possibly be explained by the large height of the trees making dispersal by animals less effective. All in all, the species studied do not only show dispersal by animals (birds, bats, and mammals) which is more common in Annonaceae because of the aromatic smell of the fruits, but 2 species have autochorous mechanisms through dehiscence of mericarps.

Apocynaceae

The Apocynaceae have syncarpous or (secondarily?) apocarpous fruits, often with follicular capsules opening at the ventral suture. *Aspidosperma* is characterized by flattened, roundish or ovoid capsular mericarps with several winged seeds and a long funicle. The mericarps open at the ventral suture, or ventrally and dorsally at the same time, separating into 2 valves. Each mericarp consists in one carpel of coriaceous or woody texture. Ridley calls the fruit a "pair of follicles".

1. *Aspidosperma album* = Hielillo negro
   Hab.: RF — Height: A
   Fruit: Capsule — Ind.: 3
   Disp.: Anemochorous — Fr. time:
2. *Aspidosperma excelsum* = Canjilón negro
   Hab.: RF — Height: A
   Fruit: Capsule — Ind.: 1760
   Disp.: Anemochorous — Fr. time:
3. *Aspidosperma marcgravianum* = Canjilón amarillo
   Hab.: RF — Height: A
   Fruit: Capsule — Ind.: 550
   Disp.: Anemochorous — Fr. time: 5
4. *Aspidosperma megalocarpon* = Hielillo blanco
   Hab.: RF — Height: A
   Fruit: Capsule — Ind.: 650
   Disp.: Anemochorous — Fr. time: 2
5. *Aspidosperma oblongum* = Olivo montañero
   Hab.: RF — Height: A
   Fruit: Capsule — Ind.:
   Disp.: Anemochorous — Fr. time:
6. *Himatanthus articulatus* = Mapolo
   Hab.: RF — Height: A
   Fruit: Capsule — Ind.: 59
   Disp.: Anemochorous — Fr. time:
7. *Parahankornia amapa* = Arbol de la vaca
   Hab.: RF — Height: a
   Fruit: Berry — Ind.: 4
   Disp.: Zoochorous — Fr. time: 2/5
8. *Tabernaemontana psychotriifolia* = Cojón de verraco
   Hab.: Savanna, RF — Height: A
   Fruit: Berry — Ind.: 10
   Disp.: Zoochorous — Fr. time:
9. Leche de burra
   Hab.: RF — Height: A
   Fruit: Drupe — Ind.: 11
   Disp.: Zoochorous — Fr. time:

1. *Aspidosperma album:* The large mericarp is flattened, of light brown colour, and covered with brown hairs. The outlines are circular to ovoid. It reaches a length of 12 cm and a width of 8,7 cm, and opens with 2 valves. The pericarp measures 0,6 cm in thickness. The endocarp inside is smooth and brittle. It contains several disc-shaped seeds of circular outlines consisting in a more or less circular seed body, up to 2 cm in diameter, which lies in the center being completely surrounded by a white papery wing. The seed is attached with a long funicle, about 2,5 cm long. The entire seed diameter reaches 7 cm. Dispersal of the seeds takes place by wind. In Surinam the tree is frequent.

2. *Aspidosperma excelsum:* The mericaprs have about circular outlines and are 8 cm long and 7,8 cm broad. They are covered with short spines, about 1 cm long (Figs. 10, 11). The pericarp is 0,6 cm in thickness. The texture is woody. The endocarp inside is sculptured. The mericarp contains several winged seeds of circular outlines. The seed body of each seed is completely surrounded by a papery wing. Dispersal of the seeds takes place by wind. In Surinam the tree is infrequent. The fundamental tissue of the pericarp consists in thin-walled parenchyma in which large groups of sclereids with layered walls are scattered (Fig. 10). As seen in t.s., the sclerenchymatous groups have more or less circular outlines. As the sclerenchyma quantitatively prevails, it makes the pericarp hard and woody. The endocarp is composed of fiber-sclereids which are frequently bent in different directions.

3. *Aspidosperma marcgravianum:* Each of the 2 mericarps is about 4,7 cm long and 4,5 cm broad exhibiting more or less circular outlines. The pericarp texture is woody and the hairy surface shows a warty or alveolate sculpturing. Frequently, only one mericarp is developed, the other one being aborted. Pericarp thickness measures about 0,3 cm. Dehiscence takes place longitudinally with 2 valves. The several seeds are flat and winged, the seed body being completely surrounded by the wing, but in an excentric manner. The diameter of the circular seed reaches 2 cm or more. Dispersal of the seeds takes place by wind. In Surinam the tree frequently occurs.

The fruit surface is rough by large warts or knobs which are covered with unicellular hairs. The fundamental tissue of the pericarp is parenchymatous, but large sclereids partly of very irregular shape are scattered within it (Fig. 10). These add the woody texture to the pericarp. The endocarp is composed of thick-walled sclerenchyma.

4. *Aspidosperma megalocarpon* (Figs. 8, 9): The fruit is composed of 2 carpels, however, often only one mericarp develops. The flattened large mericarps have ovoid to circular outlines, reaching up to 17 cm in length and 10–11 cm in width. The dark-green to yellowish mericarp surface is covered with short brown hairs. The fruit outside – with numerous lenticels – is sculptured by numerous longitudinal ribs. The pericarp wall is very thick. Each mericarp contains several seeds. The centrally located seed body of the winged seeds measures about 2,5 × 2 cm, the entire seed diameter together with the wing reaching 6,5–7,5 cm. The white to yellowish wings are very thin and papery, completely surrounding the seed body. The funicle is about 1,7 cm long and attached to the center of the seed. Dispersal of the seeds takes place by wind. The species shows a good rate of distribution also in the secondary forest. In Surinam the tree is common.

5. *Aspidosperma oblongum:* The fruit is composed of 2 mericarps. The mericarps are almost circular to ovoid, flattened, about 8 cm long and 6,7 cm broad, of woody texture, and covered with brown hairs. The fruit surface is somewhat sculptured by little knobs. The pericarp thickness measures 0,2 cm. The endocarp inside is fibrous. The capsule opens with 2 valves. The few winged seeds with a diameter of about 2 cm have circular outlines. The seed body has a velvety yellow indumentum. The long funicle reaching 3 cm in length is attached to the seed center. Dispersal of the seeds takes place by wind. The species occurs in Surinam and French Guiana too.

6. *Himatanthus articulatus:* The fruit is composed of 2 flattened mericarps of more or less cylindrical shape. The mericarps reach 29 cm in length and 4 × 2,7 cm in thickness. They are delicately longitudinally ribbed. The pericarp mea-

sures 0,3 cm in thickness. In the leathery fruit wall an outer dark-brown, a middle light-brown, and an inner light-brown transparent thin layer may be distinguished. The mericarp opens along the ventral suture. The inside of the valves is smooth and brilliant. Each mericarp contains more than 50 seeds of elliptic shape, measuring 4 × 2,5 × 0,2 cm, with a 1–2 cm broad wing completely surrounding the seed body. The seeds are dispersed by wind. The species is found in Surinam in the ridge forest and in the high savanna forest as well as in the savanna.

7. *Parahankornia amapa:* The globular fleshy berry measures 4,5 × 4 cm, but may even reach a diameter of up to 8 cm. The surface is smooth, devoid of hairs and green. The pericarp reaches about 0,1 cm in thickness. The fruit flesh is yellow and edible. The mericarp contains few flattened seeds of ovalate shape, 1,5 cm long and 1 cm borad. The berry is dispersed by animals. The species occurs in Surinam in the savanna forest and in the swampy forest too.

8. *Tabernaemontana psychotriifolia:* The mericarps of the twin-fruit are of circular shape, 4,5 cm long and 2,5 cm broad. The vernacular name of the plant is explained by the shape of the mericarps which resemble testicles of a boar. The numerous seeds are 1 cm long. The fleshy indehiscent fruits are probably dispersed by animals (birds).

9. Leche de burra: The fleshy indehiscent fruit is relatively large and contains a single seed. The drupe is dispersed by animals.

Almost all species of the Apocynaceae studied are represented by high trees of height category A, except *Parahankornia amapa.* Their fruits are twin-fruits. Three of the species studied show dispersal of their fruits by animals, including *Parahancornia* of height category a. They have an extremely low rate of distribution. All other species show dispersal of their seeds by wind. The species of *Aspidosperma* resemble one another very much in the structure of their fruits. The mericarps are more or less of circular outlines and flattened. They contain several winged seeds. The seed body is located in the seed center (with one exception) and surrounded by a papery wing. The seed body usually occupies a small part of the entire seed and the wing mainly consists of 2 epidermal layers between which an aerenchyma is expanded. The seed is, therefore, light-weighted and thin. With its large surface and light specific weight it represents a good "discflyer". Although all 6 anemochorous species of *Aspidosperma* are very much alike in their fruit and seed structure, their rate of distribution is quite different. *A. excelsum* reaches a very high rate with 1760 individuals. *A. marcgravianum* and *A. megalocarpon* have a medium distribution rate of 550 and 650 individuals, while in 3 species the number of individuals is extremely low. There is no apparent reason, why the number of individuals should be so different, if not historic-geological reasons are considered. But the Apocynaceae studied show once again with much clarity that wind dispersal of very high trees is much more advantageous than dispersal by animals.

Araliaceae

The fruits of the Araliaceae are berries or drupes. There is usually one seed in each locule. The fruits are almost exclusively dispersed by frugivorous birds, often aided by fruit-bats and mammals.

1. *Dendropanax arboreus* = Maramo blanco

Hab.: RF	Height: a
Fruit: Drupe	Ind.: 1
Disp.: Zoochorous	Fr. time: 6

2. *Didymopanax morototonii* = Sun-sun

Hab.: RF	Height: a
Fruit: Drupe	Ind.: 40
Disp.: Zoochorous	Fr. time: 6

3. *Oreopanax capitatum* = Matapalo blanco

Hab.: RF	Height:
Fruit: Berry	Ind.: 1
Disp.: Zoochorous	Fr. time:

1. *Dendropanax arboreus:* The fleshy drupes are more or less of globular shape reaching a diameter of 0,8–1 cm. They are slightly flattened and have persistent stigmas. When ripe they have a purple or dark-red brilliant colour and a slightly resinous smell due to secretory cavities. They contain up to 7 flattened seeds, 4 cm long. The fruits ripen almost throughout the year, but in Mexico preferably in March and August. They have a very wide distribution and also occur in deciduous forest. They are abundant in the secondary forest in Mexico, according to Pennington & Sarukhan (1968). The fruits are dispersed by animals, mostly birds.

2. *Didymopanax morototonii:* The fleshy drupes have a flattened globular shape reaching a length of 0,6–0,7 cm and a width of 1–1,1 cm. They are gray-green and have 2 persistent stigmas. They contain 2 monospermic stones. The flattened seeds reach 0,4 cm in diameter. On their outside they show about 10 longitudinal ribs. In Mexico the fruits ripen between January and March. The three occurs abundantly in the secondary vegetation. According to Roosmalen, it is a common tree in the rain forest and in the savanna of Surinam. The fruits are dispersed by animals, mainly birds.

3. *Oreopanax capitatum:* The fleshy indehiscent fruit probably represents a berry. It contains few seeds and is dispersed by animals, mainly birds.

The 3 species of Araliaceae studied have very few representatives in the forest studied. The trees belong probably all to height category a and their fruits are dispersed by animals, an effective dispersal mechanism of lower trees.

Bignoniaceae

The fruits are frequently capsules dehiscing with 2 valves. Dehiscence is loculicidal or septifragal. The seeds are commonly winged. The wings are papery and transparent. The seed body may be entirely surrounded by a wing. Seeds are winged on both sides in *Bignonia.* The cells in the wings are arranged in rows parallel with the long side of the wing. The embryo is characteristically heart-shaped. The seeds are dispersed by wind.

1. *Jacaranda copaia* = Simaruba
   Hab.: RF — Height: A
   Fruit: Capsule — Ind.: 49
   Disp.: Anemochorous — Fr. time: 2
2. *Jacaranda obtusifolia* = San José
   Hab.: RF — Height: A
   Fruit: Capsule — Ind.: 111
   Disp.: Anemochorous — Fr. time:
3. *Tabebuia serratifolia* = Araguaney
   Hab.: Deciduous forest, RF — Height: A
   Ind.: 18
   Fruit: Capsule — Fr. time: 5
   Disp.: Anemochorous
4. *Tabebuia stenocalyx* = Purguillo blanco
   Hab.: RF — Height: a
   Fruit: Capsule — Ind.: 704
   Disp.: Anemochorous — Fr. time: 5
5. *Tabebuia* sp. = Araguaney puig
   Hab.: RF — Height: A
   Fruit: Capsule — Ind.: 17
   Disp.: Anemochorous — Fr. time:
6. *Tabebuia* sp. = Puig
   Hab.: RF — Height: A
   Fruit: Capsule — Ind.:
   Disp.: Anemochorous — Fr. time:

1. *Jacaranda copaia* (Figs. 12, 13, 14): The large woody loculicidal capsule opens with 2 valves. It is flat, of ovoid shape and dark-brown colour, measuring 8–16,5 cm in length and 6–9 cm in width. The pericarp thickness is 0,4 cm. The fruit surface is rough. The endocarp is fibrous. Groups of fibers are arranged in different directions. The dehiscence is loculicidal and septifragal at the same time. The numerous seeds are almost circular in outline and measure about 4,5 cm in dia-

meter. The seed body is entirely surrounded by a papery wing. Dispersal of the seeds takes place by wind. Foster (1982) reports for Barro Colorado Island that on a clear September morning trees can fill the air above the canopy with millions of seeds.

The outer fruit epidermis is very small-celled and has a thin cuticle. Beneath follows a very small-celled parenchyma in which groups of sclereids are interspersed. The parenchyma cells are arranged in the form of tangential rows due to the growth in circumference of the fruit. Dilatation growth is vigorous, especially in the outermost part of the pericarp. As the numerous sclereids form an obstacle against this tangential growth being unable to extend further, the parenchyma cells separating the sclerenchyma groups considerably extend in a tangential direction. Towards the inside, the size of the parenchyma cells as well as that of the sclereids increases reaching a maximum in the vicinity of the endocarp, while the amount of stone cell groups considerably diminishes thinning out completely towards the endocarp. The sclerenchymatous endocarp shows the familiar irregular outer surface protruding into the mesocarp in the form of teeth, as seen in t.s. (Fig. 13). The endocarp is vascularized, whereby ramifications enter the mesocarp. The fiber-like sclereids of the endocarp run in different directions, they appear, consequently, partly cut transversely and partly longitudinally, as seen in transverse section. As seen in longitudinal section, the vascular bundles are surrounded by layers of parenchyma composed of roundish cells with thickened walls. The vascular bundles thus divide the endocarp into an outer stony part with irregular outlines towards the mesocarp, a looser central part composed of roundish cells in which the vascular bundles are embedded, and an inner fibrous part (Figs. 12, 13, 14). The roundish cells are sclerified parenchyma cells with thickened walls and pits. Where vascular bundles are absent, the endocarp has a more or less homogeneous structure. In Surinam, the species occurs in the swampy forest too and on ridges of the coastal plane.

2. *Jacaranda obtusifolia:* The large woody loculicidal capsule is of ovoid shape and has undulated margins. It is about 7 cm long and 3,7 cm broad, of brown colour and with a sculptured surface. It opens with 2 valves liberating numerous seeds. The seeds are about 2,7 cm long and 1,4 cm broad, flat, and of light brown colour. The seed body is entirely surrounded by a papery wing, about 1 cm in width. Dispersal of the seeds takes place by wind.

3. *Tabebuia serratifolia:* The elongated capsule measures between 31 and 65 cm in length and 1,3 to 2 cm in width and 0,2 to 1,5 cm in thickness. It is irregularly constricted between the seeds. The dehiscing capsule releases many small seeds, 1 cm long, 3,5 cm broad, and 0,1 cm in thickness. The seeds have 2 lateral wings. They are dispersed by wind. The species has a profound radical system and is resistant to drought and poor soils. It is characteristic of the deciduous forest. In Surinam, the tree is very common, also in the swampy forest and at the riverside and on ridges.

4. *Tabebuia stenocalyx:* The dry dehiscent capsule contains numerous winged seeds. The seeds are dispersed by wind.

5. *Tabebuis* sp. (Araguaney puig): The fruit is an elongated to elliptic dehiscent capsule in which the separating wall is in perpendicular position to the valves. The elongated seeds have transparent wings and are dispersed by wind.

6. *Tabebuia* sp. (Puig): The elongated capsule dehisces with 2 valves, releasing numerous seeds with transparent wings. The seeds are dispersed by wind.

Five of the six species studied are represented by high trees of height category A. All 6 species have the same fruit type, the capsule, and their seeds are dispersed by wind. These are of the "sailing flyer" type. However, the number of individuals is very low remaining below 100 in 4 species. One species has 111 individuals (*Jacaranda obtusifolia*), while *Tabebuia stenocalyx* is represented by an elevated number of individuals, 704 altogether. However, this species is of height category a. This fact is surprising, because one should assume that the highest trees are best dispersed by wind and, therefore, show the highest rates of distribution. Most of the species of Bignoniaceae are wind-dispersed. The seeds often separate from the funicle within the fruit so that they may fall out easily when the capsule splits into 2 valves. Furthermore, the seeds are arranged in the fruit like coins in a roll of money so that they may easily slip out from the fruit one after the other. The seeds with 2 lateral wings belong to the "gliding sail-plane" type. In this case, the point of gravity is in excentric position being shifted towards the front. The light seeds glide calmly to the ground describing ample helicoidal movements. As many of the Bignoniaceae are represented by high and very high trees, dispersal of their seeds by wind is advantageous for them. Families like the Bignoniaceae mainly dispersed by wind are comparatively scarce in flowering plants. The Bignoniaceae show, therefore, once more that dispersal by wind is an adaptation of high trees to the environmental conditions. Likewise, species which grow in the deciduous forest (e.g. *Tabebuia serratifolia*), have characteristically anemochorous dispersal units.

Furthermore, abundance and ecological importance of the Bignoniaceae are greatest in dry areas. However, only abundance of individuals, but not species diversity is greater in dry than in wet forest (Gentry 1976).

Bixaceae

Bixa orellana = Onoto rebalsero

Hab.: Deciduous forest, Riverside	Height: aa
Fruit: Capsule	Ind.: 34
Disp.: Autochorous and zoochorous	Fr. time:

The dry dehiscent capsule reaches up to 5 cm in length. It has an ovoid shape and is unilocular, opening with 2 valves. The dark-red fruit is densely covered with red flexible spines 0,7–0,9 cm long, and contains numerous seeds surrounded by a showy orange-red aril. The Indians use it to paint their bodies a red colour. The red principle ("achiote") is also used to dye food. The generic name *Bixa* comes from the Indian dialect "taino" in which the "x" is pronounced as the English "sh". The shrub was cultivated in former times and the seeds exported. The species uses a twofold dispersal mechanism which is autochorous by dehiscing capsules, while the seeds surrounded by an attractive aril are dispersed by animals. However, the species is not characteristic of the forest studied and the number of individuals, therefore, low. In Surinam it grows in swampy forests along rivers and on ridges, and is cultivated too.

Bombacaceae

The fruit is either a loculicidal capsule or berry-like, and the seeds are often embedded in a woolly indumentum originating from the pericarp inside, or in a pith-like tissue.

1. *Catostemma commune* = Baramán

Hab.: RF	Height: A
Fruit: Capsule	Ind.: 767
Disp.: Autochorous	Fr. time: 1

2. *Ceiba pentandra* = Ceiba

Hab.: RF	Height: A
Fruit: Capsule	Ind.: 4
Disp.: Anemochorous	Fr. time:

3. *Eriotheca* sp. = Cedro dulce

Hab.: RF	Height: A
Fruit: Capsule	Ind.: 50
Disp.: Aneomochorous	Fr. time: 8

1. *Catostemma commune* Figs. 15, 16, 17, 18): The unilocular woody capsule measures 8,5 cm in length and 4,5 cm in width. The fruit opens loculicidally, but only contains one seed. Dispersal is autochorous. Although the number of seeds is reduced to one, dispersal seems to be advantageous (with 767 individuals in the forest).

The outer pericarp epidermis is furnished with stellate hairs. Beneath follows a region of parenchyma in which roundish groups of stone cells are scattered (Fig. 15). The major part of the pericarp consists in thick-walled sclerified parenchyma. The thick walls are penetrated by numerous small pits. Less thick-walled parenchyma cells which contain tannin give a brown aspect. Numerous sclerenchymatous plates composed of stone cells extend in a radial direction of the fruit (Fig. 18). Parenchyma cells between these plates become tangentially extended and later on divide by anticlinal walls giving the aspect of expanding phloem rays. Vascular bundles are capped on the phloem side by fibers - the thickest-walled fibers lying on the bundle periphery. Slime cavities may be interspersed within the parenchyma.

The most curious characteristic of this fruit is the formation of a "rhytidome" through development of successive periderm layers (Figs. 15, 16, 17, 18). The first periderm may arise about 10–20 layers beneath the outer fruit epidermis. The cork of the periderm remains small, while the phelloderm develops considerably. It may occupy a large part comprising up to 50 layers or more. It also becomes stratified or layered by alternate formation of thin-walled cells and well-developed stone cells. Where an obstacle like a sclerenchyma plate lies in the way of the advancing periderm, this is possibly embraced in the form of a ring. In some regions, periderms profoundly penetrate into the fruit flesh (e.g. to avoid groups of sclerenchymatous cells). The cork cells may have U-shaped wall thickenings at the beginning, but later on transform into stone cells with homogeneously thickened walls. The origin of the periderm formation is discontinuous, but later on a fusion of the periderms takes place so that calottes are cut out of the pericarp. In this way, the outer pericarp regions adopt the aspect of a scaly rhytidome resembling the bark of a stem. The bark-like rind may reach 0,5 cm in thickness or more.

Lenticels on fruits are not so uncommon and occur, for example, on the apple. Periderm formations on fruits, on the other hand, are very rare. An example is *Achras sapota* (Roth & Lindorf 1972b). But the presence of a large scaly rhytidome in a pericarp has never been observed before. It is unique in *Catostemma* and its necessity or function is unclear. It may be understood as a means of protection of the capsule against mechanical injuries and desiccation.

2. *Ceiba pentandra:* The hard woody loculicidal capsule contains many seeds embedded in a cotton-like wad produced by the fruit endocarp. The fruit reaches up to 24 × 11 cm, is brown, has 5 longitudinal ribs, and opens with 5 valves. The seeds measure 0,6 × 0,5 × 0,5 cm. The calyx is persistent. The endocarp hairs are commercially used as "kapok". The seeds are dispersed by wind together with the hairs, which considerably reduce their specific weight. A single tree may cover a very large area with its woolly wad. The very high emergent tree (up to 55 m or more) is more characteristic of open regions, hence the small number of individuals in the forest studied. In Surinam the species occurs along riverbanks and in swampy forests too. The fruits ripen in April & May (Williams 1951).

3. *Eriotheca* sp. (Cedro dulce): The woody loculicidal capsule is divided into 5 locules with numerous small seeds. The fruit length measures 5,8 cm, the width 3,6 cm. The endocarp inside is covered with hairs. The seeds are dispersed by wind together with the hairs.

The 3 species of Bombacaceae studied are represented by high to very high trees and dispersal of the seeds together with a woolly mass of hairs takes place by wind in *Ceiba* and *Eriotheca*. Nevertheless, the number of individuals in the forest studied is very low. *Catostemma commune*, on the other hand, shows a large number of individuals (767), but the number of seeds per capsule is reduced to one and the dispersal method is less effective, being autochorous.

Boraginaceae

The Cordioideae are generally represented by fleshy drupes with a 1–4 locular "stone".

1. *Cordia alliodora* = Pardillo
   Hab.: RF — Height: A
   Fruit: Samara — Ind.: 16
   Disp.: Anemochorous — Fr. time:
2. *Cordia bicolor* = Caujaro blanco
   Hab.: — Height: A?
   Fruit: Drupe — Ind.:
   Disp.: Zoochrous — Fr. time: 6–7
3. *Cordia exaltata* = Alatrique blanco
   Hab.: RF — Height: A
   Fruit: Drupe — Ind.: 624
   Disp.: Zoochorous — Fr. time: 2
4. *Cordia fallax* = Alatrique negro
   Hab.: RF — Height: A
   Fruit: Drupe — Ind.: 263
   Disp.: Zoochorous — Fr. time: 6–7/12
5. *Cordia viridis* = Caujaro
   Hab.: RF — Height: A
   Fruit: Drupe — Ind.: 164
   Disp.: Zoochorous — Fr. time:
6. *Lepidocordia punctata* = Guatacare negro
   Hab.: RF — Height: A
   Fruit: Berry — Ind.: 44
   Disp.: Zoochorous — Fr. time: 9
7. *Guatacare blanco*
   Hab.: RF — Height: A
   Fruit: ? — Ind.: 3
   Disp.: Zoochorous? — Fr. time:

1. *Cordia alliodora:* The small nutlets persist with all their floral parts. The persistent scarious corolla acts as a parachute in the dispersal of the fruit by wind. The papery petals are brown. The samara contains a seed about 4–5 mm long. The tree is abundant in the secondary vegetation (dry hillside), and also in the semideciduous forest (see also Frankie, Baker & Opler 1974). Dispersal of the samara takes place during the dry season according to Janzen (1976).

2. *Cordia bicolor:* The drupe has an ovoid shape, measuring 1,2 × 0,8 cm. The surface is hairless and brilliant. The indehiscent fruitlet contains a single stone with a single seed. The stone is egg-shaped and about 1 cm long. The persistent calyx only reaches 0,2–0,4 cm, and has an indumentum. The drupe is dispersed by animals (birds). In Surinam the species is uncommon, occurring in hilly regions of the rain forest.

3. *Cordia exaltata:* The ovoid drupe, about 1,4 cm long and 1,1 cm broad contains a single stone with a single seed. The thick pericarp becomes woody. The fruit represents a derived form with a reduced number of locules and seeds. The fruit flesh is possibly edible. Dispersal of the drupe takes place by animals (birds and mammals).

The fruit has a small-celled outer epidermis with thick outer walls which are not cutinized, but covered with a cuticle (Fig. 19). Beneath follows a subepidermal layer of small cells. Almost the entire pericarp consists in a cartilaginous collenchyma (Fig. 20) the cells of which contain small chloroplasts. Due to the irregular wall thickenings, the thickned angles of the walls and the enlarged pits, the cells adopt an irregular shape (Figs. 20, 21). Towards the inside the cells become compressed in a radial direction. The cell size diminishes somewhat towards the inside. The vascular bundles mainly lie on the inside of the hard pericarp part. The pericarp owes its woody texture to the tough cartilaginous tissue. Towards the inside follow large tracheidal cells with spiral wall thickenings. Dispersal of the fruit takes place by animals. The rate of distribution of this species is high with 624 individuals.

4. *Cordia fallax:* The dry indehiscent drupe contains a single seed. Dispersal takes place by animals (e.g. birds).

5. *Cordia viridis:* The dry indehiscent drupe contains a single seed. Dispersal takes place by animals (e.g. birds).

6. *Lepidocordia punctata:* The fleshy indehiscent fruit is small and contains only few seeds. The berry is dispersed by animals.

7. Guatacare blanco: The fruit is unknown.

The 7 species studied of Boraginaceae belong all to height category A. However, most species (6) have drupaceous fruits which are dispersed by animals. Their rates of distribution are relatively low, except in *Cordia exaltata* (with 624 individuals). This species is followed by *Cordia fallax* with 263 individuals, and by *C. viridis* with 164 individuals. The only species that is dispersed by wind having a samara with a papery parachute of petals is represented by very few individuals (16). According to Opler, Baker & Frankie (1975), the ancestors of *Cordia* species had fleshy fruits. Animal dispersal is common in shurbs and understory trees, while the higher trees are dispersed by wind. Species represented by shrubs use birds as their dispersers, species of medium height are dispersed by mammalia. The fleshy fruits of small trees and shrubs are often mucilaginous and adstringent. Sticky *Cordia* fruits (with slimy contents) show epizoochory, usually ornitho-epizoochory.

The fruits of the species studied here are drupaceous belonging to trees of height category A. These species may possibly be considered still in a stage of progressing development. Only *C. alliodora* is adapted to wind dispersal.

Burseraceae

The species of Burseraceae considered in these studies have drupaceous fruits with 1–5 stones each containing a single seed. However, at maturity the fruits dehisce representing capsules from this point of view. But generally only the outer fruit part dehisces, while the stone remains closed. In the Protieae, the stones are not united, but separate at maturity. The pericarp is characterized by secretory cavities which contain resinous and oily substances. Oil-birds (*Steatornis*) eat and collect the fruits (Van der Pijl 1972). The fruits are also eaten by fish.

1. *Protium decandrum* = Azucarito
   Hab.: RF — Height: A
   Fruit: Drupe — Ind.: 2669
   Disp.: Zoochrous (autochorous) — Fr. time: 2/5–6/11
2. *Protium heptaphyllum* = Tacamajaca negra
   Hab.: RF — Height: a
   Fruit: Drupe — Ind.: 1
   Disp.: Zoochorous (autochorous) — Fr. time:
3. *Protium neglectum* = Azucarito blanco
   Hab.: RF — Height: A
   Fruit: Drupe — Ind.: 293
   Disp.: Zoochorous (autochorous) — Fr. time: 2
4. *Protium neglectum* = Caraño
   Hab.: RF — Height: a
   Fruit: Drupe — Ind.: 1157
   Disp.: Zoochorous (autochorous) — Fr. time:
5. *Protium* sp. = Azucarito amarillo
   Hab.: RF — Height: A
   Fruit: Drupe — Ind.:
   Disp.: Zoochorous (autochorous) — Fr. time:
6. *Protium* sp. = Caraño blanco
   Hab.: RF — Height: a
   Fruit: Drupe — Ind.: 26
   Disp.: Zoochorous (autochorous) — Fr. time: 2/8
7. *Protium* sp. = Sipuede
   Hab.: RF — Height: A
   Fruit: Drupe — Ind.: 155
   Disp.: Zoochorous (autochorous) — Fr. time:
8. *Tetragastris panamensis* = Aracho
   Hab.: RF — Height: A
   Fruit: Drupe — Ind.: 698
   Disp.: Zoochorous (autochorous) — Fr. time:
9. *Tetragastris panamensis* = Caraño negro
   Hab.: RF — Height: a
   Fruit: Drupe — Ind.: 408
   Disp.: Zoochorous (autochorous) — Fr. time: 2/5
10. *Tetragastris?* or *Dacryodes?* = Aracho blanco
   Hab.: RF — Height: a
   Fruit: Drupe — Ind.:

Disp.: Zoochorous (autochorous) Fr. time: 11–12

11. *Trattinickia* sp. = Maro
Hab.: RF Height: a
Fruit: Drupe Ind.: 33
Disp.: Zoochorous (autochorous) Fr. time:

12. Maramo
Hab.: RF Height: A
Fruit: Drupe Ind.: 9
Disp.: Zoochorous (autochorous) Fr. time:

1. *Protium decandrum:* The globular drupaceous capsule shows 5 ribs and is about 3 cm long and 2,3 cm broad. It is 5-locular and contains 5 seeds. The pericarp is 0,2 cm in thickness. Dispersal seems to be very effective, as the species is represented by 2669 individuals. The species is typical of Guiana. It is dispersed by birds.

2. *Protium heptaphyllum:* The drupaceous capsule is either asymmetrically ovoid and contains a single stone or is of globular shape and contains 2–3 stones. It measures about 1,3 × 1,5 × 0,8 – 1,5 cm. The fruit is dehiscing. According to Pittier, the fruit is a resinous drupe which abundantly contains a perfumed oleo-resin similar to the "elemi" which is of economic importance. Dispersal takes place by birds. This tree is not characteristic of the rain forest, but has its main distribution all over the coastal plains of Venezuela and the plains of the Orinoco river, hence the small number of individuals in the forest studied. It may also be dispersed by fish. In Surinam, the species mainly occurs in the savanna forest and in the ridge forest, and also along the riverside.

3. *Protium neglectum* (Azucarito blanco): The drupe is either asymmetrically ovoid with a single stone or ovoid and 2–4 (5) locular with 2–4 (5) seeds. Apparently, a reduction of locules and seeds is taking place here. The fruit size measures 1,8 × 1,5 – 2,4 × 1 – 1,5 cm. Dispersal is mainly zoochorous (e.g. by birds). In Surinam the species is common.

4. *Protium neglectum* (Caraño): The drupaceous capsule dehisces and only contains few seeds. Dispersal mainly takes place by animals (birds). The 2 types of trees designated as *Protium neglectum* probably correspond to 2 different species (vernacular names!) and also differ in the height categories. The smaller trees show a far higher number of individuals (1157, as compared with 293). Roosmalen describes P. neglectum as a shrub or small tree up to 20 m. In Aamzonian forest 1–3 individuals occur on one hectare (Black et al. 1950).

5. *Protium* sp. (Azucarito amarillo): The drupaceous capsule is furnished with 3 ribs and has a woody endocarp which is 3-locular each locule containing a single seed. The fruit measures 2,5 cm in length. It is dispersed by birds.

6. *Protium* sp. (Caraño blanco): The 3-locular globular drupaceous capsule is flattened in an axial direction, measuring 1,5 cm in length and 2,9 cm in diameter. It opens loculicidally. The pericarp thickness reaches 0,2 cm. The inner fruit surface corresponding to the endocarp inside is reticulate. Each locule contains a single triangular seed. Dispersal takes place by birds.

7. *Protium* sp. (Sipuede): The drupaceous capsule is dehiscing and contains only few seeds. It is dispersed by animals (birds).

8. *Tetragastris panamensis:* The drupaceous capsule opens with 2–5 valves and contains 2–5 stones. The fruit is very variable in its shape depending on the number of stones it contains. It is globular to obliquely ovoid and then contains only one stone. It measures about 3,5 × 3,5 cm and has a smooth red surface. The arillate seeds are dispersed by mammals (e.g. monkeys) and by birds. In Surinam the species is very common.

9. *Tetragastris (panamensis?)* (Caraño negro): The drupaceous capsule opens with 2–5 valves and only contains few seeds. Each seed is enclosed in its own stone. The fruit is either globular and contains 2–5 stones or of ovoid shape and then contains only a single seed. It measures about 4 × 3,5 × 3,5 cm, and has a smooth surface. The fruit may be dispersed by birds.

10. *Tetragastris* or *Dacryodes?:* The drupaceous capsule is dehiscent and contains only few seeds. Dispersal takes place by birds and monkeys.

11. *Trattinickia* sp.: The drupaceous capsule

opens with 2 valves, separating into 2 stones. The 2 stones are separated by a very small layer of fruit flesh. The endocarp is woody and has a wrinkled surface. The drupe may be dispersed by birds.

12. Maramo: The dehiscent drupaceous capsule contains only few seeds. Dispersal may take place by birds.

It seem to be possible that the Burseraceae studied had originally capsular fruits which are progressively transforming into drupes – the mesocarp becoming fleshy and the seeds becoming reduced in number. Most of the species studied are represented by very high (height category A) or high trees (height category a). Their original dispersal mechanism is autochorous. The fruits of all species studied are more or less alike being capsular drupes or drupaceous capsules, often separating at maturity into separate stones, each single-seeded. Their pericarp is resinous and oily through abundance of secretory cavities. However, the 12 species differ enormously in the number of individuals present in the forest studied. This circumstance can certainly not be attributed to the dispersal mechanism which is similar in all species studied, although there may be certain differences, and some strategies may be more advantageous than others. Dispersal may still be partly autochorous through separating stones, but definitely also takes place via animals (birds, mammals, fish). Comparing the 2 types of trees identified as *Protium neglectum* one can observe that the zoochorous dispersal mechanism is more advantageous in the smaller trees (a: with 1157 individuals), as compared with the higher trees (A: with 293 individuals). Species which are not characteristic of the rain forest naturally show a lower number of individuals (e.g. *Protium heptophyllum*). But the unequal distribution of the species, i.e. the low number of individuals in some species, and the high number in others calls for another way of explanation which may be "historic-geological" (see p. 169/170).

Capparidaceae

The fruit often consists in 2 carpels. The pistil borne on a gynophore not seldom develops into a capsule dehiscing by valves, a berry or a drupe.

1. *Capparis amplissima* = Burro muerto

Hab.: RF	Height: A
Fruit: Drupe	Ind.: 3
Disp.: Zoochorous	Fr. time: 8

2. *Capparis guaguaensis* = Toco negro

Hab.: RF	Height: a
Fruit: Berry	Ind.: 7
Disp.: Zoochorous	Fr. time: 2/5

3. *Crataeva tapia* = Toco blanco

Hab.: RF	Height: aa
Fruit: Berry	Ind.: 1
Disp.: Zoochorous	Fr. time:

1. *Capparis amplissima:* The one-seeded dry indehiscent fruit should rather be considered a drupe. It may have a creamy endocarp. Dispersal probably takes place by animals (e.g. birds).

2. *Capparis guaguaensis:* The fleshy indehiscent fruit contains numerous seeds and represents a berry. It is dispersed by animals (e.g. birds).

3. *Crataeva tapia:* The globular to elliptic fruit dangles from a 2,5–6 cm long gynophore and contains numerous kidney-shaped seeds. The berry reaches a diameter of 2–3,5 (7) cm, and has a smooth green, yellow or orange surface. The edible juicy pulp has a white colour and a sweet taste. The rind has a repulsive odour. The berry is dispersed dy animals. In Surinam the species is common on ridges at the coast line. The Capparidaceae are of no importance in the rain-forest studied, concerning their number of individuals. Noteworthy is that the highest tree species (*Capparis amplissima*) shows a deduced fruit form with only one seed being modified to a drupe, while the other 2 species have fleshy berries.

Caryocaraceae

Caryocar nuciferum = Almendra

Hab.: RF	Height: A
Fruit: Drupe	Ind.: 10
Disp.: Zoochorous	Fr. time: 2/6

This heavy globular giant drupe reaches 15 cm in diameter or more. It is gray-brown and has a thick fleshy shell containing 1–2 (–4) stones. Each stone is kidney-shaped and contains one seed. Occasionally, the fruit encloses only one functional seed and 3 abortive ones, each in its own pyrene. The endocarp measures about 8 mm in thickness. Its inside is smooth, while its outside shows a tuberculate appearance. The large seeds, about 3×4,8 cm up to 7,5 cm, are kidney-shaped and nut-like. The seed coat measures 1 mm in thickness. The large embryo is edible and oily, filling most of the seed cavity with its hypocotyl.

The endocarp is stone-hard and fibrous in its inner part. The fiber bundles run in different directions. Those of the outer layers are partly radially (longitudinally) arranged, while those of the inner layers run tangentially to the fruit surface. The endocarp outside has a very conspicuous tuberculate appearance, being covered with innumerable knobs. These protuberances are composed of thick-walled roundish cells with pits and a brown content (Figs. 22, 23). The exocarp is said to be oily. The mesocarp of this giant fruit is probably eaten by animals. But the type of dispersal seems to be disadvantageous, as the fruit is too large for the number of seeds it contains (1–2). Owing to the size of the fruit, there are relatively few fruits on the tree. This is probably a very primitive form of dispersal (original barochory?) apparently representing a "dead end" judging by the number of individuals in the forest. Germination of the seeds is at first slow, but rapid on emergency (Corner 1976). Guiana represents a refuge area of this species (Prance 1973). In Surinam the species is uncommon, occurring in hilly regions of the rain forest in the West.

Celastraceae

The fruits are usually capsules, drupes, samaras or berries. The seeds often have a colourful aril. Capsules may contain winged seeds.

1. *Goupia glabra* = Congrio blanco

Hab.: RF	Height: A
Fruit: Berry	Ind.: 18
Disp.: Zoochorous	Fr. time:

2. *Maytenus* sp. = Pilón negro

Hab.: RF	Height: A
Fruit: Capsule	Ind.: 237
Disp.: Autochorous	Fr. time: 12

3. *Zinowiewia* aff. *australis* = Guatacare montañero

Hab.: RF	Height: A
Fruit: Samara	Ind.: 39
Disp.: Anemochorous	Fr. time:

1. *Goupia glabra:* The fleshy indehiscent berry contains only few seeds. Dispersal takes place by animals.

2. *Maytenus* sp.: The small dry and dehiscent loculicidal capsule contains few seeds which are possibly furnished with an aril. Dispersal is autochorous and, probably also takes place by animals (aril).

3. *Zinowiewia* aff. *australis:* The dry indehiscent fruit is winged and has a light wight. It contains a single seed. Dispersal of the samara takes place by wind.

The 3 species of Celastraceae studied are all of height category A. Nontheless, all 3 have different dispersal mechanisms. One is dispersed by animals and has the lowest rate of distribution. Another is dispersed by wind and shows a little higher number of representatives. The highest number of individuals has the species with autochorous dispersal mechanism by dehiscing capsules. Possibly, the seeds are also dispersed by animals and this is the reason for the higher dispersal rate. Pure animal dispersal in high trees seems less advantageous than dispersal by wind or autochorous dispersal.

Cochlospermaceae

The loculicidal capsules are often furnished with a skin-line endocarp that separates tangentially from the rest of the pericarp. Dehiscence takes place with 3 valves. The capsules contain numerous seeds with a curved embryo and an oily endosperm.

Cochlospermum sp. (orinocense?) = Carnestolendo

Hab.: Semideciduous and tropophilous forest	Height: a
	Ind.:
	Fr. time:
Fruit: Capsule	
Disp.: Anemochorous	

The large capsule, up to 7 cm long and 5 cm wide, of parchment-like consistence, opens loculicidally with 3 valves, but additionally the outer pericarp part separates tangentially from the endocarp (Figs. 24, 25, 26, 27, 28, 29). However, the separation from the endocarp takes place first - otherwise the endocarp would also open loculicidally. In fact, the endocarp opens only on top so that the seeds may be released, and only partly splits from the partition walls in a septicidal way. The septs result partly exposed, first being covered by a tiny parenchymatous skin which later decays so that the septs finally consist in a strong network of veins. In the completely opened fruit, the partial septicidal separation of the endocarp may become more obvious. We thus discover here 3 different types of dehiscence in one and the same fruit: first tangential splitting of the endocarp from the exo- and mesocarp, then loculicidal dehiscence of the united exo- and mesocarp, and finally partial septicidal opening of the endocarp. The seeds are set free through the partial septicidal opening of the endocarp. The fruit surface is velvety through gold-brown hairs (Fig. 28, 29), and very delicately ribbed (Fig. 27). The elongated small seeds are helicoidally bent and covered with about 8 mm long haires which spread out in one plane (Fig. 24). Owing to the presence of the long hairs the seeds have a low specific weight and may be carried by wind over large distances.

The only species of Cochlospermaceae present in the rain-forest studied is of little importance being more characteristic of drier habitats such as the semideciduous or deciduous forest. The relatively small tree of height category a shows dispersal by wind, although this type of dissemination is more restricted to high trees. However, the species is characteristic of drier habitats where wind dispersal is more common. The tree, therefore, abundantly appears in secondary vegetation, being a pioneer after felling or fire. In Surinam, it occurs more frequently on granite and in rocky areas.

Combretaceae

The fruit of the Combretaceae is commonly one-seeded and has a leathery texture. It is frequently alate, but seldom a drupe. Dispersal often takes place by water (fruits with floating tissue) or by wind, but more seldom by animals.

1. *Buchenavia capitata* = Pata de danto redondo

Hab.: RF	Height: A
Fruit: Drupe	Ind.:
Disp.: Zoochorous	Fr. time: 8

2. *Terminalia amazonia* = Pata de danto (blanco)

Hab.: RF	Height: A
Fruit: Samara	Ind.: 390
Disp.: Anemochorous	Fr. time: 5

3. *Terminalia guayanensis* = Pata de danto amarillo

Hab.: RF	Height: A
Fruit: Samara	Ind.: 39
Disp.: Anemochorous	Fr. time: 2/5

1. *Buchenavia capitata:* The elliptic to roundish fruit reaches 2–3 up to 10 cm in diameter and shows 5 more or less prominent ribs. Its surface is yellow-brownish. The juicy fruit flesh encloses a single seed. The drupe is dispersed by animals. In Surinam the species is found scattered in the rain forest.

2. *Terminalia amazonia* (Figs. 30, 31, 32): The dry samara is 5-angular, as seen in t.s. (Fig. 30). Three of these angles form small wings, while 2 of them extend to about 1 cm long wings which end with about 4 indentations. The yellow samara measures about 2,2 cm from end to end of the wings. The wings are papery and their surface is silky. With the 2 long wings, the samara somewhat resembles a butterfly when seen in front view. Dispersal of the samara takes place by wind. The wings mainly consist in thin-walled prosenchymatous cells elongated parallel to the surface of the wings. The wings are strengthened on their outside by several subepidermal layers of collenchymatous cells with thickened and partly wavy walls (Fig. 31, 32). The vascular bundles which alternate with the wings in their position are surrounded by sclerenchymatous plates (Fig. 32).

The well-developed main body in the center (endocarp) is composed of radial cell rows which arose from the activity of the ventral meristem in a way comparable to that of the cacao fruit, for example (Fig. 31) (Roth 1977). This tissue consists in thin-walled parenchyma cells with numerous pits and very small druses (Fig. 32). The single seed is 2 mm long. In Surinam, the tree grows in the ridge forest and in the savanna forest too. The samara is dispersed by wind.

3. *Terminalia guayanensis* (Fig. 30): The 2-winged samara of the "sailing flyer" type, is about 6,5 cm long and 2,2 cm broad. The wing lenght is 3 cm. The wings flank both sides of the fruit, while the single seed lies in the center. The wings are papery and have a smooth and silky surface. The locule extends on both sides to the wings. The samara is flat and light in weight. The wings composed of fibers show a delicate longitudinal ribbing. Dispersal of the dry samara takes place by wind.

Both species of *Terminalia* are represented by high trees (height category A) and both show dispersal of their samaras by wind. In Mexico *T. amazonia* reaches up to 70 m in height with a diameter of 3 m. *T. amazonia* with the smaller samara shows a rate of distribution which is 10 times higher than that of *T. guayanensis.* Is the smallness and the lighter weight of the fruit responsible for this phenomenon or is the ability of the tree to reproduce well by suckers the reason of its more successful distribution? However, *T. guayanensis* may possibly have suckers too. *Terminalia* is well known for its ramification in the form of stories. Both species show the same ramification of their branches. The flowers of all 3 species are condensed in inflorescences. From the pantropic genus *Terminalia* (about 200 species) savanna species have originated.

Buchenavia capitata, on the other hand, although a high tree too, develops a completely different type of fruit in the form of a drupe dispersed by animals. Interesting are the different fruiting times: The zoochorous *Buchenavia* bears fruits in August, while *T. guayanensis* and *T. amazonia* fruit in February/May and May respectively. In this case, we may assume an adaptation to the different dispersal agents. Anemochorous dispersal units are preferentially dispersed during the dry season.

Dichapetalaceae

Tapura guianensis = Jabón

Hab.: RF	Height: A
Fruit: Drupe	Ind.: 1527
Disp.: Zoochorous	Fr. time: 6/8/10

The dry leathery drupe has an elliptic shape, measuring 1,4 × 1,7 cm, and is of a velvety light-green colour. The calyx is persistent. The 1–2 stones have a woody endocarp. The fruit contains a single seed. It is dispersed by animals. Although animal dispersal in high trees is not always advantageous, this species shows a high number of individuals in the forest studied. According to Prance (1973) it is a typical widespread forest species of Amazonia with considerable variation in leaf size and shape and in other variable characters, in this way perhaps reflecting isolation of various populations during drier periods. According to Roosmalen, the species is common in Surinam in the rain forest as well as in the savanna forest and in the secondary forest. Imataca is the refuge area of *Tapura guianansis*, which fact may explain the high number of individuals.

Ebenaceae

The fruits of the Ebenaceae are almost exclusively dispersed by frugivorous birds, often aided by fruit bats and mammals. Almost all species of the genus *Diospyros* are bird-dispersed (Ridley 1905).

1. *Diospyros guianensis* = Carbón

Hab.: RF	Height: A
Fruit: Berry	Ind.: 176
Disp.: Zoochorous? Anemochorous?	Fr. time: 2

2. *Diospyros ierensis* = Barabá

Hab.: RF	Height: A
Fruit: Berry	Ind.: 1
Disp.: Zoochorous	Fr. time: 12

3. *Diospyros melinonii* = Moradito

Hab.: RF	Height: A?
Fruit: Berry	Ind.: 8
Disp.: Zoochorous	Fr. time: 6

1. *Diospyros guianensis:* The hard globular berry, about 5 cm long and 4,5 cm in diameter, has a persistent calyx, and contains several (6–7) seeds. The seeds are 2,8 cm long and 2 cm broad. They are very hard and of woody texture. Their colour is black. The seed has a wing which is dentate on the apical side. The ripe fruit has a smooth and brilliant blackish surface. It is poisonous being used as a fish poison. Dispersal of the fruit by birds is, therefore, very doubtful. It would be more likely that the seeds are dispersed by wind, if the fruit would not be indehiscent.

The outer fruit epidermis has very thick outer and anticlinal walls which are completely cutinized (Fig. 33). The outer part of the pericarp consists in pure parenchyma. In the inner part, groups of sclereids almost entirely replace the parenchyma which becomes scarce between the sclerenchyma (Fig. 33).

In Surinam, the species mainly occurs in the swampy forest along rivers and creeks.

2. *Diospyros ierensis:* The fleshy indehiscent berry contains several small seeds. The fruit is dispersed by animals, probably birds.

3. *Diospyros melinonii:* The dry indehiscent elliptic berry measuring 3 × 2,5 up to 4 × 4 cm has a hairy brown-green to yellow surface. It contains several small dark-brown seeds, measuring 1,6 × 0,7 × 0,6 cm, with a slimy surface. The fruit may be dispersed by animals, probably birds. In Surinam the tree is very common.

Although the species of the genus *Diospyros* usually have berries which are eaten by birds, while the seeds are inedible and hard so that they are dispersed in an endozoochorous way, *Diospyros guianensis* has poisonous fruits of mimetic character. The berries are black and brilliant, attracting birds by their colour. The seeds have a wing which, however, may not be very effective in wind dispersal. Furthermore, the fruit is indehiscent and the seeds are set free only when the fruit rots or when it is opened mechanically. *Diospyros guianensis* is of height category A and has the highest number of individuals, while the other 2 species of *Diospyros* show very low numbers. Surprising are the different fruiting times of the 3 species. The fruiting time in June may be adapted to bird dispersal, while December is a period when less fruits are offered. The fruiting time of *D. guianensis* in February, however, is a period which is very typical of wind-dispersed species. It coincides with the dry period and may indicate that wind dispersal in this species is – at least – in development. On the other hand, February is a month when food is very scarce.

Elaeocarpaceae

The fruits are generally capsules and only seldom drupes are found. Reduction of the seed number may be observed in the genus *Sloanea.*

1. *Sloanea grandiflora* = Cabeza de araguato
   Hab.: RF Height: a
   Fruit: Capsule Ind.: 168
   Disp.: Autochorous and zoochorous Fr. time: 2/5
2. *Sloanea guianensis* = Aletón
   Hab.: RF Height: A
   Fruit: Capsule Ind.: 167
   Disp.: Aotochorous and zoochorous Fr. time: 3
3. *Sloanea laurifolia* = Almidón
   Hab.: RF Height: A
   Fruit: Capsule Ind.: 8
   Disp.: Autochorous Fr. time: 6
4. *Sloanea megacarpa* = Onoto montañero
   Hab.: RF Height: a
   Fruit: Capsule Ind.: 66
   Disp.: Autochorous Fr. time: 6
5. *Sloanea robusta* = Manira
   Hab.: RF Height: A
   Fruit: Capsule Ind.: 47
   Disp.: Autochorous and zoochorous Fr. time: 8
6. *Sloanea* sp. = Espina de erizo
   Hab.: RF Height: A
   Fruit: Capsule Ind.: 1
   Disp.: Autochorous Fr. time: 8–9/12

1. *Sloanea grandiflora* (Figs. 34): The woody globular capsule is densely covered with bristles up to 2,2 cm long. Dehiscence takes place loculicidally with 4–5 valves. The capsule diameter measures up to 5 cm. Pericarp thickness is 0,2 cm. The mesocarp is parenchymatous, but the endocarp is very fibrous. The fruit contains only 4–5 seeds with a red and fleshy aril. Dispersal may take place by animals which eat the aril. The bristles may also stick to the fur of animals. In Surinam, the species occurs in the high savanna forest and in the creek forest.

2. *Sloanea guianensis:* The woody capsule is globular to elliptic and measures 1,5–3 cm in length. It is dark-brown and covered with numerous woody bristles about 0,7 cm long. It dehisces with 4 valves. The calyx is persistent. Although being a capsule it contains only a single seed (1,7 × 0,7 cm) surrounded by a dark-red fleshy aril. Apart from the autochorous dispersal mechanism the seeds may be disseminated by animals which eat the aril. In Surinam, the species is common and also occurs in the savanna forest.

3. *Sloanea laurifolia:* The woody ovoid capsule opens with 3 valves. Fruit length is 4,6 cm, width 3,5 cm. The pericarp measures 0,8 cm in thickness. The calyx is persistent. The capsule surface is rough and very hairy.

The outer pericarp epidermis is densely covered with unicellular hairs. Beneath follow several layers of small-celled parenchyma. Towards the inside the parenchyma becomes large-celled. Scattered in between the parenchyma occur radially elongated plates of fiber-like sclereids. The elongated sclereids are arranged in a radial direction. Due to the growth in thickness (dilatation growth) of the fruit, the parenchyma cells lying between the radial "rays" of sclerenchyma are extended tangentially and partly divided anticlinally in the same way as the parenchyma cells of a dilating phloem ray (Fig. 35). The dry capsule contains several light seeds. Dispersal is autochorous.

4. *Sloanea megacarpa:* The dry capsule contains few light-weighted seeds. The dispersal mechanism is autochorous.

5. *Sloanea robusta:* The woody capsule is elliptic to globular, measuring 2,1 × 2 cm, and opens loculicidally with 4 valves. The fruit surface is covered with bristles up to 0,8 cm long. When open, the capsule offers a star-like appearance as seen in front view. The pericarp measures 0,4 cm in thickness. The one to several seeds are furnished with an orange aril. Besides the autochorous dispersal mechanism the plant is disseminated by animals (birds) which eat the aril. In Surinam, this species is very rare, but also occurs on slopes.

6. *Sloanea* sp. (Espina de erizo): The woody capsule is ovoid, its surface being densely covered with prickles which reach 0,4 cm in length. Capsule length is 1,5 cm, width 1,1 cm. Dehiscence takes place loculicidally with 2 valves.

The Elaeocarpaceae studied are of little importance in the forest concerning the numer of individuals. Their fruit type, a capsule with a reduced number of seeds, may be advantageous for high trees, and most species are represented by high trees. The bristles and prickles on the fruit surface are secondary formations, but their function is uncertain. Dispersal is autochorous, on the one hand, but also may take place by animals (birds) which eat the aril of the seeds. A combination of 2 dispersal types may be advantageous for the "presence of individuals". But the differing number of individuals of the species studied cannot be explained by the dispersal mechanism alone. However, an autochorous dispersal mechanism of capsules is characteristic of high trees, at least in the area studied, being also realized in the genus *Sloanea*.

Euphorbiaceae

The mostly 3-locular fruits consist in 3 carpels, whereas numerous carpels are characteristic of *Hura crepitans* (about 14). The fruit is often a capsule dividing into 3 coccae which separate from the central column. Each locule (coccus) usually opens with 2 valves and contains 1–2 seeds. *Hymenocardia* has winged fruits. The pericarp of the capsules has often an outer fleshy layer. Stone fruits and berries are seldom. The seeds are often furnished with a caruncle developing from the outer integument in the micropylar region. Myrmecochory is frequent. The testa of the seeds is hard, but possibly an outer fleshy layer is present (*Sapium*). In certain species, the testa has a slimy epidermis transformed into a palisade layer. The well-developed endosperm is oily.

1. *Amanoa guianensis* = Hicaquillo
   Hab.: River side — Height: a
   Fruit: Capsule — Ind.:
   Disp.: Autochorous — Fr. time: 5
2. *Aparisthmium cordatum* = Tabalí morado
   Hab.: RF — Height: a
   Fruit: Capsule — Ind.: 5
   Disp.: Autochorous — Fr. time:
3. *Chaetocarpus schomburgkianus* = Cacho
   Hab.: RF — Height: A
   Fruit: Capsule — Ind.: 3679
   Disp.: Autochorous — Fr. time: 2/6/12
4. *Conceveiba guianensis* = Nicolás
   Hab.: RF — Height: a
   Fruit: Capsule — Ind.: 14
   Disp.: Autochorous and myrmecochorous — Fr. time: 2/7/10
5. *Croton gossypifolia* = Sangre de drago
   Hab.: Secondary forest — Height: a
   Fruit: Capsule — Ind.: 14
   Disp.: Autochorous — Fr. time:
6. *Croton matourensis* = Canelo
   Hab.: RF — Height: A
   Fruit: Capsule — Ind.: 10

Disp.: Autochorous Fr. time:

7. *Croton xanthochloros* = Canelón
   Hab.: RF Height: A
   Fruit: Capsule Ind.: 180
   Disp.: Autochorous Fr. time:
8. *Drypetes variabilis* = Kerosen
   Hab.: RF Height: A
   Fruit: Drupe Ind.: 423
   Disp.: Zoochorous Fr. time: 3/7/10
9. *Drypetes variabilis?* = Kerosen negro
   Hab.: RF Height:
   Fruit: Drupe Ind.: 2
   Disp.: Zoochorous Fr. time: 3/5/12
10. *Hieronyma laxiflora* = Aguacatillo
    Hab.: RF Height: A
    Fruit: Drupe Ind.: 43
    Disp.: Zoochorous Fr. time: 8/11
11. *Mabea piriri* = Pata de paují
    Hab.: RF Height: a
    Fruit: Capsule Ind.: 1706
    Disp.: Autochorous Fr. time: 5/8/12
12. *Mabea taquiri* = Pata de paují lacreado
    Hab.: Deciduous forest Height:
    Ind.:
    Fruit: Capsule Fr. time: 10–11
    Disp.: Autochorous
13. *Margaritaria nobilis* = Ojo de grulla
    Hab.: RF Height: a
    Fruit: Capsule Ind.: 40
    Disp.: Autochorous Fr. time:
14. *Pera schomburgkiana* = Pilón rosado
    Hab.: RF Height: A
    Fruit: Capsule Ind.: 95
    Disp.: Autochorous Fr. time: 2/5–6
15. *Phyllanthus* sp. = Grosella montañera
    Hab.: RF Height: a
    Fruit: Capsule Ind.: 5
    Disp.: Autochorous Fr. time:
16. *Piranhea longepedunculata* = Caramacate
    Hab.: RF Height: A
    Fruit: Capsule Ind.: 33
    Disp.: Autochorous Fr. time: 7
17. *Pogonophora sagotii* = Flor de mayo
    Hab.: RF Height: A
    Fruit: Capsule Ind.: 190
    Disp.: Autochorous Fr. time: 10
18. *Sandwithia guianensis* = Gaspadillo
    Hab.: RF Height: a
    Fruit: Capsule Ind.:
    Disp.: Autochorous Fr. time:
19. *Sapium aubletianum* = Caucho blanco
    Hab.: RF Height: A
    Fruit: Capsule Ind.: 5
    Disp.: Autochorous Fr. time: 8/10
20. *Sapium* sp. = Caucho
    Hab.: RF Height: A
    Fruit: Capsule Ind.: 39
    Disp.: Autochorous Fr. time:
21. *Sapium* sp. = Caucho morado
    Hab.: RF Height: a
    Fruit: Capsule Ind.: 20
    Disp.: Autochorous Fr. time:
22. *Sapium* sp. = Lechero
    Hab.: RF Height: A
    Fruit: Capsule Ind.: 9
    Disp.: Autochorous Fr. time:
23. *Sapium* sp. = Lechero blanco
    Hab.: RF Height: A
    Fruit: Capsule Ind.: 12
    Disp.: Autochorous Fr. time: 7
24. Bizcochito
    Hab.: RF Height: aa
    Fruit: Capsule Ind.: 349
    Disp.: Autochorous Fr. time:

1. *Amanoa guianensis* (Fig. 39): The dark-green to bluey capsular fruit of globular shape separates into 3 coccae. Each of these coccae is about 3,3 cm long and 1,4 cm broad and contains a single seed. The coccae dehisce loculicidally opening with 2 valves. Exo- and mesocarp are hard, while the endocarp is brittle. The fruit surface is knobby. The endocarp partly breaks off the mesocarp in a tangential direction (Fig. 39). The seed is 2,2 cm long and 1,7 cm broad. Its colour is brown with light-coloured patches (maculated). Hilum and raphe are prominent. A depression opposite to the micropylar region makes te seed heart-shaped.

The ground tissue of the fruit mesocarp consists in parenchyma which clearly demonstrates the influence of the dilatation growth of the fruit, the parenchyma cells being arranged in precise tangential rows (Fig. 39). Interspersed within the parenchymatous tissue are sclerenchymatous plates, elongated in a radial direction and reflecting the growth of the pericarp in a radial direction. The sclereids of the endocarp show a "T"-shaped arrangement in so far as the sclereids of the outer layers are extended in a radial direction, those of the innermost layers in a tangential direction. The dispersal of the seeds is autochorous, but the rate of distribution is very low. In Surinam the species occurs in the swampy forest and in the humid savanna forest.

2. *Aparisthmium cordatum:* The dry dehiscent capsule measures about 0,7 × 1 × 1 cm. It is yellow-green to red-brown with a delicate short indumentum or without hairs and a smooth surface. Each of the 3 locules opens with 2 valves and contains a single seed of a elliptic shape and with a smooth surface, measuring 0,45 × 0,4 cm. The seed outside is somewhat longitudinally striped. Dispersal is autochorous. In Surinam the small tree grows at the margin of the rain forest, in the savanna forest, and at the river-side.

3. *Chaetocarpus schomburgkianus:* The dry dehiscent and woody capsule measures 1,7 × 1,2 cm. The surface is covered with numerous spines about 0,15 cm long. The fruit is first green, later yellow-green, and red-brown at maturity. Each of the 3 locules opens with 2 valves and contains a single black and brilliant seed, 0,4 × 0,3 cm. The species also grows in the savanna. In Surinam, the species is common in the rain forest and in the high savanna forest as well.

4. *Conceveiba guianensis:* The 3-angular fruit is about 2,5 cm long and 2 cm broad, of ovoid shape and of a hard and woody texture. The surface is somewhat rough and knobby. Pericarp thickness is about 0,3 cm. The endocarp has palisade structure. The fruit is a capsule and each of the 3 locules opens with 2 valves. However, in our material the indehiscent fruit had one large locule with a well-developed seed and 2 small aborted locules with an aborted seed each. The seed is 0,8 cm long and 1 cm broad, more or les of globular shape and has a smooth surface covered with very small knobs. The seed coat is papery and brittle. A caruncle is being developed. Roosmalen indicates a creme-coloured to whitish edible aril on the seed.

The outer epidermis of the pericarp is small-celled. The mesocarp consists in a collenchymatous ground tissue in which the vascular bundles are dispersed. The fiber-sclereids of the endocarp are arranged in 3 different directions: those of the outermost layers are cut transversely in a transverse section, while in the following central layer, which occupies the major part of the endocarp, the fiber-sclereids follow a radial direction. In the innermost part of the endocarp, on the other hand, the fibers take a tangential direction. Through this arrangement, the endocarp attains the most possible strength withstanding tearing and straining forces.

Dispersal of the seeds is probably myrmecochorous. In Surinam, the tree is common and grows in the swampy forest, mainly on the river side, too.

5. *Croton gossypifolia:* The dry dehiscent capsule contains only few seeds. The dispersal mechanism is autorchorous.

6. *Croton matourensis:* The woody globular capsule measures about 0,6 cm in diameter, is green-brown, with a scaly surface and a persistent style. Each of the 3 locules opens with 2 valves to release a single elliptic seed, 0,35 × 0,25 cm, with a brilliant surface. The calyx is persistent. The dispersal mechanism is autochorous. In Surinam, the species is common and also grows in the savanna forest.

7. *Croton xanthochloros:* The dry dehiscent capsule contains only few seeds. The dispersal mechanism is autochorous.

8. *Drypetes variabilis* (Figs. 37, 38): The pear-shaped drupe is about 2,2–2,5 cm long and 1,7 cm broad. The pericarp is 0,3 cm thick, the endocarp measures 0,1 cm in thickness. The yellow-green fruit is first fleshy, but later hard. The only seed is 1,5 cm long and 1 cm broad, brown, and has a large groove on one side (caruncle?). The surface is deeply furrowed, giving the impression of a brain-like structure. The raphe is very prominent and the seed coat is strongly vascularized.

The fruit has a small-celled outer epidermis. Beneath follow a few layers of small-celled parenchyma. In the following layers, stone cell groups are very densely packed. Towards the inside the parenchyma cells enlarge in size and often become radially extended (Fig. 37). Stone cell groups are scattered all over the mesocarp. The outer surface of the endocarp shows very irregular outlines with protrusions penetrating into the mesocarp. This is partly due to vascular bundles lying in the vicinity of the endocarp which leave imprints in the form of concavities on the endocarp surface. Rhomboid crystals are found on the endocarp outside. In the outer endocarp part the sclerenchymatous elements are partly extended in a radial direction and partly in a tangential direction. In the inner endocarp part, on the other hand, most of the fibrous elements take a tangential direction (Fig. 38).

The indehiscent fruit, a drupe with a reduced number of seeds, is probably dispersed by animals. The rate of distribution of the species is medium. In Surinam, the species is preferentially found on the hill side.

9. *Drypetes variabilis?* (Kerosen negro): The indehiscent drupe contains only one seed. Dispersal takes place by animals.

10. *Hieronyma laxiflora:* The fleshy globular and indehiscent drupe, 0,3–0,5 cm in diameter, contains 1–2 seeds and is dispersed by animals (birds). It has a red-brown to blackish colour. The calyx is persistent. The high tree grows in Surinam in the rain forest as well as in the savanna forest and in the secondary forest.

11. *Mabea piriri:* The woody capsule measures about 1,9–2 cm in length and 1,1–1,8 cm in width, is ovoid to globular and has a rough gray-green to yellow-brown surface. Each of the 3 locules opens loculicidally releasing a single, more or less globular seed, 1 cm long and 0,7 cm broad, of black-brown colour, hard texture and a smooth surface. The calyx is persistent. Dispersal of the seeds takes place in an autochorous way. The rate of distribution is very high (with 1706 individuals). In Surinam, the species occurs in the swampy forest and at the river side too.

12. *Mabea taquiri:* The 3-locular woody capsule is more or less globular, and measures about 1,2 cm in diameter. The surface is red-brown and velvety. Each locule opens with 2 valves and releases a single seed. The calyx is persistent. The dispersal of the seeds is autochorous. In Surinam, the species grows at the river side and in the savanna too. *Mabea taquiri* occurs as a liana too, according to Hallé, Oldemann and Tomlinson (1978).

13. *Margaritaria nobilis:* The 5-locular globular capsule measures about 0,6 × 0,9 × 1,2 cm, being somewhat flattened in the long axis. It is first fleshy, but later becomes dry. The surface is somewhat vascularized, smooth and yellow-green. Each of the 5 locules opens with 2 valves and contains a single 3-angular seed with a brilliant blue surface. The calyx is persistent. The dispersal of the seeds is autochorous. The species is common in Surinam, preferentially occurring in the swampy forest and on sandy ridges.

14. Pera schomburgkiana: The ovoid woody capsule has 3 ribs and measures about 1,2–1,7 cm in length. It has a brown velvety surface and opens loculicidally and septicidally with 6 valves presenting a star-shaped appearance when open. The texture is shell-like and the surface netty. Each locule contains a single flattened seed, 0,6 × 0,5 × 0,3 cm, with a black brilliant surface. Dispersal of the seeds takes place in an autochorous way. In Surinam, the species is very common in the high savanna forest and in the savanna creek forest.

15. *Phyllathus* sp.: The dry dehiscent capsule contains only few seeds. The dispersal mechanism is autochorous.

16. *Piranhea longepedunculata* (Fig. 36): The ovoid capsule separates into 3 cocci and a central column. Each of the cocci contains one seed. The cocci are about 1 cm long. The pericarp texture is shell-like and the surface is covered with small light-coloured knobs (lenticels?). The seed is 0,4 cm long and 0,3 cm broad, blackish, of globular shape and of hard texture. The surface is smooth. The raphe is prominent. The seeds are dispersed in an autochorous way.

17. *Pogonophora sagotii:* The dry dehiscent capsule is woody and contains only few seeds which are dispersed in an autochorous way.

18. *Sandwithia guianensis:* The dry dehiscent capsule contains only few seeds which are dispersed in an autochorous way.

19. *Sapium aubletianum:* The dry globular cap-

sule is about 0,6 cm long and 0,5 cm broad and has a smoothish surface. It opens loculicidally only on one side with 2 valves to release a single globular seed, about 0,4 cm long and 0,3 cm broad, of light-brown colour and hard texture and with a somewhat sculptured surface. Opposite to the dehiscence line a longitudinal rib occurs on the fruit. Dispersal of the seed is autochorous. The rate of distribution is extremely low. In Surinam, the species is very common in the ridge forest, in the swampy forest, and on the hill side too. In open areas it occurs as a shrub.

20. *Sapium* sp. (Caucho): The dry dehiscent capsule contains only few seeds. Dispersal of the seeds is autochorous.

21. *Sapium* sp. (Caucho morado): The dry dehiscent capsule contains only few seeds which are dispersed in an autochorous way.

22. *Sapium* sp. (Lechero): The dry dehiscent capsule contains only few seeds which are dispersed in an autochorous way.

23. *Sapium* sp. (Lechero blanco): The dry dehiscent capsule contains only few seeds which are dispersed in an autochorous way.

24. Bizcochito: The dry dehiscent capsule contains only few seeds which are dispersed in an autochorous way.

Of the 24 species of Euphorbiaceae studied 12 belong to height catagory A, 9 to height catagory a, and only one to height category aa. Most of these species have a very low rate of distribution below 100 individuals. Only 4 species show a number of individuals between 180 and 423. Two species show a high number of individuals: *Mabea piriri* of height category a with 1706 individuals, and *Chaetocarpus schomburgkianus* of height category A with 3679 individuals. Most of the species are characteristic of the rain forest including the 2 species with the highest number of individuals. The majority of the species (21) use an autochorous dispersal mechanism through dehiscing capsules. Only a few species (3) have drupaceous fruits which are dispersed by animals. It is, however, very likely that also several of the autochorous species additionally use animal dispersal by arilloids on the seeds. According to van der Pijl, saurochory is an original feature in Euphorbiaceae. Many seeds of Euphorbiaceae have a caruncle near the micropyle and are dispersed by ants, being myrmecochorous (e.g. *Conceveiba guianensis*), which is a more recent type of dispersal. The throwing mechanism of the capsules is, therefore, often combined with myrmecochory. It is not impossible that the 2 species with the highest numbers of representatives use such a combination of dispersal mechanisms. Otherwise the high rate of distribution could not be explained or only by other, e.g. historic-geologic facts. However, in the genus *Sapium* the seed is devoid of a caruncle. The example of the Euphorbiaceae shows once more that in high trees of the rain forest autochorous dispersal is more advantageous than in lower ones and that animal dispersal is less advantageous in high trees.

Flacourtiaceae

The unilocular ovary comprises 2–10 carpels. Placentation is parietal. The capsules or berries have usually numerous seeds. These are often conspicuously arillate. The baccate fruits, especially the smaller ones, are bird-dispersed, often aided by fruit-bats and mammals.

1. *Banara guianensis* = Pan de acure
   Hab.: RF — Height: a
   Fruit: Berry — Ind.: 125
   Disp.: Zoochorous (birds) — Fr. time:
2. *Banara nitida* = Cayenito
   Hab.: RF — Height: A
   Fruit: Berry — Ind.: 5
   Disp.: Zoochorous — Fr. time: 5–7
3. *Casearia guianensis* = Rastrojero
   Hab.: RF — Height: aa
   Fruit: Berry — Ind.: 29

Disp.: Zoochorous (birds) — Fr. time: 5–6

4. *Homalium racemosum* = Nicua

Hab.: Lowland, Riverside	Height: a
	Ind.:
Fruit: Capsule	Fr. time:
Disp.: Autochorous	

5. *Laetia procera* = Jobo macho

Hab.: RF	Height: A
Fruit: Capsule	Ind.: 44
Disp.: Autochorous and zoochorous	Fr. time: 2/8

6. *Xylosma* sp. = Barba de acure

Hab.: RF	Height: a
Fruit: Berry	Ind.: 5
Disp.: Zoochorous (birds)	Fr. time:

7. *Zuelania guidonia* = Vela de barco

Hab.: RF	Height: A
Fruit: Capsule	Ind.: 2
Disp.: Autochorous and zoochorous	Fr. time:

8. Maspara

Hab.: RF	Height: a
Fruit: Berry	Ind.:
Disp.: Zoochorous (birds)	Fr. time: 1/11

1. *Banara guianensis:* The black globular berry has only little fruit flesh, but contains numerous brilliant brown seeds. The fruit diameter measures 0,7–0,8 cm. Dispersal takes place by birds, probably aided by fruit-bats and mammals. In Surinam, the tree grows at the riverside, at forest margins, in the secondary forest, and on ridges.

2. *Banara nitida:* The small dry berry contains numerous seeds. Dispersal takes probably place by animals.

3. *Casearia guianensis:* Almost all species of *Casearia* are bird-dispersed. The fruits open with 3 valves and a showy aril of orange, red or yellow colour is frequently attached to the seed. The fleshy fruit of *C. guianensis* contains numerous seeds and is dispersed by animals (birds).

4. *Homalium racemosum:* The small top-shaped capsule has a hairy surface and contains only one seed. It is about 0,5 cm long. The perigone is persistent.

The outer fruit epidermis is covered by a thin cuticle. Pluricellular hairs are numerous and the stomata elevated above the epidermis level. The majority of the pericarp is composed of thin-walled parenchyma cells. Several layers beneath the outer epidermis are small-celled. Then follows a larger-celled ground parenchyma in which groups of stone cells are irregularly dispersed. Rows of parenchyma cells radiating from the stone cell groups form a characteristic pattern. Small druses occur here and there in the ground tissue. The small persistent perigone does not play any role in dissemination. Dispersal is autochorous. In Surinam, the tree grows at the riverside, and on granitic islets in valleys.

5. *Laetia procera:* The elliptic to pear-shaped capsule is berry-like and opens with 3 valves each with a broad rib. It measures 2,7 × 1,5 cm and has a greenish or pink surface. The seeds are glued together by their fleshy arils which have a milky aspect and texture and a sweet taste. The seeds are dispersed by birds, fruit-bats and mammals.

The outer fruit epidermis is very small-celled. Beneath the epidermis lies a cartilaginous collenchyma. Very characteristic of the outer pericarp part are secretory cavities of an enormous size, surrounded by an epithelium. Towards the inside follows a large zone of dead radially elongated parenchyma cells with somewhat thickened walls and pits (Fig. 41). They may serve as a water-storing tissue. This region occupies more than one half of the entire pericarp. The inner part of the pericarp consists of roundish parenchyma cells. A cork cambium may develop in the outer pericarp part cutting out small calottes of parenchymatous tissue (Fig. 41). Periderm formation in fruits has earlier been observed (see also Catostemma). In Surinam, the tree grows principally on sandy soil.

6. *Xylosma* sp. (Barba de acure): The fleshy indehiscent fruit is small and contains few seeds. The berry is dispersed by animals.

7. *Zuelania guidonia:* The fleshy dehiscent capsule reaches 8 cm in diameter and has a wrinkled yellow-green surface. The numerous angular seeds are about 5 mm long. They are surrounded by a

yellow pulp. Calyx and stamens are persistent. The capsule shows delayed dehiscence. Dispersal of the seeds takes place by animals.

8. Maspara: The fleshy fruit is indehiscent and contains numerous small seeds. Dispersal of the berry takes place by animals.

The species of Flacourtiaceae present in the rain-forest studied are of little importance concerning their number of individuals. Most of the fruits are dispersed by animals. Either the fruits themselves are fleshy or the seeds have a fleshy and often colourful aril. According to Ridley most of the species of Flacourtiaceae are dispersed by birds, often aided by fruit-bats and mammals. Some of the species show an autochorous dispersal mechanism, as the capsules are dehiscent – however, some capsules have a delayed dehiscence. A transitional stage towards the berry seems to be present in these cases. The species seem to be adapted to animal dispersal and the autochorous dispersal seems to be of little importance. Therefore, we can only speak of vestiges of a diplochorous dispersal mechanism. In *Homalium racemosum* the number of seeds is reduced to one. Reduction of the capsules and their transformation into berries or possibly drupes (with only one seed) seems to be on the way here. The fleshyness of the capsules and the delayed dehiscence point in this direction, as well as the partial reduction of seeds. The very high trees (height category A) show a very small number of individuals, whereas the smaller tree *Banara guianensis* of height category a contributes 125 individuals. This fact proves again that animal dispersal is more advantageous in lower trees than in very high ones.

Guttiferae

The ovary is composed of one to many carpels. The capsules open septicidally or septifragally. Berries or drupes sometimes occur. The ovules per carpel are one to many. The seeds are often arillate.

1. *Calophyllum brasiliense* = Cachicamo
   Hab.: RF — Height: A
   Fruit: Drupe — Ind.: 9
   Disp.: Zoochorous (birds) — Fr. time: 2
2. *Caraipa richardiana* = Hicaco
   Hab.: RF — Height: a
   Fruit: Capcule — Ind.: 1893
   Disp.: Autochorous — Fr. time: 1/8/12
3. *Clusia* sp. = Copey
   Hab.: RF — Height: A
   Fruit: Capsule — Ind.:
   Disp.: Autochorous and zoochorous — Fr. time:
4. *Rheedia* aff. *spruceana* = Cozoiba rebalsera
   Hab.: RF — Height: A
   Fruit: Berry — Ind.: 53
   Disp.: Zoochorous (birds) — Fr. time:
5. *Rheedia* sp. = Cozoiba blanca
   Hab.: RF — Height:
   Fruit: Berry — Ind.: 1
   Disp.: Zoochorous (birds) — Fr. time:
6. *Rheedia* sp. = Cozoiba negra
   Hab.: RF — Height: A
   Fruit: Drupe — Ind.: 25
   Disp.: Zoochorous (birds) — Fr. time: 7
7. *Rheedia* sp. = Cozoiba picuda
   Hab.: RF — Height: A
   Fruit: Drupe — Ind.: 13
   Disp.: Zoochorous (birds) — Fr. time:
8. *Symphonia globulifera* = Copeicillo
   Hab.: Lowland, swamps — Height: A
   Ind.: 30
   Fruit: Drupe — Fr. time: time
   Disp.: Zoochorous (birds)

9. *Tovomita brevistamina* = Coloradito
   Hab.: RF — Height: a
   Fruit: Capsule — Ind.: 264
   Disp.: Autochorous — Fr. time: 3
10. *Tovomita* cf. *calodictyos* = Mangle negro
   Hab.: RF — Height: a
   Fruit: Capsule — Ind.: 76
   Disp.: Autochorous — Fr. time: 6
11. *Vismia guianensis* = Lacre
   Hab.: RF — Height: a
   Fruit: Berry — Ind.: 41
   Disp.: Zoochorous (birds) — Fr. time:
12. *Vismia macrophylla* = Lacre negro
   Hab.: RF — Height: a
   Fruit: Berry — Ind.: 50
   Disp.: Zoochorous (birds) — Fr. time: 6/11–12
13. *Vismia* sp. = Lacre amarillo
   Hab.: RF — Height: a
   Fruit: Berry — Ind.: 21
   Disp.: Zoochorous (birds) — Fr. time:
14. Peramancillo
   Hab.: RF — Height: A
   Fruit: Berry? — Ind.: 37
   Disp.: Zoochorous? — Fr. time:

1. *Calophyllum brasiliense:* The fleshy drupe is of globular to ovoid shape and measures 2,2–3 cm in length. The yellow-green surface shows a short and delicate indumentum. In a dry state, the surface presents a network of wrinkles. There is only one locule and a single seed (1,5 × 1,3 cm). The endocarp is hard. Secretory canals with a resinous liquid add a strong aromatic odour to the fruit. According to Roosmalen, the tree is frequent in savannas, in the humid savanna forest and in swampy forests. The edible part of the fruit is thin. According to Ridley, the fruit is dispersed by rolling when fallen, but on the tree birds and bats often eat and disperse the fruits. Although the fruits are dull coloured and not very showy, monkeys like to eat them. The seeds are rich in oil which is used commercially as a remedy against diseases of the skin. In Surinam the tree is common, principally occurring in savannas, humid savanna forest and marsh forest.

2. *Caraipa richardiana:* The asymmetric spindle-shaped green fruit measures about 3,7–8 cm in length and 1,8–2,5 cm in width. The elongated 3-angular capsule has 3 locules each with a single seed. It opens septicidally as well as septifragally in so far as the pericarp wall breaks off the central column in a tangential direction so that a triangular column remains in the center. Autochorous dispersal takes place by liberation of the seeds. Although the seeds are devoid of wings or an aril, the dispersal mechanism of this species seems to be better than that of the other Guttiferae studied, as 1893 individuals were counted in the forest, unless other reasons count for the presence of the tree. Possibly, the seeds with 2 large cotyledons attract animals. In Surinam the three grows along the Maratakka river.

The fundamental tissue of the pericarp consists in thin-walled parenchyma. Large secretory canals occur scattered in this tissue, each canal being surrounded by an epithelium. Sclereids often border the canals or are embedded within the parenchyma. The endocarp consists in few layers of tangentially elongated sclerenchymatous elements with somewhat thickened walls.

3. *Clusia* sp. (Copey): The large fruit is a plurilocular septicidal capsule with 11–15 carpels in which the endocarp separates tangentially from the mesocarp, each valve being crowned by a stigma. The endocarp additionally opens loculicidally, otherwise the seeds could not escape. The "calyx bracts" are persistent. A sticky secretion in secretory cavities is characteristic of the fleshy fruit wall. The seeds have a yellow to orange aril. Dispersal of the seeds takes place by animals (e.g. birds).

The outer fruit epidermis is small-celled, but has a relatively thick cuticle. Beneath lies a small-celled parenchyma in which secretory cavities with a resinous oily content are embedded (orange coloration with Sudan III). A cartilaginous collenchyma already develops in the outer pericarp region consisting in large cells with thick wavy walls (Fig. 46). Vascular bundles are embedded in this tissue. The relatively small endocarp is composed of tangentially elongated cells. The inner epidermis of the pericarp is small-celled. Most species of the genus *Clusia* also occur in the savanna forest, according to Roosmalen. The dispersal of the arillate seeds is zoochorous after dehiscence of the capsule.

4. *Rheedia* aff. *spruceana:* The fleshy berry only contains few seeds. Dispersal takes place by animals.

5. *Rheedia* sp. (Cozoiba blanca): The fleshy berry only contains few seeds. Dispersal takes place by animals.

6. *Rheedia* sp. (Cozoiba negra), (Figs. 45, 49): The fleshy ovoid greenish to light-brown fruit (3,5 cm long and 2,8 cm in width), densely covered with protuberances, is drupaceous, having a woody endocarp. Each of the 3 locules contains one large seed.

The outer pericarp epidermis is furnished with thick cutinized outer walls which appear horseshoe-shaped in t.s., some cells extending into unicellular hairs (Fig. 49). The large excrescences (humps) are formed by the outer mesocarp part. The fundamental tissue of the mesocarp in which large secretory cavities are irregularly dispersed, is collenchymatous. Towards the inside, the mesocarp cells become more and more compressed. Dispersal of the fruit with its edible mesocarp takes place by animals.

7. *Rheedia* sp. (Cozoiba picuda) (Figs. 45, 47, 48): The almost globular fleshy drupaceous fruit is 3-locular, each locule being furnished with a single seed. The length of the fruit measures 7,2 cm, the width 5,5 cm. The fruit surface is rough. The endocarp is papery and reaches 0,5 mm in thickness. On drying out, the endocarp separates tangentially from the mesocarp (Fig. 45).

The outer pericarp epidermis is palisade-shaped resembling the palisade layer of certain seeds, e.g. of Leguminosae. In contrast to the epidermis of the "Cozoiba negra", where the epidermis walls are horseshoe-shaped as seen in transverse section, the epidermis cells of Cozoiba picuda have thick anticlinal walls and are throughout cutinized. The cell lumen becomes narrower towards the cell center (Fig. 47). The ground tissue of the mesocarp appears in the form of an angular collechyma. Abundantly scattered in the ground tissue are elongated secretory cells or tubes with a yellowish content. The endocarp consists in thinner-walled parenchyma in its outer part and in compressed collenchyma in its inner part. It separates from the mesocarp along the borderline of the parenchyma.

8. *Symphonia globulifera:* The ovoid to globular drupe (3–4 × 2 cm) has a brownish to violet colour and contains 1–3 large seeds. Apparently, the reduction of the seeds to one is not stabilized yet. The pericarp contains a yellow latex in secretory cavities. The calyx is persistent. The fleshy fruits are softish and the seeds become tough and flexible with age. As *S. globulifera* prefers swampy grounds, the fruit is often dispersed by water, the seeds being ready to germinate when the fruit floats. However, the pulp of the fruits attracts animals. In Amazonian forests 13–16 individuals occur on one hectare (Black et al. 1950). In Surinam the species is common growing in the swampy forest and in the marsh forest.

9. *Tovomita brevistamina:* The dry capsule opens septicidally with 4–6 valves. A 4–6 winged column remains in the center, as the valves separate tangentially from the column. The few seeds are devoid of an aril. Dispersal is autochorous.

10. *Tovomita* cf. *calodictyos:* The dry capsule has a light weight and contains few seeds without aril. Dispersal is autochorous.

11. *Vismia guianensis:* The fruit is first ovoid, but later globular, and reaches 0,8–1,2 cm. The surface is dark-green or brown. The fleshy berry contains numerous small seeds. The fruit is dispersed by animals (birds, rodents). In Surinam, the tree is very common mainly occuring in the secondary forest, but in the savanna forest and in the rain forest too.

12. *Vismia macrophylla:* The fruit is a globular berry (0,8–1,3 cm in diameter), green, and crowned by the 0,4–0,5 cm long pistil. The fleshy berry contains numerous seeds and is dispersed by animals (birds). The tree is very common in Surinam.

13. *Vismia* sp. (Lacre amarillo): The fleshy berry contains numerous seeds and is dispersed by animals (e.g. birds).

14. "Peramancillo": The fruit is a berry?

The species of Guttiferae studied have partly capsular fruits, partly berries and partly drupes. In the Clusieae the fruits are all capsular and the seeds arillate. The fruits of *Rheedia* are either berries or - by reduction of the number of seeds - drupes (3 to one-seeded). *Caraipa richardiana* as well as the species of *Tovomita* studied develop capsules. *Calophyllum brasiliense* and *Symphonia globulifera* produce drupes. *Symphonia* fruits are usually one-seeded, according to Corner. And the species studied of *Vismia*, finally, have berries. As Ridley points out, most of the species of Guttiferae are dispersed by birds aided by fruit-bats and mammals. Of the capsular fruits studied only *Clusia* has arillate seeds, being diplochorous. The other species with capsules (species of *Tovomita* and *Caraipa richardiana*) show only autochorous dispersal by fruit dehiscence. The high rate of occurrence of *Caraipa richardiana* cannot be explained only by the autochorous dispersal mechanism. Possibly the seeds with their large cotyledons attract animals which eat some and disperse others. The genus *Caraipa* is now placed within the Bonnetiaceae. However, the species dispersed by animals have the lowest numbers of individuals in the forest studied, while the 3 autochorous species, *Caraipa richardiana, Tovomita brevistamina,* and *T.* cf. *calodictyos*, show the highest numbers of individuals (1893, 264 and 76). All three are of height category a.

Hernandiaceae

Hernandia guianensis = Cojoro

Hab.: RF	Height: A
Fruit: Nut	Ind.: 9
Disp.: Zoochorous	Fr. time:

The black ovoid hard nut is about 2,5–3 cm long and furnished with 6–8 ribs. The nut is completely surrounded by a cupule with a small opening on top. Together with the cupule the fruit measures 5,5 cm in length and 4–6 cm in diameter. The cupule is of papery texture. The scutate stigma on top of the nut is lobed. The number of species present in the forest studied is very low. The fruit is probably dispersed by animals. The low rate of distribution of this species seems to confirm the theory that animal dispersal in high trees is not so advantageous. In Surinam, the species also occurs in the swampy forest as well as in the secondary forest.

Hippocrateaceae

Melocotoncito

Hab.: RF	Height: aa
Fruit:	Ind.: 43
Disp.:	Fr. time:

The fruit is unknown and the species is of little importance in the forest studied.

Humiriaceae

Saccoglottis cydonioides = Ponsigue montañero

Hab.: RF	Height: A
Fruit: Drupe	Ind.: 37
Disp.: Zoochorous	Fr. time: 2

The almost globular to pear-shaped 5-locular drupe (about 2,7 × 2,5 cm) is of orange-red colour and contains a single seed which is dark brown and ovoid, whereas all the other ovules abort. The woody and fibrous endocarp shows 5 ribs on its outside. Resinous cavities are characteristic of the endocarp.

The outer pericarp epidermis is small-celled and has thickened outer walls. Below follows a layer of thin-walled palisade cells with collapsed (wavy) walls, when the fruit is dried out. These cells are devoid of starch and probably function as a water reservoir. The fundamental tissue is represented by thin-walled roundish parenchyma cells densely filled with large starch grains (Fig. 50). These are very regularly arranged along the walls giving a wreath-like aspect, as seen in t.s. Interspersed within the ground parenchyma are groups of sclereids. The endocarp consists in fiber bundles which cross one another in different directions (Fig. 50). Large secretory cavities ("resinous" cavities = "Harzlücken" of Engler) surrounded by an epithelium occur in the endocarp.

The family is of very little importance in the forest studied. The fruit is dispersed as a whole, since the mesocarp is fleshy and edible. The endocarp, on the other hand, is hard and indigestible. The fruit is dispersed by animals, but the dispersal mechanism of this high tree seems not to be very advantageous. In Surinam the species is uncommon.

Lacistemaceae

Lacistema aggregatum = Don Juán

Hab.: RF — Height: a
Fruit: Capsule — Ind.:
Disp.: Autochorous & zoochorous — Fr. time:

The ovoid red capsule measuring 1 × 0,6 cm with a short indumentum has a bitter-sweet tasting fruit flesh. It opens with 3 valves and contains a single seed. It shows a twofold dispersal mechanism which is first autochorous by the dehiscing capsule and later takes place by animals (birds) which eat the sweet pulp surrounding the seed. In Surinam the species is represented by a low shrub (3–4 m) or tree (up to 20 m) and grows in the swampy forest as well as in the secondary forest and in savannas.

Lauraceae

The pseudomonomerous ovary composed of 3 carpels is unilocular and contains a single seed. The fruit is a berry or drupe, partly or completely surrounded by an enlarged fleshy or woody "cupule" (receptacle or perigone?).

1. *Aniba excelsa* = Laurel Rollet

Hab.: RF — Height: A
Fruit: Drupe — Ind.: 134
Disp.: Zoochorous — Fr. time: 2–3/6

2. *Aniba riparia* = Laurel amarillo

Hab.: RF — Height: A
Fruit: Drupe — Ind.: 34
Disp.: Zoochorous — Fr. time: 8/12

3. *Beilschmiedia curviramea* = Aguacatillo moises

Hab.: RF — Height: a
Fruit: Drupe — Ind.: 139
Disp.: Zoochorous — Fr. time: 12

4. *Endlicheria cocuirey* = Laurel negro

Hab.: RF — Height: a
Fruit: Drupe — Ind.: 300
Disp.: Zoochorous — Fr. time:

5. *Nectandra grandis* = Laurel

Hab.: RF — Height: A
Fruit: Drupe — Ind.: 20
Disp.: Zoochorous — Fr. time: 6

6. *Nectandra pichurin* = Laurel eucalipto
Hab.: RF — Height: a
Fruit: Drupe — Ind.: 30
Disp.: Zoochorous — Fr. time: 12

7. *Ocotea duotincta* = Laurel verde
Hab.: RF — Height: a
Fruit: Drupe — Ind.: 207
Disp.: Zoochorous — Fr. time: 6

8. *Ocotea martiana* = Laurel baboso
Hab.: RF — Height: A
Fruit: Drupe — Ind.: 7
Disp.: Zoochorous — Fr. time:

9. *Ocotea nicaraguensis* = Laurel blanco
Hab.: RF — Height: A
Fruit: Drupe — Ind.: 43
Disp.: Zoochorous — Fr. time: 2/11

10. *Ocotea* cf. *subalveolata* = Laurel paraguito
Hab.: RF — Height: a
Fruit: Drupe — Ind.: 39
Disp.: Zoochorous — Fr. time:

11. Laurel benitez
Hab.: RF — Height: A
Fruit: Drupe — Ind.: 10
Disp.: Zoochorous — Fr. time:

12. Laurel canelo
Hab.: — Height:
Fruit: Drupe — Ind.:
Disp.: Zoochorous — Fr. time:

13. Laurel marrón
Hab.: RF — Height:
Fruit: Drupe — Ind.: 5
Disp.: Zoochorous — Fr. time:

14. Laurel rastrojero
Hab.: RF, cultivated — Height:
Fruit: Drupe — Ind.:
Disp.: Zoochorous — Fr. time: 6

1. *Aniba excelsa:* The fleshy elliptic drupe (berry) contains a single seed and is surrounded by a cupule. The fruit is dispersed by animals (birds).

2. *Aniba riparia:* The yellow fleshy drupe (a berry according to Roosmalen) surrounded by a cup-shaped (half-globular) cupule contains a single seed. It measures 2–2,5 cm in length and 1,5–1,7 cm in diameter, and has a flattened top. The cupule is covered with little knobs. The fruit is dispersed by animals (birds). The species is very rare in Surinam.

3. *Beilschmiedia curviramea:* The drupe contains a single seed and is dispersed by animals, mainly birds.

4. *Endlicheria cocuirey:* The fleshy indehiscent drupe contains a single seed. It is surrounded by a cupule. The fruit is dispersed by animals.

5. *Nectandra grandis:* The ovoid fleshy fruit is about 4,2 cm long and 2 cm in diameter. The cup-shaped cupule measures about 3,5 cm in length and 2,3 cm in diameter. It is of reddish colour, with a somewhat sculptured surface by knob-like lenticels, and has longitudinal ribs. The black fruit is smooth and has a short pointed tip. The only seed is 3 cm long and 1,6 cm broad. The integument is papery. The 2 fleshy cotyledons fill the entire seed. The drupe resembling an "acorn" is dispersed by animals. The species is common in Surinam.

The outer epidermis cells of the pericarp have extremely thickened outer walls so that the cell lumen becomes considerably reduced. The ground tissue of the pericarp is represented by parenchyma. In the outer pericarp part close to the epidermis, very large plates of sclerenchyma are found (Fig. 52). In the inner part of the pericarp, the conducting tissue mainly consisting in tracheids is arranged in the form of layers which alternate with layers of compressed parenchyma. The outer epidermis of the seed coat – in contact with the cutinized inner pericarp epidermis – is furnished with heavily thickned anticlinal cell walls (Fig. 52).

6. *Nectandra pichurin:* The fleshy drupe contains a single seed. It is surrounded by a cupule. The fruits have a penetrating smell, probably due to eteric oil cells, and the seeds are very adstringent, a natural defense against predators. They are used in popular medicine against diarrhea, dysentery, and nervous irritations. The fruits are dispersed by animals (birds).

7. *Ocotea duotincta* (Fig. 51): The fleshy drupe has a lenght of 1,5 cm and a width of 0,7 cm including the cupule. Cupule length is about 0,9 cm. The cupule surrounds the fruit proper about half-ways. The fruit endocarp corresponding to the inner pericarp epidermis is palisade-shaped.

The single seed is about 0,6 cm long and 0,4 cm broad. The fruit is dispersed by animals (birds).

8. *Ocotea martiana:* The fleshy drupe contains a single seed. It is surrounded by a cupule. The fruit is dispersed by animals (birds).

9. *Ocotea nicaraguensis:* The fleshy drupe contains a single seed. It is surrounded by a cupule. The fruit is dispersed by animals (birds).

10. *Ocotea* cf. *subalveolata:* The fleshy drupe contains a single seed. It is surrounded by a cupule. The fruit is dispersed by animals (birds).

11. Laurel benitez: The fleshy drupe contains a single seed. It is dispersed by animals.

12. Laurel canelo: The ovoid drupe is about 1,5 cm long being surrounded by a cupule of 0,9 cm length and 0,7 cm width. The endocarp is palisade-shaped. The fruit contains a single seed and is dispersed by animals.

13. Laurel marrón (Fig. 51): The fleshy ovoid drupe is enclosed within a large cupule, 1,5 cm long and 1,8 cm broad with a thickness of 0,5 cm. The fruit proper is about 0,8 cm broad und surpasses the cupule in length by about 1 cm. The fruit which is dispersed by animals encloses a single seed.

The pericarp has a thick-walled outer epidermis with especially thickened walls. Beneath follow several layers of very small cells. The ground tissue consists in parenchyma in which groups of stone cells are scattered. Due to the growth in circumference of the fruit the parenchyma cells are partly extended tangentially.

14. Laurel rastrojero: The fleshy drupe contains a single seed. It is dispersed by animals.

Six of the species of Lauraceae studied belong to height category A, while five are of height category a. The fruits of all species studied resemble one another very much, representing single-seeded fleshy drupes surrounded by a cupule. In their outer appearance they resemble therefore an "acorn", the fruit of *Quercus*. The fruit pericarp consists in a parenchymatous ground tissue in which stone cell groups are scattered. These make the pericarp somewhat hard and brittle. All fruits are dispersed by animals, probably mainly birds, often aided by fruit bats and mammals. At least some of the seeds seem to be protected against destruction by predators either by bitter-tasting substances (smell and taste like an almond) or by an adstringent taste. The number of individuals present in the forest studied is very low in most species. However, the lower trees of height category a seem to have a somewhat better rate of distribution. The highest number of individuals shows *Endlicheria cocuirey* of height category a with 300 representatives. This fact proves again that animal dispersal is more advantageous in the lower forest layers.

Lecythidaceae

The fruit of the Lecythidaceae frequently corresponds to a woody pyxidium, either septate or with more or less reduced septs. The ovary is united with the receptacle. A disc is present in the neighbourhood of the calyx. Sepals are persistent. The seeds are without endosperm. The majority of the Lecythidaceae show biennial fruiting.

According to Prance & Mori (1978), the funicle of most Lecythidaceae is surrounded by a fleshy aril. The wing of the anemochorous seeds of Couratari is interpreted by the authors as a flattened aril.

1. *Couratari multiflora* = Tampipio
   Hab.: RF — Height: A
   Fruit: Capsule (pyxidium) — Ind.: 499
   Fr. time: 5
   Disp.: Anemochorous

2. *Couratari pulchra* = Capa de tabaco
   Hab.: RF — Height: A
   Fruit: Capsule (pyxidium) — Ind.: 170
   Fr. time: 2/5
   Disp.: Anemochorous

3. *Eschweilera chartacea* = Guacharaco amarillo
   Hab.: RF — Height: A
   Fruit: Capsule (pyxidium) — Ind.: 100
   Fr. time: 2/5–6
   Disp.: Autochorous (and zoochorous)

4. *Eschweilera corrugata* = Guacharaco rosado
Hab.: RF | Height: A
Fruit: Capsule (pyxidium) | Ind.: 6 | Fr. time: 2/5/9
Disp.: Autochorous (and zoochorous)

5. *Eschweilera grata* = Cacaito
Hab.: RF | Height: A
Fruit: Capsule (pyxidium) | Ind.: 3446 | Fr. time: 2/5
Disp.: Autochorous (and zoochorous)

6. *Eschweilera odora* = Cacao
Hab.: RF | Height: A
Fruit: Capsule (pyxidium) | Ind.: 1025 | Fr. time: 2/8–9
Disp.: Autochorous (and zoochorous)

7. *Eschweilera subglandulosa* = Majaguillo
Hab.: RF | Height: A
Fruit: Capsule (pyxidium) | Ind.: 2374 | Fr. time: 3/6
Disp.: Autochorous and zoochorous

8. *Eschweilera* cf. *trinitensis* = Majaguillo negro
Hab.: RF | Height: A
Fruit: Capsule (pyxidium) | Ind.: 803 | Fr. time: 8/10
Disp.: Autochorous (and zoochorous)

9. *Eschweilera* sp. = Majaguillo erizado
Hab.: RF | Height: A
Fruit: Capsule (pyxidium) | Ind.: 1 | Fr. time:
Disp.: Autochorous (and zoochorous)

10. *Gustavia augusta* = Merguo
Hab.: Lowland | Height: aa
Fruit: Capsule (pyxidium) | Ind.: 153 | Fr. time: 8?
Disp.: zoochorous

11. *Lecythis davisii* = Tinajito
Hab.: RF | Height: A
Fruit: Capsule (pyxidium) | Ind.: 247 | Fr. time: 1
Disp.: Autochorous (and zoochorous)

1. *Couratari multiflora* Figs. 55, 60): The cylindrical dehiscent pyxidium is 5,5–6 cm long and 2,5–3 cm in diameter, has a slightly triangular shape, and resembles a pipe-bowl (fig. 55). The colour is brown and the fruit surface rough. The pericarp width measures 0,2 cm. The lid, 2,5–3 cm in diameter, dehisces together with the central colum which is 5 cm long. Below the lid margin lies the calyx ring. The fruit texture is leathery and fibrous. The fruit surface is slightly ribbed by vascular bundles.

The pericarp consists in layers of fiber bundles alternating with layers of parenchymatous cells (Fig. 60). The numerous seeds reach a length of 3,8 cm and a width of 1,1 cm. They are flat and have an ovoid shape (Fig. 55). The seed body is surrounded by a smooth papery wing. Dispersal of the seeds takes place by wind. In Surinam, the tree mainly grows in the swampy forest on the river side. It flowers at the beginning of the large dry periode after the shedding of the leaves.

2. *Couratari pulchra* (Figs. 56, 57, 59): The cylindrical urn-shaped dehiscent pyxidium resembles a long pipe-bowl. It is 10–22 cm long and 3,5–5 cm in diameter. Pericarp thickness is 0,4 cm. The pericarp consists in alternating layers of fiber bundles and layers of roundish parenchyma cells. The fruit is brown to dark-red, woody, and has a rough surface through numerous lenticels which appear as light dots. The lid dehisces together with the central column, resembling a thunder-wedge of 10 cm length. A calyx ring remains below the lid. The numerous elongated seeds are 7,8 cm long and 2,2 cm broad. The seed body is completely surrounded by a flattened papery wing, 0,7 cm in thickness. The wings are densely vascularized. Dispersal of the seeds takes place by wind.

3. *Eschweilera chartacea:* The dehiscent cup-shaped pyxidium measures about 5,3 cm in length and 4,6 cm in diameter, somehow resembles a pipe, and has a woody texture. It shows a rusty brown colour and has a rough and corky surface covered with yellowish lenticels. The lid is flat. The pericarp measures 0,5 cm in thickness. Septs are absent. The fruit contains 1–3 triangular seeds, measuring 2,8 × 1,7 × 1,3 cm. Dispersal of the seeds is autochorous by the dehiscing capsule, but possibly also takes place by animals. The tree occurs scattered in the rain forest of Surinam.

4. *Eschweilera corrugata:* The cone-shaped dehiscent pyxidium measures about 1,5–2 cm in length and 2,4–3 cm in diameter and shows 6 calyx tips below the lid margin. The few seeds reach 2,2 × 1,2 × 1,2 cm and are ribbed on their outside. Dispersal of the seeds is autochorous, but may also take place by animals. In Surinam, the tree is very frequent in the humid as well as in the swampy forest.

5. *Eschweilera grata* (Fig. 53): The short and small urn-shaped pyxidium is dehiscent and measures 2,5 cm in length and 2,2 cm in diameter. Its texture is woody and its colour brown. The surface is rough and wrinkled. The pericarp reaches 0,1 cm in thickness. The lid is devoid of a central column and septs are absent. The calyx bases surround the fruit in a ring-like manner. There are few laterally flattened seeds. Dispersal of the seeds is autochorous, but these may also be dispersed by animals. The rate of distribution of this species is enormously high and actually the highest among all species of the Lecythidaceae studied.

6. *Eschweilera odora:* The woody urn-shaped dehiscent pyxidium is about 4,5 cm long and 7 cm in diameter. Pericarp thickness measures 0,3 cm. The colour of the fruit is light-brown and the surface is rough due to the presence of lenticels. There are 3 locules separated from one another by 3 septs and 2 seeds lie in each locule. The seeds are 3 cm long and 2,1 cm broad, of triangular shape and light-brown. The seed coat is brittle and has a smooth and shiny outside. The seed coat is highly vascularized and the funicle is thick. This description does not completely coincide with that of Roosmalen for *E. odora.* Roosmalen's description rather coincides with that of the "Majaguillo erizado". Dispersal of the seeds is autochorous, but may possibly also take place by animals. The species is common in Surinam.

7. *Eschweilera subglandulosa:* The woody dehiscent pyxidium is about 2,2 cm long and measures 4 cm in diameter. The entire pericarp thickness is 0,5 cm, of this the endocarp occupies 0,2 cm. The fruit is urn-shaped and dark-brown and has a wrinkled surface, rough by lenticels. The calyx ring consists in 5–6 pointed sepals. The capsule encloses a single seed (2,5 × 1,9 × 1,4 cm) With a cream-coloured aril. Dispersal of the seeds is first autochorous, but also takes place by animals (arils!). They are much eaten by agoutis. In Surinam, the tree grows in the swampy forest, mainly along rivers and creeks.

8. *Eschweilera* cf. *trinitensis* (Fig. 53): The small woody urn-shaped pyxidium is short (3 cm long) and 6 cm broad. It is dehiscent. The pericarp measures 0,6 cm in thickness. The fruit is brown and has a rough surface. The woody sepals surround the urn in the form of a curved ornament. The lid is very short. Septs are not present. The fruit is autochorous, but the few seeds are possibly also dispersed by animals.

9. *Eschweilera* sp. ("Majaguillo erizado") (Fig. 55): The short woody deshiscent pyxidium measures up to 4,4 cm in length and 4 cm in diameter and is cup-shaped. The 6 persistent calyx tips are pointed. The colour of the fruit is dark-brown and the surface rough. The lid is devoid of a column. The pericarp thickness is 0,3 cm. There is usually only one (to few) seed, 2,8 cm long and 1,6 cm broad with a light-brown colour. The seed has a peanut-like ribbed surface with 5–6 ribs which correspond to the vascularization of the seed coat. The raphe is prominent and perhaps fleshy in fresh material. Dispersal of the seeds is autochorous, but possibly also takes place by animals.

10. *Gustavia augusta* (Figs. 57, 60): The urn-shaped indehiscent pyxidium with nearly globular outlines measures about 3,5–4 × 4–6 cm. Its top is flattened, the lid is flat and has a little central peg. Beneath the lid lies the padded calyx ring. The fruit wall has a fibrous and brittle texture. The fruit is brown when ripe and has a rough surface.

The pericarp width is about 0,1 cm. The seeds with a fleshy funicle are embedded in fruit flesh. The seeds are eaten and dispersed by animals, e.g. squirrels and monkeys. In the Botanical Garden in Caracas I have never seen dehisced fruits, but only fruits opened by animals (squirrels). Trees have significantly more branches where *Cebus* monkeys feed on the terminal buds (Oppenheimer & Lang 1969). The young fruit mainly consists in parenchyma in which the vascular bundles are scattered. In the region of the virtual dehiscence line a unilateral meristem (resembling a ventral meristem) is acting (Fig. 60). *Gustavia* is the most primitive genus of tropical Lecythidaceae having the simplest type of androphore (Prance 1976), while *Eschweilera* and *Couratari* are more derived. *G. augusta* is flowering and fruiting every year. In Surinam the small tree mainly grows in the swampy forest where it is very common.

11. *Lecythis davisii* (Figs. 54, 58): The very big urn-shaped pyxidium measures 19 cm in length and 15 cm in diameter and has a woody texture. It is brown and has a corky surface with lenticels. The lid dehisces together with the short pointed column. The numerous seeds are 3 cm long and 1 cm broad. They are furnished with 5 main ribs, of brown colour and woody texture and have a slightly rough surface. The edible seeds which are embedded in a swet edible fruit flesh have a nutlike taste. They are attached with a large fleshy funicle. The pericarp is mainly composed of fiberlike sclereid bundles running in various directions and of little parenchyma. Dispersal is first autochorous, but later certainly assisted by animals, e.g. agoutis, squirrels, pacas, pecaries. In Surinam, the tree is rare in the humid tropical forest.

The species of *Couratari* studied have a very characteristic pyxidium in the form of an elongated pipe-bowl. The lid dehisces together with the central column (and 3 septs). The seeds are flat, elongated and the seed body is surrounded by a papery wing. The species of *Eschweilera* studied form a woody pyxidium often without septs and with several relatively large seeds. The lid generally dehisces without central column. The seeds are devoid of wings. The species of *Lecythis* studied has a large woody pyxidium with a broad calyx ring. The lid is relatively small and attached to it remains the central column which is relatively short and pointed. The seeds are devoid of wings. The species of *Gustavia* studied has a nearly globular pyxidium with a flattened top, a padded calyx ring and a flat lid with a small central peg on top. The kidney-shaped seeds are devoid of a wing.

The genus *Lecythis*, studied by Mori (1970), concerning ecology and uses, produces edible "nuts" which are more palatable than those of *Bertholletia excelsa*. Only the seeds of *L. minor* are poisonous by their high content of selenium which causes loss of fingernails and hair, nausea, vomiting, diarrhea, muscle pain, convulsions, and dizziness. Although the woody urn-shaped pods dehisce with a lid, the seeds are dispersed either by bats or tree rodents (squirrels) or by ground animals such as deer, pacas, agoutis and peccaries. The woody capsule is considered a protection of the seeds from insect predation and damage of the unripe seeds, especially by rodents. The fleshy funicle is eaten by bats and the seed dropped in flight or at the rooting place. There are about 45 species of the genus represented by high or emergent trees of the tropical rain forest of the New World with the center of distribution in the Brazilian Amazon and in Guiana. Mori emphasizes that the frequency of individuals of *Lecythis* in the tropical forest is generally very low. Individuals of *L. costaricensis* are, for example, separated by 1–2 kilometers.

Ten of the eleven species studied are represented by high trees of height category A and only one species (*Gustavia augusta*) is represented by a low tree or shrub. The fruits of almost all species studied are dehiscent pyxidia. Only 2 species have winged seeds (in the genus *Couratari*) which are dispersed by wind. However, the rate of distribution is not very high in these cases. The other species are likewise autochorous, but the dispersal of their seeds is probably assisted by animals. In 2 species (*Lecythis davisii* and *Gustavia augusta*) the seeds are surrounded by fruit flesh, but the number of individuals of these species in the forest is not very high. *Gustavia augusta* of the lowest story (aa) is definitely dispersed by animals (e.g. squirrels). The highest numbers of individuals are found in the genus *Eschweilera*. However, 3 species of *Eschweilera* have a low to very low rate of distribution. (*E. chartacea* with 100 individuals, *E. corrugata* with 6 individuals, and *E.* sp. with only one individual). The largest number of individuals is found in *E. grata* with 3.446 individuals, followed by *E. subglandulosa* with 2.374 individuals, *E. odora* with 1.025 individuals, and *E.* cf. *trinitensis* with 803 individuals. All species studied are typical of the rain forest with the exception of *Gustavia augusta* which is characteristic of lowlands and swampy forests. There is no obvious reason visible why certain species of *Eschweilera* are considerably better represented than others. The autochorous dispersal seems, however, to be advantageous for high trees. Once again it becomes obvious that high trees in the forest are either autochorous or use dispersal by wind. The large numbers of individuals in 4 species of Eschweilera can possibly be explained by historic-geologic facts.

Gustavia augusta is concerned the most primitive genus of tropical Lecythidaceae, while *Eschweilera* and *Couratari* are more derived. It is perhaps not by accident that *Gustavia augusta* is a very small tree or shrub and that it is dispersed by animals. The fleshy funicle of the seeds is also a relatively primitive character, according to van der Pijl.

Leguminosae

The fruit of the Leguminosae has probably developed from the many-seeded arillate legume to the one-seeded drupe, nut or samara. The arillate seeds are dispersed by animals. Pods with mimetic seeds that are dispersed by animals are derived forms. The autochorous method with pods dehiscing into 2 valves and with dry hard seeds devoid of an aril seems to represent a derived form (Corner, van der Pijl). Autochorous dispersal preferably occurs in high trees. Concerning their fruit type, the Mimosaceae seem to include more primitive forms, while the Papilionaceae show the most derived fruit forms e.g. with anemochorous dispersal mechanisms.

The fruit length oscillates between a few millimeters and one meter (e.g. in *Vigna sesquipedalis*). Burkart (1952) distinguishes a variety of fruit types in the Leguminosae, according to their morphological structure: e.g. follicles opening at the ventral suture only, transversely septate loments, drupaceous loments, cryptoloments with a septate endocarp, nutlet, geocarpium, samara, drupe and others.

Mimosaceae

The Mimosaceae produce a variety of fruit forms, although the pod is the basic fruit type. Pods are partly dehiscent and partly indehiscent. Seed dispersal frequently occurs by animals, as the seeds produce fleshy parts (sarcotesta, aril) or the fruit itself produces a fleshy endocarp. Dispersal by fish and reptiles, still occurring in the genus *Inga*, is considered relatively primitive.

1. *Enterolobium cyclocarpum* = Caro negro
   Hab.: Deciduous forest, RF — Height: A
   Ind.: 4
   Fruit: Indehiscent pod — Fr. time:
   Disp.: Zoochorous
2. *Enterolobium* sp. = Caro blanco
   Hab.: Deciduous forest, RF — Height: A
   Ind.: 1
   Fruit: Indehiscent pod — Fr. time: 7
   Disp.: Zoochorous
3. *Inga alba* = Guamo colorado
   Hab.: RF — Height: A
   Fruit: Indehiscent pod — Ind.: 409
   Disp.: Zoochorous — Fr. time: 2

4. *Inga capitata* = Guamo negro
   Hab.: RF — Height: A
   Fruit: Indehiscent pod — Ind.: 404
   Disp.: Zoochorous — Fr. time: 3
5. *Inga fastuosa* = Guamo peludo
   Hab.: RF — Height: a
   Fruit: Indehiscent pod — Ind.: 43
   Disp.: Zoochorous — Fr. time:
6. *Inga heterophylla* = Guamo pata de morocoy
   Hab.: RF — Height: A
   Fruit: Indehiscent pod — Ind.: 54
   Disp.: Zoochorous — Fr. time:
7. *Inga lateriflora* = Guamo cinta
   Hab.: RF — Height: a
   Fruit: Indehiscent pod — Ind.: 144
   Disp.: Zoochorous — Fr. time: 3
8. *Inga leiocalycina* = Guamo pompeya
   Hab.: RF — Height: a
   Fruit: Indehiscent pod — Ind.: 2
   Disp.: Zoochorous — Fr. time: 3
9. *Inga rubiginosa* = Guamo terciopelo
   Hab.: RF — Height: aa
   Fruit: Indehiscent pod — Ind.: 240
   Disp.: Zoochorous — Fr. time: 2/8
10. *Inga ruiziana* = Guamo balbino
    Hab.: RF — Height: aa
    Fruit: Indehiscent pod — Ind.: 43
    Disp.: Zoochorous — Fr. time: 5
11. *Inga scabriuscula* = Guamo rabo de mono
    Hab.: Savanna, RF, Riverside — Height: a
    Ind.: 287
    Fruit: Indehiscent pod — Fr. time: 2–3
    Disp.: Zoochorous
12. *Inga splendens* = Guamo cambur
    Hab.: RF — Height: a
    Fruit: Indehiscent pod — Ind.:
    Disp.: Zoochorous — Fr. time: 5
13. *Inga* sp. = Guamo caraota
    Hab.: RF — Height: A
    Fruit: Indehiscent pod — Ind.: 2
    Disp.: Zoochorous — Fr. time:
14. *Inga* sp. = Guamo liso
    Hab.: Savanna, Deciduous forest — Height: a
    Ind.: 14
    Fruit: Indehiscent pod — Fr. time: 2/5
    Disp.: Zoochorous
15. *Inga* sp. = Guamo macho
    Hab.: RF — Height: a
    Fruit: Indehiscent pod — Ind.: 29
    Disp.: Zoochorous — Fr. time:
16. *Inga* sp. = Guamo mora
    Hab.: RF — Height: A
    Fruit: Indehiscent pod — Ind.:
    Disp.: Zoochorous — Fr. time:
17. *Inga* sp. = Guamo rastrojero
    Hab.: Secondary forest, deciduous forest — Height: a
    Ind.: 6
    Fr. time:
    Fruit: Indehiscent pod
    Disp.: Zoochorous
18. *Inga* sp. = Guamo verde
    Hab.: RF — Height: A
    Fruit: Indehiscent pod — Ind.: 179
    Disp.: Zoochorous — Fr. time:
19. *Parkia oppositifolia* = Caro montañero
    Hab.: RF — Height: A
    Fruit: Indehiscent pod — Ind.: 42
    Disp.: Zoochorous — Fr. time: 2/12
20. *Parkia pendula* = Cascarón
    Hab.: RF — Height: A
    Fruit: Dehiscent pod — Ind.: 19
    Disp.: Autochorous — Fr. time: 3
21. *Parkia ulei* = Clavellina
    Hab.: RF — Height: A
    Fruit: Indehiscent pod — Ind.: 26
    Disp.: Zoochorous — Fr. time:
22. *Parkia* sp. = Tamarindo monetañro
    Hab.: RF — Height: A
    Fruit: Dehiscent pod — Ind.:
    Disp.: Autochorous — Fr. time:
23. *Pentaclethra macroloba* = Clavellino
    Hab.: RF — Height: a
    Fruit: Dehiscent pod — Ind.: 2534
    Disp.: Autochorous and zoochorous — Fr. time: 2
24. *Piptadenia psilostachya* = Yiguire
    Hab.: RF — Height: A
    Fruit: Dehiscent pod — Ind.: 185
    Disp.: Anemochorous — Fr. time: 2
25. *Pithecellobium basijugum* = Curarina chiquita
    Hab.: RF — Height: aa
    Fruit: Dehiscent pod — Ind.: 561

Disp.: Autochorous and zoochorous Fr. time:

26. *Pithecellobium* cf. *claviflorum* = Hueso de pescado blanco
   Hab.: RF Height: A
   Fruit: Dehiscent pod Ind.:
   Disp.: Autochorous and zoochorous Fr. time: 2
27. *Pithecellobium jupunba* = Samán montañero
   Hab.: RF Height: A
   Fruit: Dehiscent pod Ind.: 142
   Disp.: Autochorous and zoochorous Fr. time:
28. *Pithecellobium pedicellare* = Hueso de pescado (amarillo)
   Hab.: RF Height: A
   Fruit: Dehiscent pod Ind.: 91
   Disp.: Autochorous and zoochorous Fr. time: 1/3
29. *Pithecellobium* sp. = Guamo rebalsero
   Hab.: Lowland, Riverside Height: a
   Ind.: 44
   Fruit: Indehiscent pod Fr. time:
   Disp.: Zoochorous
30. *Pethecellobium* sp. = Samán
   Hab.: RF Height: A
   Fruit: Indehiscent pod Ind.: 12
   Disp.: Zoochorous Fr. time: 2
31. *Stryphnodendron polystachyum* = Masaguaro
   Hab.: RF Height: A
   Fruit: Dehiscent pod Ind.: 73
   Disp.: Autochorous Fr. time: 9/12
32. *Stryphnodendron purpureum* = Josefina
   Hab.: RF Height: a
   Fruit: Dehiscent pod Ind.: 26
   Disp.: Autochorous Fr. time: 8/10
33. *Stryphnodendron* sp.? = Palo blanco
   Hab.: Deciduous forest Height: A
   Ind.:
   Fruit: Indehiscent pod Fr. time:
   Disp.: Zoochorous

1. *Enterolobium cyclocarpum* (Fig. 62): The woody dark brown indehiscent pod is helicoidally bent in one plane so as to form a complete circle and has a smooth shiny surface. The surface humps are formed by the seeds, about 11–20 in number. The fruit circle measures 7–12 cm in diameter and is flattened in the median plane. The seeds are likewise flattened, brown and brilliant, about 2,3 × 1,5 cm measuring. They are embedded in a sweet fibrous fruit flesh which is edible and originates from the so-called "seed cushion" corresponding to a kind of ventral meristem originating from the endocarp (see Roth 1977). The light line is conspicuous in the testa of the seeds. The seeds become free after rotting of the fruit shell. The large indehiscent pods fall to the ground where they are eaten or carried about by a variety of mammals attracted by the sweet flavour of the fruit flesh. Normal dispersal takes place in the intestinal tract of large vertebrates and mechanical transport by squirrels and other smaller vertebrates. Pecaries (*Dicotyles tajacu*) crush some seeds and spit out others. They are good dispersal agents. Even cattle like the fruit and grow stout with the harvest. Chemical defenses in the seeds prevent attacks by bruchids (Janzen 1969). The light weight of the fruit tissue and the air spaces around the seeds give the fruit enough buoyancy to be carried by rivers to the ocean and drift there. However, ocean currents have played little role in the spread of this species (Gunn and Dennis 1976).

The tree is characteristic of open forests and even savannas and grows well in secondary forest. It produces fruit during the dry season.

2. *Enterolobium* sp. (Caro blanco) (Fig. 63): The woody indehiscent pod is somewhat bent helicoidally, but less than in *E. cyclocarpum*, and shows no constrictions between the seeds. It has a brown colour. The seeds are embedded in a sweet edible fruit flesh originating from the "seed cushion" (endocarp). Liberation of the seeds occurs after rotting of the fruit shell and dispersal takes place by animals (rodents, deer, and others).

The genus *Inga* comprises primitive Mimosoideae dispersed by animals such as fish, reptiles, birds, mammals. The seeds are generally surrounded by a sweet pulp corresponding to the "old sarcotesta", according to van der Pijl (1966, 1972) or to the endocarp (van der Pijl 1955). The integument of *Inga* is of spongy consistency and seems to adhere inseparably to the fruit endocarp, but separates with facility from the embryo (Nadelmann 1889, Boelcke 1946). However, in the species studied the origin of the pulpa was not investigated. The entire fruit drops at maturity and does not dehisce, as a rule, or has delayed dehiscence. The seeds become free by rotting of the fruit shell. In *I. feuillei* the integument adopts a spongy consistency in its outer part (mainly the outer epidermis) with juicy cells rich in glucose, while the embryo becomes hard. The cotyledons are protected against insolation (anthocyanin layer), desiccation and heat (thick cuticle, lack of stomata), predation (tannin). When the sarcotesta is eaten by birds, the embryo remains nude, but well protected (Borzi 1903).

3. *Inga alba:* The indehiscent pod is flattened measuring about 15–20 cm in length, 1,6 cm–1,8 cm in width, and 0,8 cm in thickness. The fruit surface is smooth and has a somewhat veined appearance. The pod which contains up to 10 seeds is somewhat constricted between the seeds. Dispersal takes place by animals (mammals). In Surinam, the high tree grows in the high savanna forest too.

4. *Inga capitata:* The flattened pod is 9–10 cm long, 2–3 cm broad, and 0,8–1 cm thick. The surface is devoid of hairs, but shows numerous lenticels. The fruit margins are undulated. The indehiscent fruit contains many small seeds and is dispersed by animals. The species is common in hilly regions in Surinam.

5. *Inga fastuosa* (Fig. 61): The large flattened pod with a strengthened ventral suture has a hairy surface resembling a fur. It is longitudinally helocoidally bent and about 15 cm long. The several seeds are embedded in a fleshy edible pulpa (endocarp?). The pod remains indehiscent. Dispersal takes place by animals.

6. *Inga heterophylla:* The flattened indehiscent pod is about 8–16 cm long, 1–2 cm broad, and 0,5–0,7 cm thick, straight or somewhat bent, and of green to yellowish colour. The surface is smooth, the margin thickened. The fruit is constricted between the seeds. Up to 14 seeds are embedded in a sweet edible pulp. Dispersal takes place by animals. The species is common in the savanna forest of Surinam.

7. *Inga lateriflora:* The flattened pod is 10–30 cm long, 3 cm broad, and 1 cm thick being irregularly constricted between the seeds. The fruit margin is strongly thickened. Up to 11 seeds occur in each pod which are surrounded by a sweet edible pulp. The pod remains indehiscent. Dispersal takes place by animals. In Surinam, the tree also occurs in the savanna forest.

8. *Inga leiocalycina:* The very flat indehiscent pod reaches 15–25 cm in length, 2 cm in width, and 1 cm in thickness. It is somewhat constricted between the seeds. The fruit margin is somewhat thickened. The surface is hairless. Up to 12 seeds are embedded in a sweet edible pulp (endocarp?). Dispersal takes place by animals. In Surinam, the tree is found in the marsh forest, mainly at the river side, and in the rain forest preferably on sandy soil.

9. *Inga rubiginosa:* The indehiscent pod is 20–30 cm long, 2,5–3 cm broad, and 1,6 cm thick, somewhat bent, and has a brown velvety indumentum, hence the vulgar name "terciopelo" which means velvet. The fruit is somwhat constricted between the seeds. Dispersal takes place by animals. In Surinam, the species is found mainly on slopes, preferably on bauxit.

10. *Inga ruiziana:* The dry indehiscent pod contains numerous seeds. The fruit is dispersed by animals.

11. *Inga scabriuscula* = *I. edulis* (Fig. 64): The dark-green elongated pod is more or less circular in transection and has numerous longitudinal ribs. It is about 20–30 cm long, and 1–2 cm in diameter. The indehiscent pod has a leathery texture and its surface is covered with a brown velvety indumentum (Fig. 64). About 5–6 seeds are embedded in a sweet edible white pulp (endocarp?). Dispersal takes place by animals. In Surinam, the tree frequently occurs along creeks and in hilly regions.

12. *Inga splendens:* The inhediscent pod is very flat, somewhat to strongly bent, 15–21 cm long, 3 cm broad, and 1–2 cm thick. In a young stage, the fruit has a delicate indumentum, but later becomes hairless. Between the seeds the pod is somewhat constricted. Dispersal takes place by

animals. The species is common in Surinam.

13. *Inga* sp. (Guamo caraota): The indehiscent dry pod contains numerous small seeds. Dispersal takes place by animals.

14. *Inga* sp. (Guamo liso): The indehiscent dry pod contains numerous small seeds. Dispersal takes place by animals.

15. *Inga* sp. (Guamo macho): The dry indehiscent pod contains numerous small seeds. Dispersal takes place by animals.

16. *Inga* sp.? (Guamo mora): The dry indehiscent pod contains numerous small seeds. Dispersal takes place by animals.

17. *Inga* sp. (Guamo rastrojero): The dry indehiscent pod contains numerous small seeds. Dispersal takes place by animals.

18. *Inga* sp. (Guamo verde): The dry indehiscent pod contains numerous small seeds. Dispersal takes place by animals.

Species of *Parkia* may develop a pod with a sweet endocarp pulpa which is eaten by birds and bats. The seeds are rich in fat and protein.

19. *Parkia oppositifolia:* The indehiscent pod is 20–25 cm long and 5 cm broad, bent, and in a young stage it is furnished with a velvety indumentum, but later becomes hairless. The seeds are arranged in one row. Dispersal takes place by animals. The tree is rare in Surinam.

20. *Parkia pendula:* The dry dehiscent pod is flat and somewhat bent. It measures 7,5–20 cm in length, 2–2,5 cm in width, and 0,2–0,3 cm in thickness. At maturity it has a reddish colour. The pod is somewhat constricted between the seeds and has a thickened margin. It contains about 20 seeds. The fruit opens with 2 valves. The fruits ripen during the dry period, when the tree is leafless. The dispersal mechanism is autochorous. In Surinam the species is very common in the rain forest. In the Amazonian forest 0–2 individuals occur on one hectare (Black et al. 1950).

21. *Parkia ulei:* The dry indehiscent pod is flat, and measures about 25 cm in length, 3–4 cm in width, and 0,7 cm in thickness. It is straigth or sligthly bent and has a rusty-brown indumentum. The fruit is somewhat constricted between the seeds. The seeds are arranged in a single rown and possibly embedded in an endocarp pulpa. Dispersal takes place by animals. In Surinam, the species frequently occurs in the rain forest.

22. *Parkia* sp. (Tamarindo montañero): The large flat pod opens with 2 valves. It contains numerous seeds arranged in 1–2 rows. The dry pod ripens during the dry season. The dispersal mechanism is autochorous.

23. *Pentaclethra macroloba:* The flattened linear-subpatulate pod is bent, about 30–38 cm long, 3,3–6 cm broad, and 0,8–1,1 cm thick. The ventral suture is strengthened. The explosive dry woody pod opens with 2 valves, and the valves are veined longitudinally. The seeds are actively thrown out by curling of the valves and may be ejected up to 10 meters from the crown edge. However, dispersal by floatation and scatterhoarding through rodents is more effective. Fruit development takes 3–4 months. Rapid elongation and lignification of the fruit delays seed development so that the seeds are protected from sucking insects. Only one legume matures in each inflorescence. The seeds are brown, rhomboid, 3 cm in diameter, and have a moderately soft permeable seed coat. Seedlings of *P. macroloba* are generally abundant. Germination is rapid and occurs within 3 weeks after dehiscence. The seeds contain an alkaloid and free amino acids which serve as a nitrogen storage for the seedling. The toxic nitrogenous compounds render the seeds virtually immune to vertebrate and invertebrate predators. According to Janzen, toxicity evolved in response to host specific insect predation. Parrots and squirrels can eat a few seeds without damage. Hartshorn (1972) suggests that the toxicity of the seeds represents an important factor for the abundance of *P. macroloba* in the forest. Toxicity of seeds together with an excellent germination power leads to abundance of seedlings. Furthermore, the ability of *P. macroloba* to grow well on the older alluvial soils of low fertility reduces interspecific competition. *P. macroloba*, therefore, comes to be the dominant tree in the tropical wet forest of La Selva, Costa Rica, where it reaches a height of up to 40 m.

Pentaclethra macroloba has a disjunt distribution in Latin America. It extends from Northeastern Venezuela across the coastal lowlands of the

Guianas into the Eastern Amozan Basin. A second population occurs in Panamá and Western Colombia and a third one in the Eastern lowlands of Nicaragua, Costa Rica and Western Panamá. Dry forest and savanna separate the disjunct areas. The explanation of the disjunct populations is that they were once part of a larger continuous population extending from Nicaragua to Brazil during better conditions, according to Hartshorn (refuge theory). In contrast to the observations in Venezuelan Guiana, *P. macroloba* shows a fruiting peak from July to October in Costa Rica. In Surinam, the tree grows in the marsh- and riparian forest.

24. *Piptadenia psilostachya:* The dry fruit is dehiscent and liberates winged anemochorous seeds. The fruit shell is rich in tannins used for tanning. The delicate wing of the seed originates from the integument. Dispersal of the winged seeds takes place by wind.

25. *Pithecellobium basijugum:* The flat pod is about 20 cm long, 1–1,5 cm broad, and has the shape of a stick. The surface is hairless and reticulately veined. The fruit is swollen above the seeds and irregularly constricted between them. It dehisces along the dorsal suture. Dispersal is autochorous, but probably also takes place by animals, e.g. birds. After falling, fruits are eaten by ruminants. The tree is rare in Surinam.

26. *Pithecellobium* cf. *claviflorum:* The dry flat pod is bent and opens along the dorsal suture. The valves bend helicoidally when the fruit dehisces. The seeds are devoid of an aril. The dispersal mechanism is autochorous, but may also be zoochorous (by birds, ruminants). The species is rare in Surinam.

27. *Pithecellobium jupunba:* The dry pod is bent in a circle, measuring about 10 cm in length and 0,8–1,6 cm in width. It is green to dark-brown and swollen above the seeds. It opens along the dorsal suture, while the ventral suture is somewhat thickened and strengthened. The numerous seeds (10–12) measure 0,7 × 0,5 × 0,4 cm. They are white with a blue-green spot. The dispersal is autochorous, but may also be zoochorous (by means of birds) when the fruit is dehisced. In Surinam, the tree is very frequent in the rain forest as well as in the marsh forest and in the high savanna forest. It flowers and fruits several times a year, according to Roosmalen.

28. *Pithecellobium pedicellare:* The dry dehiscent pod is flat, straight, of leathery consistency, about 10–14 cm long, 2–2,5 cm broad, and 0,5 cm thick, and has a dark-brown colour. It dehisces along the dorsal suture and contains up to 18 bean-shaped seeds, measuring 0,7 × 0,3 × 0,2 cm. The dispersal mechanism is autochorous, but may be assisted by animals (birds). The species occurs scattered in the rain forest of Surinam.

29. *Pithecellobium* sp. (Samán): The dry cylindric pod is indehiscent and contains numerous small seeds. The dispersal mechanism is probably zoochorous (by birds).

30. *Pithecellobium* sp. (Guamo rebalsero): The dry and indehiscent pod contains numerous seeds. Dispersal takes possibly place by animals.

Although *Pithecellobium* seeds have no chemical defenses and are heavily attacked by bruchid beetles, according to Janzen, the species of *Pithecellobium* studied in this paper show relatively high numbers of individuals. As Janzen (1980) emphasizes, *Pithecellobium* compensates the damage of predators by larger crops.

31. *Stryphnodendron polystachyum:* The dry dehiscent pods are assembled in large conglomerations. Each pod is strongly bent, flat, about 2,5–3 cm long and 1,5 cm broad. The seeds are egg-shaped and measure 0,6 × 0,4 × 0,25 cm. They are separated by partition walls. The dispersal mechanism is autochorous. In Surinam, the tree occurs very frequently, mainly in hilly areas.

32. *Stryphnodendron purpureum:* The dry dehiscent pod is somewhat bent, very thick, and has a rusty-brown velvety indumentum. It measures 14 × 1,2 × 0,6 cm. It is swollen above the seeds and opens along the dorsal suture, releasing 12–14 flattened seeds, about 0,75 × 0,6 × 0,35 cm in size. The seeds are separated by partition walls. The dispersal mechanism is autochorous. The tree is rare, even in Surinam where it occurs along the Surinam river.

33. *Stryphnodendron* sp.? (Palo blanco): The dry pod is indehiscent and contains numerous seeds. Dispersal takes possibly place by animals.

In contrast to *Pentaclethra macroloba* seeds, the

seeds of *Stryphnodendron excelsum* and possibly of other species of *Stryphnodendron* are attacked by beetle larvae. This may be the reason for the low number of individuals in the forest studied. Furthermore, the genus is centered in the Amazon Basin (Hartshorn 1972).

The Mimosaceae studied play a very important part in the composition of the forest concerning their number of species and individuals (33 species with 5.616 individuals altogether). However, most of the species are represented by relatively few individuals. This may partly be explained by the fact that some species belong to different habitats, but are not characteristic of the forest studied (e.g. *Enterolobium cyclocarpum, Enterolobium* sp., *Inga scabriuscula, Inga* sp. = Guamo liso, *Inga* sp. = Guamo rastrojero, *Pithecellobium* sp. = Guamo rebalsero, *Stryphnodendron* sp. = Palo blanco). The fruits of all species studied of *Inga* can be considered indehiscent pods which fall to the ground and are eaten by animals for the sweet pulp surrounding the seeds, no matter to which height category the trees belong, as dispersal takes place on the ground. The genus *Inga* is primitive and the ancient way of saurochory is still being observed within the genus. The number of individuals is, however, not very high in any species studied of *Inga*, and only six species reach an elevated number, such as *Inga alba, I. capitata, I. scabriuscula, I. rubiginosa, I.* sp. = Guamo verde, and *I. lateriflora.* The 2 species of *Enterolobium* studied definitely belong to the deciduous forest and are weakly represented in the rain forest for this reason. All 4 species of *Parkia* are very weakly represented, whether their fruits are dispersed by animals or have an autochorous dispersal mechanism. All 4 species are of height category A. The species which reaches the highest number of individuals of all Mimosaceae studied is *Pentaclethra macroloba.* The tree is of height category a and contributes 2.534 individuals. The fruits are explosive pods which possibly throw the seeds many meters (up to 10 meters). This dispersal mechanism combined with floatation and scatterhoarding by animals is advantageous, as the tree is high and may reach up to 40 m. Futhermore, germination of the seeds seems to be very effective, as seedlings of this species are abundant in the forest. The species is resistant and grows on soils of low fertility reducing interspecific competition in this way. However, in prehistoric times it had a much larger extension (refuge theory). The above mentioned facts may explain the frequency of individuals of this species. *Piptadenia psilostachya* has a good dispersal mechanism by dehiscent pods which release winged seeds. As the tree is of height category A, wind dispersal is advantageous, but nevertheless, the number of individuals remains moderate. The genus *Pithecellobium* is related to *Inga.* According to van der Pijl, the evolution of *Pithecellobium* starts with ornithochory of sarcotesta seeds dangling on the dehiscing valves which are often coloured. Through the torsion of the valves caused by crossing fiber layers the presentation of "berry-seeds" becomes effective. Arilloid seeds may become replaced by mimetic seeds, also dispersed by animals. Pods which drop to the ground are eaten by ruminants and the arilloids have disappeared in this case. We may therefore assume that even the autochorous species are partly dispersed by animals. This fact would explain that *Pithecellobium basijugum* of the lowest height category (aa) has the highest rate of distribution within the species of *Pithecellobium* studied with 561 individuals. The lowest tree species may have the best chance to be dispersed well by animals. The rate of distribution of *Stryphnodendron polystachyum* and *S. purpureum* is low and the dispersal mechanism is autochorous. The lower autochorous species (*S. purpureum*) has less individuals.

Janzen (1977a) indicates that the total seed production of *Pithecellocium* is much greater than that of *Enterolobium*, but that probably no more than 1% of *Pitchellobium* seeds survive, before the pods are shed. *Pitchellocium* seeds are heavily attacked by larvae of bruchid beetles. Seeds of *Enterolobium*, on the other hand, are not attacked possibly because of the presence of unpalatable chemicals. Pods of both species are eaten by herbivorous mammals, including cattle, deer, and several rodents (Hatheway & Baker 1970).

Of the 33 species of Mimosaceae studied, 19 belong to height category A, 11 to height category a, and 3 to the lowest height category (aa). Of these 33 species one is anemochorous, 4 are autochorous only, 5 species are autochorous and zoochorous at the same time, and 23 species are zoochorous. Zoochory thus predominates. Of the 3 lowest species (aa) 2 are zoochorous and one is auto-zoochorous. Of the 4 autochorous species 3 are of height category A and one is of height category a. The anemochorous species belongs to the highest height category (A). Of the 5 auto-zoochorous species 3 belong to height category A, one to a and another one to aa. Of the 23 zoochorous species 2 are of height category aa, 9 of height category a, and 12 of A. It thus becomes obvious again that anemochorous and autochorous species are preferably high trees of height category A, while very low trees or shrubs of height category aa are usually zoochorous.

Of the 33 species of Mimosaceae studied 23 have indehiscent pods, and only 10 have dehiscent pods. In the genus *Inga*, the seeds are often surrounded by a sugary fruit flesh. The indehiscent pods of *Stryphnodendron* may be furnished with a fleshy mesocarp.

Caesalpiniaceae

The pods of the Caesalpiniaceae studied undergo a great variety of metaporphoses, from dehiscent several-seeded forms to one-seeded indehiscent types, such as the drupe and the samara.

1. *Bauhinia* sp. = Pata de vaca
   Hab.: RF — Height: a
   Fruit: Dihiscent pod — Ind.: 18
   Disp.: Autochorous — Fr. time:
2. *Brownea latifolia* = Rosa de montaña
   Hab.: Lowland, RF, deciduous forest — Height: a
   Fruit: Dehiscent pod — Ind.: 434
   Disp.: Autochorous — Fr. time:
3. *Cassia grandis* = Cañafistola montañera
   Hab.: RF, Deciduous & riparian forests — Height: A
   Fruit: Indehiscent pod — Ind.: 3
   Disp.: Zoochorous — Fr. time:
4. *Cassia multijuga* = San Francisco marimari
   Hab.: RF — Height: a
   Fruit: Samara — Ind.: 14
   Disp.: Anemochorous — Fr. time: 1
5. *Cassia* sp. = Cañafistola (sabanera)
   Hab.: Savanna, RF — Height: a
   Fruit: Indehiscent pod — Ind.: 2
   Disp.: Zoochorous — Fr. time: 8/11
6. *Copaifera (pubiflora?)* = Aceitillo
   Hab.: Semiarid regions, RF — Height: A
   Fruit: Dehiscent pod — Ind.: 2
   Disp.: Autochorous and zoochorous — Fr. time:
7. *Crudia glaberrima* = Arepito rebalsero
   Hab.: Lowland, riverside — Height: A
   Fruit: Dehiscent pod — Ind.:
   Disp.: Autochorous — Fr. time:
8. *Crudia oblonga* = Algarrobo rebalsero
   Hab.: Lowland — Height: A
   Fruit: Dehiscent pod — Ind.: 1
   Disp.: Autochorous — Fr. time: 5/9
9. *Dialium guianense* = Gateado
   Hab.: RF — Height: A
   Fruit: Drupe — Ind.:
   Disp.: Zoochorous — Fr. time:
10. *Dimorphandra gonggrijpii* = Mora
    Hab.: RF — Height: A
    Fruit: Dehiscent pod — Ind.: 2
    Disp.: Autochorous — Fr. time:
11. *Eperua falcata* = Caraota
    Hab.: RF — Height: A
    Fruit: Dehiscent pod — Ind.:
    Disp.: Autochorous — Fr. time:

12. *Eperua jenmani* = Concha de suela
Hab.: RF Height: A
Fruit: Dehiscent pod Ind.:
Disp.: Autochorous Fr. time:
13. *Hymenaea courbaril* = Algarrobo
Hab.: RF Height: A
Fruit: Indehiscent pod Ind.: 12
Disp.: Zoochorous Fr. time:
14. *Lecointea amazonica* = Nispero
Hab.: RF Height: A
Fruit: Indehiscent pod Ind.: 127
Disp.: Zoochorous Fr. time:
15. *Macrolobium acaciaefolium* = Arepito negro
Hab.: Lowland, Riverside Height: A
Ind.:
Fruit: Dehiscent pod Fr. time: 2
Disp.: Autochorous and zoochorous
16. *Mora excelsa* = Mora rebalsera
Hab.: RF Height: A
Fruit: Dehiscent pod Ind.:
Disp.: Autochorous, achorous Fr. time:
17. *Peltogyne porphyrocardia* = Zapatero
Hab.: Deciduous forest, RF, riverside Height: A
Ind.: 217
Fr. time: 2/6/10
Fruit: Dehiscent pod
Disp.: Autochorous and zoochorous
18. *Peltogyne pubescens* = Nazareno
Hab.: RF Height: A?
Fruit: Dehiscent pod Ind.: 1
Disp.: Autochorous and zoochorous Fr. time:
19. *Peltogyne* sp. = Zapatero negro
Hab.: RF Height: A
Fruit: Dehiscent pod Ind.: 16
Disp.: Autochorous and zoochorous Fr. time: 10
20. *Sclerolobium paniculatum* = Guamillo
Hab.: RF Height: A
Fruit: Samara Ind.: 421
Disp.: Anemochorous Fr. time: 1–2/5
21. *Sclerolobium* sp. = Guamillo blanco
Hab.: RF Height: A
Fruit: Samara Ind.: 22
Disp.: Anemochorous Fr. time:
22. *Sclerolobium* sp. = Guamillo rojo
Hab.: RF Height: A
Fruit: Samara Ind.: 174
Disp.: Anemochorous Fr. time:
23. *Swartzia schomburgkii* = Olivero
Hab.: RF Height: A
Fruit: Dehiscent pod Ind.:
Disp.: Zoochorous Fr. time:
24. *Swartzia* sp. = Macho
Hab.: RF Height: a
Fruit: Indehiscent pod? Ind.: 226
Fr. time: 2/5
Disp.: Zoochorous
25. *Tachigalia paniculata* = Palo de Maria
Hab.: RF Height: A
Fruit: Indehiscent pod Ind.: 139
Disp.: Zoochorous Fr. time:

1. *Bauhinia* sp.: The dry dehiscent pod contains only few seeds. Dispersal takes place by explosion whereby the seeds are thrown out of the fruit. The dispersal mechanism is autochorous.

2. *Brownea latifolia:* The dry dehiscent pod contains few seeds. The dispersal mechanism is autochorous.

3. *Cassia grandis:* The dry indehiscent blackish-brown pod is woody and measures about 25–60 (80) cm in length, 4 cm in width, and 3 cm in thickness weighing a fraction of one gram up to 0,25 kg. The surface is delicately veined. On the ventral suture are 2 prominent longitudinal ribs. The large hard (1,25 cm long) seeds weighing 247–904 mg, are embedded in a swet pulp, possibly originating from the endocarp. Dispersal mainly takes place by animals. The sweet odiferous, molasses-like, water-soluble material surrounding the seeds hardens with age. It attracts dispersal agents such as deer, which swallow the seeds and pass them intact, furthermore squirrels, agoutis, tapirs, peccaries, and large rodents such as pacas (*Cuniculus paca*). A fully mature tree bears about 300 (up to 700) pods in one crop each of which contains 10–90 seeds (as an average 40–55). This makes about 15.000 seeds per tree. The seed coat protects the seed so well that it must be scarified either by passage through a mammal's intestine or

otherwise to germinate. Unscarified seeds can live at least for 8 years. The pods hang on the tree for 2–5 months until they are blown down or removed by dispersal agents. The tree bears fruit every other year. It has to produce a large quantity of seeds, as most of the seeds are killed by larvae of two bruchid beatles (Janzen 1971, 1977d). Guppy (1917) notes that *Cassia grandis* fruits support prolonged flotation without injury. They are regular constituents of beach drift in the West Indies. The tree has a profound radical system. It has a wide natural circum-Caribbean distribution in deciduous and riparian forests. In Surinam it is not very common. The fruits mature in the dry season (February-April).

4. *Cassia multijuga:* The dry indehiscent fruit is winged and has a light weight, but contains numerous seeds. According to Roosmalen *C. multijuga* is not winged, about 10–15 (–21,5) cm long, 1–1,5 (–2) cm broad, and 3 cm thick. The surface is smooth and hairless. About 50 transverse ribs indicate the position of the underlying seeds. Folowing the notes of Rollet, the fruit should be dispersed by wind, representing a samara. This dispersal type is very characteristic of plants growing in the savanna. In Surinam the small tree (12 m) or shrub is very common mainly occurring in riparian forests and on rocks.

5. *Cassia* sp. (Cañafistola sabanera): The dry fruit is indehiscent and contains numerous seeds. It is probably dispersed by animals.

6. *Copaifera* (*pubiflora?*): The dry fruit is dehiscent and contains a single large seed with an aril. Seed dispersal takes place by animals (e.g. birds). As the tree is characteristic of semi-arid regions, the low number of individuals in the forest studied explains itself.

7. *Crudia glaberrima:* The dry pod is of leathery consistency, about 10–18 cm long, 5–10 cm broad, and 0,2–0,3 cm thick and contains few (1–2) seeds, which become free by dehiscence of the fruit. The fruit surface is brown and has a velvety indumentum. The venation of the pod is prominent on the surface. The dispersal takes place in an autochorous way. The species occurs in the marsh forest of Surinam.

8. *Crudia oblonga:* The dry pod is dehiscent and contains only few seeds. Dispersal is autochorous.

9. *Dialium guianense:* The fleshy fruit is indehiscent and contains only one seed (rarely two). The seed is surrounded by the pithy soursweet endocarp-pulp which has a taste similar to that of tamarind. The drupe has an ovoid to globular shape, is about 1,5–2,5 cm long and 1,5 cm in diameter, dark-green to brown with a smooth hairless surface. According to Roosmalen, the fruit opens with 2 valves. The seeds are dispersed by animals, tree mammals, e.g. by monkeys in Africa. In Surinam the tree is common in the rain forest.

10. *Dimorphandra gonggrijpii:* The large and dry dehiscent pod contains only few hard seeds. Dispersal is autochorous.

11. *Eperua falcata:* The dry and woody dehiscent pod somehow imitates the shape of the sheath of a cutlass. In Spanish it has, therefore, occasionally the name "Palo machete". The fruit is about 20–30 cm long, 6–8 cm broad, and 0,5 cm thick, and is somewhat curved. It has a red to red-brown velvety indumentum and contains 2–5 seeds which are of ovalate shape and flattened, measuring 4 × 2,5 × 1 cm. Dispersal is autochorous. The inflorescences (racemes) are pendulous on 0,5–2 meter long peduncles (penduliflorae) and are pollinated by bats. In Surinam, the tree occurs scattered in the rain forest, but may become more frequent along creeks and in the high savanna forest, especially of the West. The species is indigenous of Guiana and the Amazonas.

12. *Eperua jenmani:* The flat woody pod is dehiscent and contains few large seeds. The shape of the fruit is sword-like. Dispersal is autochorous.

13. *Hymenaea courbaril:* The woody flattened pod is indehiscent, about 9–17,5 (20) cm long, 4–6,5 cm broad, and 1–3,5 cm thick, weighing 25–125 gr. The surface is green to brownish and covered with numerous lenticels. The woody pericarp wall is about 0,4–0,5 cm thick and somewhat constricted between the seeds. The fruit remains 4–11 months on the tree and drops in the middle of the dry season. The resin content in the pericarp deters squirrels and parrots (Janzen 1978). The dark-brown to red-brown seeds measure 2,2 × 1,5 cm

and are somewhat flattened. The pod contains few seeds (1) 2–4 (6). The hard-shelled seeds may remain in good condition for several months in the indehiscent pod (Marrero 1942). The seeds are embedded in a dry and sweet, edible, mealy tasting pulpa-powder of an unpleasant odour and of yellowish colour, the endocarp pulpa. The pods are rich in resin ("copal") which is used in the churches as in incense. Pounded when young the fruits are applied in the curation of bone fractures by the natives. The tannin content of the pericarp is high. The dry powder around the seeds is said to help against pectoral diseases. It is furthermore eaten toasted.

The seedlings need shade first, but later get adapted to insolation. Their growth is slow. The radical system of the tree is profound. The species is widely distributed in tropical regions. However, it is more characteristic of open forests and even savannas, according to Pittier, hence the low rate of distribution in the forest studied. Seed dispersal mainly takes place by animals (tree mammals), the hard-shelled seeds being well protected against injuries. However, Guppy (1917) observed seeds as beach drift in Jamaica. The seeds being able to float may be dispersed by water too. Predispersal predation takes place by weevils, postdispersal predation by agoutis. Seed predation is the major force for deviation from annual fruiting, and also for setting the timing and variance of seeding distribution with the fruiting year (Janzen 1978). In Surinam, the tree is frequent in the rain and marsh forest, mainly on sandy ridges and in the riparian forest.

14. *Lecointea amazonica:* The dry indehiscent fruit is relatively large and heavy and contains only few seeds. Dispersal is zoochorous.

15. *Macrolobium acaciaefolium* (Fig. 66): The pod has ovoid to almost circular outlines, is of yellowish colour and measures 6 × 4,6–5 × 0,8 cm. The surface is hairless and smooth, and delicately veined in a transverse direction. The dorsal suture is strengthened. The dehiscent somewhat woody fruit releases a single seed. Dispersal is autochorous, but ichthyochory is also recorded. In Surinam, the tree is common, especially in the marsh and riparian forest.

16. *Mora excelsa:* The dry dehiscent fruit contains a single kidney-shaped seed. The pod measures 10-20 cm in length, 5–7 cm in width, and 4–5,5 cm in thickness. It is brown and hairless. The seed measures 7–12 × 4–7 × 3–4 cm. *M. excelsa* flowers at intervals of ca. 18 months or 2 years and normally has a good crop of seed each time (Richards 1952). Seeds of the same crop germinate together immediately after falling, having a high germination percentage. Although the fruit is dehiscent, the species is not considered to have an autochorous dispersal mechanism, because of the weight of the seed. With a size of 12,5 × 7 cm, it is one of the largest existing seeds, weighing up to 0,4 kg. Van der Pijl (1972), therefore, speaks of achory. The heavy seeds just fall to the ground (barochory) germinating immediately and the "local" seedling has to wait for an opportunity to continue growing. The slow spread of the species is partly attributed to the heavy weight and the large size of the seed (Marshall 1934, Richards 1952). The species also grows in swamps and flooded areas. An intercotyledonary cavity makes the seed buoyant (Gunn & Dennis 1976). Although being an aggressive species the spread of the tree is low, possibly owing to the ineffective and primitive dispersal mechanism. In Surinam, the up to 50 m high tree with a stem diameter of 2 m, is occasionally dominant in all stories of the riparian forest in the West.

17. *Peltogyne porphyrocardia:* The dry leathery dehiscent pod (about 3 cm in diameter) is flattened, roundish and contains a single seed with a small aril. The dispersal is autochorous and possibly also takes place by animals (aril). However, the species is more characteristic of the deciduous forest. Nonetheless it is represented by 217 individuals.

18. *Peltogyne pubescens:* The deshiscent pod is slightly triangular, measuring 2–2,5 × 3–4,5 × 0,2–0,6 cm. It is delicately veined, yellow or brown. At maturity only the dorsal suture and the pedicel have a thick indumentum. The only seed is flat, measuring 2,5 × 1,4 × 0,4 cm. It is black, has fine hairs on its surface and an aril on top in the form of a little cap. The dispersal mechanism is autochorous, but may be assisted by animals (aril).

The tree is very common, also in the high savanna forest of Surinam.

19. *Peltogyne* sp. (Zapatero negro): The dry leathery fruit is dehiscent and contains a single seed with a small aril. Dispersal is autochorous and probably also takes place by animals (aril).

20. *Sclerolobium paniculatum:* The elongated pod is flat and winged and contains few seeds. It remains indehiscent. Dispersal of the samara takes place by wind.

21. *Sclerolobium* sp. (Guamillo blanco): The elongated dry pod is winged and contains few seeds. It remains indehiscent. Dispersal of the samara takes place by wind.

22. *Sclerolobium* sp. (Guamillo rojo): The elongated dry pod is fat and winged and only contains few seeds. It remains indehiscent. Dispersal of the samara takes place by wind.

23. *Swartzia schomburgkii* (Figs. 65, 66): The elongated pod reaches up to 7 cm in length and 2–3 cm in diameter. The pericarp texture is leathery and the colour brownish. The fruit surface shows a dense reticulate venation. The pod opens ventricidally as well as dorsicidally (Fig. 65), and contains 1–2 seeds. A constriction is visible between the seeds (Fig. 65). The blackish seeds are relatively large, about 1,5–2,5 cm long (including the aril) and have a smooth and very shiny surface. The fleshy aril is white to yellowish-white, somewhat transparent and horny and the funicle is long (about 1–2 cm long) and folded in the pod so that the seeds hang out of the dehisced pod (Fig. 65) in a typical pendulous position to attract bats and birds. The tree is cauliflorous (Fig. 66) and bears fruit every year. In Surinam it is rarely growing in hilly and mountainous regions of the rain forest.

24. *Swartzia* sp. (Macho): The dry pod is indehiscent and contains numerous small seeds. Dispersal takes possibly place by animals.

25. *Tachigalia paniculata:* The dry pod is of papery or leathery consistency, flattened and elongated, measuring 6–12 × 1,5–3 × 0,1 cm. The surface is smooth and longitudinally veined, at first with a velvety indumentum, but later hairless. The indehiscent pod contains one to several seeds. Dispersal takes possibly place by animals. In Surinam, the tree is frequent especially in the riparian forest.

The 25 species of Caesalpiniaceae studied have a relatively low distribution in the forest studied. Most species have a number of individuals between 1 and 22. Five species contribute 100 or more than 100 individuals (between 100 and 226), and only 2 species show an elevated number of individuals: *Brownea latifolia* with 434 species, and *Sclerolobium paniculatum* with 421 species. The fruit of *Sclerolobium paniculatum*, a samara, is dispersed by wind. The tree belongs to the highest story (A). The fruit of *Brownea latifolia*, on the other hand, is dehiscent and the dispersal mechanims is autochorous. However, the species belongs to the a-story. The dispersal mechanism of *Sclerolobium paniculatum* may be quite effective, as dispersal by wind seems to be an adaptation of high-story trees.

Concerning the height category of the trees we may say that most species belong to height category A, the highest story, 20 altogether when we include the one of which it is doubtful to which category it belongs. Of the 5 species belonging to height category a, 2 are probably dispersed by animals, and one of them has 226 individuals. One species of height category a is dispersed by wind, as the fruit is winged, according to Rollet. This species has only 14 individuals. And 2 species of height category a have dehiscent pods and their dispersal mechanism is autochorous. One of them has 18 individuals, and the other one 434. According to our statistics, the autochorous dispersal mechanism seems to be most successful in high trees.

Of the 20 species belonging to height category A, 7 are probably mainly dispersed by animals, and 2 of these have a higher number of individuals (127 and 139 respectively). The other species remain at or below 12 individuals. Three of the species belonging to height category A are autochorous and are probably additionally dispersed by animals. One of them reaches up to 217 individuals. Seven of the species belonging to height category A are autochorous only having dehiscent pods. They have the lowest numbers of individuals. And the resting 3 species of height category A are dispersed by wind (species of *Sclerolobium* having winged pods or samaras). One of them is represented by 421 individuals, one by 174, and the third by 22 individuals. A wind-dispersed species (*S. paniculatum*), therefore, shows the highest number of individuals in the A-story, a further contribution to the assumption that wind-dispersal is most advantageous in the uppermost stories of the forest. A diplochorous dispersal mechanism (e.g. autochorous combined with zoochorous) is more successful than a haplochorous one which seems to be logical. The autochorous dispersal mechanism is most frequent in high trees. However, in the Caesalpiniaceae studied, this type of dispersal seems not to be very advantageous considering the low number of individuals in the forest studied. However, some of the species actually do not belong to the rain forest formation. But, as we shall see later, the number of individuals present in the forest depends on many factors besides the dispersal mechanism.

Worth mentioning is the fact that the 2 species of *Cassia*, i.e. *Cassia multijuga*, on the one hand, and *Cassia* sp. (Cañafistola), on the other, show very different fruiting times. While *Cassia* sp. fruits during the wet season, i.e. in August and November, *Cassia multijuga* shows a very different fruiting time in January, when the dry season starts. This difference in the fruiting times coincides with the observations of Janzen that wind-dispersed fruits or seeds are dispersed during the dry season. The samara of *C. multijuga* follows this dispersal pattern, while the zoochorous indehiscent fruits of *C.* sp. are better dispersed during the wet season when animal life is very active.

Papilionaceae

The Papilionaceae show an enormous variety of fruit forms, partly dehiscent and partly indehiscent, with numerous seeds or with a reduced number of seeds even reduced to one, comprising alate forms or pods with spines, and septate pods. This large variation in shape is derived from the basic fruit type: the legume.

1. *Alexa imperatricis* = Leche de cochino

Hab.: RF	Height: A
Fruit: Dehiscent pod	Ind.: 1262
Disp.: Autochorous	Fr. time: 3/5

2. *Andira retusa* = Pilón rebalsero

Hab.: RF	Height: A
Fruit: Drupe	Ind.: 3
Disp.: Zoochorous	Fr. time:

. *Andira* sp. = Canelito negro

Hab.: RF	Height: A
Fruit: Drupe	Ind.: 14
Disp.: Zoochorous	Fr. time:

4. *Centrolobium paraense* = Cartán

Hab.: RF, Deciduous forest	Height: A Ind.:
Fruit: Samara	Fr. time: 11
Disp.: Anemochorous	

5. *Clathrotropis brachypetala* = Caicareño montañero

Hab.: RF	Height: A
Fruit: Dehiscent pod	Ind.: 188
Disp.: Autochorous	Fr. time:

6. *Clathrotropis* sp. = Caicareño amarillo

Hab.: RF	Height: A
Fruit: Dehiscent pod	Ind.:
Disp.: Autochorous	Fr. time:

7. *Diplotropis purpurea* = Congrio

Hab.: RF	Height: A
Fruit: Samara	Ind.: 206
Disp.: Anemochorous	Fr. time: 11

8. *Dipteryx odorata* = Sarrapia

Hab.: Deciduous forest, RF	Height: A Ind.: 13
Fruit: Drupe	Fr. time: 11
Disp.: Zoochorous	

9. *Hymenolobium* sp. = Alcornoque montañero

Hab.: RF	Height: A
Fruit: Indehiscent pod	Ind.: 9
Disp.: Anemochorous?	Fr. time:

10. *Lonchocarpus sericeus* = Jebe
Hab.: Savanna, RF — Height: A
Fruit: Samara — Ind.: 251
Disp.: Anemochorous — Fr. time:

11. *Lonchocarpus* sp. = Majomo blanco
Hab.: Deciduous forest — Height: A
Ind.:
Fruit: Samara — Fr. time:
Disp.: Anemochorous

12. *Lonchocarpus* sp. = Tocorito blanco
Hab.: RF — Height: A?
Fruit: Samara — Ind.:
Disp.: Anemochorous — Fr. time: 6

13. *Machaerium* sp. = Vainespa
Hab.: Semideciduous and deciduous forest, RF — Height: A
Ind.: 14
Fr. time:
Fruit: Samara
Disp.: Anemochorous

14. *Ormosia paraensis* = Pionio
Hab.: RF — Height: A
Fruit: Dehiscent pod — Ind.: 89
Disp.: Autochorous and zoochorous — Fr. time:

15. *Ormosia* sp. = Pericoca
Hab.: RF — Height: A
Fruit: Dehiscent pod — Ind.: 17
Disp.: Autochorous and zoochorous — Fr. time: 1/8

16. *Platymiscium pinnatum* = Roble
Hab.: Deciduous forest — Height: A
Ind.: 3
Fruit: Samara — Fr. time:
Disp.: Anemochorous

17. *Pterocarpus officinalis* = Cacú
Hab.: Lowland, swampy forest — Height: A
Ind.: 61
Fruit: Samara — Fr. time: 3
Disp.: Anemochorous, (hydrochorous?)

18. *Pterocarpus rohrii* = Sangrito alado
Hab.: RF — Height: A
Fruit: Samara — Ind.:
Disp.: Anemochorous — Fr. time: 5

19. *Pterocarpus* sp. = Sangrito
Hab.: RF — Height: A
Fruit: Samara — Ind.: 15
Disp.: Anemochorous — Fr. time:

29. Roble blanco
Hab.: Deciduous forest, RF — Height: A
Ind.:
Fruit: Samara — Fr. time:
Disp.: Anemochorous

1. *Alexa imperatricis:* The dry pod is dehiscent and contains many large seeds. Dispersal is autochorous. The species occurs on slopes in Surinam.

2. *Andira retusa:* The name "Andira" means "bat" in a Brazilian Indian language. Bats scatter fruits and seeds far from the pasture tree. The fruit is indehiscent and drupaceous. It has a more or less globular shape with a very thick woody (fibrous) fruit wall and contains only a single seed. The drupe is dispersed by bats and other animals (birds and, on the ground, rodents).

3. *Andira* sp. (Canelito negro): The drupaceous pod is indehiscent and contains a single seed. It has a globular shape and a fibrous woody fruit wall. It is dispersed by bats and other animals.

4. *Centrolobium paraense* (Fig. 67): The very large samara (10–20 cm long) has one large papery wing with delicate parallel ribbing. The single seed is hidden in the fruit base which is armed with a great number of thin spines protecting the embryo. A single strong spine is inserted above the spiny base projecting from the dorsal suture. The indehiscent samara is a "screw flyer". Dispersal of the fruit takes place by wind. The fruit type is very derived considering the large wing and the reduction of the seed number. Apparently, this type of dispersal is not very successful, as the wing is too large to guarantee dispersal over larger areas. Furthermore, the species is more characteristic of the deciduous forest. Species of *Centrolobium* belong to the screw-flyer type (Ulbrich 1928). The point of gravity is excentric. A usually laterally attached wing is characterized by rotating movements during the fall.

C. paraense occurs in Brasil and Guiana. It is deciduous and has pink flowers in panicles. The wood is orange-red. The vulgar name "Cartán" is derived from the Amerindian language.

5. *Clathrotropis brachypetala:* The dry woody

legume is elongated and flattened in the median plane. It measures about 15–20 cm in length, 5–7 cm in width, and 3 cm in thickness. The colour is brown. The dorsal suture is strengthened and up to 0,3 cm broad, while the thickened ventral suture opens at dehiscence. The dehiscent pod contains few (1–2) large seeds embedded in a white spongy parenchyma. Each seed measures 4–5 × 3–4 × 1–2,5 cm. The species is autochorous, but birds may also intervene in its dispersal. In Surinam, the tree reaching 10 (-20) meters mainly occurs in the swampy forest and in the riparian forest.

6. *Clathrotropis* sp. (Caicareño amarillo): The dry pod is dehiscent and contains only few seeds. Dispersal is autochorous.

7. *Diplotropis purpurea:* The elongated pod is flat and has a papery consistency. It measures about 10–18 cm in length, 2,5–4,5 cm in width, and 0,05 cm in thickness. It is light-green and shows a reticulate venation. The dorsal suture is somewhat winged, the wing reaching only 0,2 cm in width. The pod remains indehiscent and contains 1–2 seeds, 2,5 × 0,7 × 0,2 cm. The calyx is persistent. Dispersal takes place by wind. Although moderately large, the fruit is relatively light and apt for wind dispersal by its flattened shape. It is on the way to transform into a samara. The species is represented by 206 individuals. In Surinam, the high tree (up to 40 m) is common.

8. *Dipteryx odorata:* The plum-like indehiscent pod is egg-shaped to somewhat elongated, measuring 5–6,5 × 3 × 2,5 cm, dark-green, with a smooth surface. The only seed is elongated and flattened, measuring about 3,3 × 1 × 0,9 cm. The indehiscent legume may be dispersed by bats, birds and other animals. The endocarp is hard and woody. The seeds ("tonka beans") have an aromatic perfume due to the content of coumarin. They are used in the preparation of perfumes and in the processing of tobacco to add aroma to the leaves. The species also grows in the Llanos. Along the Rio Orinoco the tree grows very well reaching enormous dimensions. The pink flowers occur in panicles. The fruits may reach the size of a Philippine mango. Associations of *D. odorata* are called "Sarrapiales". In former times, the seeds were exported from Ciudad Bolivar in Venezuela. André (1964) mentions that *Dipteryx odorata* prefers granitic soils. The harvest extends from February to May or June. The scarce pulp of the fruit is sticky and tasteless. From October to November when the fruit is still small, green, and immature, it is largely visited by parrots (e.g. *Ara macao*) which taste and destroy a quantity of the harvest. After a year with a rich harvest, one or two years with almost no fruits regularly follow. The species is indigenous of Guiana and Brazil. In Surinam it occurs scattered.

9. *Hymenolobium* sp. (Alcornoque montañero): The dry indehiscent pod is elongated, flattened, of membranaceous consistency, with reticulate venation, and contains 1–2 seeds. It may possibly be dispersed by wind (or animals?).

10. *Lonchocarpus sericeus:* The dry indehiscent fruit, about 3–12 cm long, and 2–3 cm broad, is winged and contains only few seeds. Sometimes it is constricted between the seeds. The samara of membranaceous consistency is dispersed by wind. The tree is resistant against drought and is long-lived. The racemes of violet flowers appear before the coriaceous leaves. The legume is brown and covered with hairs. Parts of the tree are used to stun fish and capture it.

11. *Lonchocarpus* sp. (Majomo blanco): The dry indehiscent pod is winged and contains only few seeds. The samara is dispersed by wind. The flowers occur in racemes. The tree is deciduous and grows at the margins of the Llanos.

12. *Lonchocarpus* sp. (Tocorito blanco): The dry indehiscent fruit is winged and contains only few seeds. The samara is flattened in the median plane and has a persistent calyx. Dispersal takes place by wind.

13. *Machaerium* sp. (Vainespa): The dry indehiscent fruit is winged and contains a single seed. The wing shows reticulate venation. The samara is dispersed by wind.

A profound study of the genus *Ormosia* has been presented by Rudd (1965). About 100 tropical species, 50% of which are endemic of the New World, are known. All species are woody, some are timber trees and reach a height up to 60 meters, others are small and shrubby, and 2 are scandent. The red or red and black seeds are used for necklaces which gave the genus its name (from the Greek "hormos" = necklace). The seeds of some species contain the alcaloids ormosine and ormosinone with a physiological effect resembling that of morphine or the alcaloid panamine which exercises hypotensive actions.

Most species are components of the tropical rain forest or subtropical forest. The inflorescences are racemose and many-flowered. The fruits contain 1-6 seeds and are dehiscent in most species. The fruit colour is fulvous to black-brown, when the pod is glabrous, but light to dark-brown or greyisch, when the fruit is pubescent. The seeds of many American species are bicoloured red and black. The marking of some species is very constant, while in others it is characteristically variable from all red to almost all black, even within the same pod. The seeds are commonly elliptic in shape. The chromosome number is 2n = 16. Genera most related are *Clathrotropis, Diplotropis*, and *Sophora.*

14. *Ormosia paraensis:* The dry dehiscent pod is of woody consistency, pear-shaped, about 3–6,5 cm long, 2–3,5 cm broad, and 1,2–2,2 cm thick. It has first a velvety indumentum, but later adopts a smooth and brilliant black or dark-brown surface. The pericarp is relatively thick (2–5 mm). The fruit contains 1–2 (3) seeds. 2-seeded fruits develop a constriction between the seeds. The seeds are more or less of circular outlines, flattened, measuring 1,2–1,5 cm in length, 1–1,3 cm in width, and 0,8–1 cm in thickness. The hilum is 2 mm long and 1 mm broad. The hard-shelled seeds are bicoloured red with a black spot. Occasionally, the showy seeds are used for necklaces, hence the greek name "hormos" modified into *Ormosia.* According to van der Pijl, the seeds are mimetic. The plant contains the alkaloids ormosine and ormosinone which have a physiological effect resembling morphine. The alkaloid panamine (e.g. in *O. panamensis*) has a hypotensive action. The cooked seeds of *Ormosia* species are used against pains of the heart or for sore of the throat (Rudd 1965). The "coral-seeds" are characteristic of the genus *Ormosia* and simulate an edible aril. Dispersal is autochorous, but probably also takes place by birds. In Surinam, the tree is common. The species is indigenous in French Guiana, Surinam, Southern Venezuela, and Brazil. It grows on sandy or clay soil in the lowland, the high jungle of the Guianas, Venezuela and the middle to lower Amazonian region of Brazil, at elevations up to 800 meters.

15. *Ormosia* sp. (Pericoca): The dry dehiscent pod contains only few seeds. These are stone-hard, a characteristic of the genus *Ormosia,* but some birds may split and eat them, dropping some for dispersal. Dispersal is autochorous, but probably also takes place by birds.

16. *Platymiscium pinnatum:* The dry indehiscent pod is winged and of light weight. It contains a large flattened kidney-shaped seed. The pod is elongated, of membranaceous consistency and reticulately veined. The tree is deciduous. Dispersal of the samara takes place by wind.

17. *Pterocarpus officinalis* (Fig. 68): The light-green almost circular pod ("disc-flyer") measures 3–5 × 2,5–5 × 0,4 cm and has a prominent reticulate venation. The dorsal suture is furnished with an up to 1 cm broad membranaceous wing, almost completely surrounding the seed which forms the center of gravity. The indehiscent pod with its single seed corresponds to a samara. Dispersal partially takes place by wind. As the pericarp has a leathery to woody (corky) texture and since the wing is small, wind dispersal may not be very effective. Besides, due to the fibrous pericarp with a corky surface and due to an empty space around the seed, the fruit has buoyancy and may be dispersed by water, too. The species is characteristic of lowlands, river banks, and swamps. Lying on the ground, the pod dehisces partially, so as to permit the seed to germinate. Seedlings of *P. officinalis* are abundant in the rain forest, according to Richards. The species is common in Surinam, in the swampy as well as in the riparian forest.

18. *Pterocarpus rohrii:* The flattened pod has almost circular outlines, measuring 5–8 cm in diameter and 0,3 cm in thickness (Fig. 68). It is light-yellow to light-brown. The single seed causes a swelling in the fruit center being completely surrounded by a 1,5–2 cm broad papery wing with a streaky venation. The winged samara has a low specific weight and is dispersed by wind. Howev-

er, the rate of distribution of this species is very low. In Surinam, it is common. In Amazonian forests one individual occurs on one hectare (Black et al. 1950).

19. *Pterocarpus* sp. (Sangrito) (Fig. 69): The flattened pod has almost circular outlines and shows a somewhat rough and reticulately veined surface. Its colour is brownish and the texture woody. One half of the fruit is surrounded by a reduced woody wing. The fruit is indehiscent and contains a single seed. The fruit type is very derived and the means of dispersal are doubtful, as the wing seems to be more or less functionless.

20. Roble blanco: The dry indehiscent fruit is winged and contains a single seed. Possibly this is a species of *Pterocarpus*. Dispersal takes place by wind.

All species of Papilionaceae studied are of height category A. It is therefore not surprising that quite a few of them are dispersed by wind. Furthermore, several species with samaras are more characteristic of the semideciduous forest or of the savanna. Most of the species contribute only few individuals. Seven species do actually not belong to the humid tropical forest, 6 pertaining to the semideciduous forest or even to the savanna, while one comes from swampy areas. It is quite clear that these species have only few representatives in the humid tropical forest. Only 3 species have between 150 and 250 representatives. And only a single species, *Alexa imperatricis*, contributes a high number of individuals, 1.262 in total. This species has a dehiscent fruit and is autochorous. The seeds are relatively large. Besides *A. imperatricis*, there are only 4 other species autochorous. Of these, one species has 188 individuals, another 89 (also dispersed by birds). Four species are probably dispersed by animals. These have very low numbers of individuals. Eleven (12) species are dispersed by wind. This is a very high rate of wind-dispersed species which fact may be related to the height of the trees. Of the wind-dispersed species 2 have somewhat elevated numbers of individuals: *Diplotropis purpurea* with 206 individuals, and *Lonchocarpus sericeus* with 251 representatives. It thus results that wind-dispersal or autochorous dispersal is more advantageous for high trees than dispersal by animals. It is furthermore obvious that almost all species occurring in the semideciduous or deciduous forest or in the savanna are wind-dispersed. Their dispersal units are carried further. They may also be more apt to occupy more open habitats. *Pterocarpus officinalis* growing in swamps is probably also dispersed by water. It is, however, worth mentioning that the anemochorous samara of *P. officinalis* is dispersed during the dry season.

In his studies on the dispersal mechanisms of the Leguminosae Buchwald (1895) distinguishes autochorous types in which the fruits open mechanically after drying out. An example is *Pentaclethra macroloba* often found along rivers. Buchwald suggests that the autochorous mechanism is not apt to acquire new areas, as dissemination takes place in the immediate vicinity of the mother plant, while in the already occupied regions the number of individuals can easily be increased. Autochorous species characteristically occur in thickets, steppes, savannas and deserts, but also occupy mountainous areas, river banks, swampy regions, and the coast line.

Within the anemochorous types Buchwald recognizes granular seeds which remain in the fruit until the wind takes them away, or winged seeds such as of *Piptadenia*. This type is closely related to the autochorous one, as the fruits are likewise dehiscent and the seeds are drawn away by the wind. It is found in the high forest, the mountainous forest, at the river side, and in the savanna as well. The second type is represented by the samara such as of *Pterocarpus*. Representatives are found in the savanna, in thickets, steppes, mountainous areas, swamps, at the coastal line, and along rivers.

Hydrochory is realized in those fruits or seeds which develop cavities filled with air or have cotyledons with a spongy tissue.

The zoochorous species protect their seeds either mechanically with a hard testa composed of palisade cells (e.g. *Cassia*) or chemically. The fleshy edible parts are represented by the meso- or endocarp or by an arillus. The hard seeds of *Parkia* are embedded in a soft endocarp pulpa eaten by birds and bats. The endocarp becomes juicy and pulpy in *Dialium guianense.* It is preferred by tree mammals. In *Andira inermis* and *Dipteryx odorata* the mesocarp becomes fleshy and is eaten by birds and bats. The fleshy mesocarp of *Hymenaea courbaril* is eaten by tree mammals. The mimetic seeds of *Ormosia* are dispersed by birds (van der Pijl 1956).

While zoochory is a primary phenomenon, anemochory as well as autochory of the explosive pod are considered derived by van der Pijl (1956). According to this author, the prototype of the legume is a semifleshy follicle dehiscing at one suture, with twisting valves and often dangling seeds provided with an edible layer. Dehiscence is primarily directed towards the exposure of the seeds. Later, an abscission layer is developed on the seeds so that an autochorous mechanism of the explosive pod resulted.

Comparing all 3 families of Leguminosae with one another we come to the following results. The Mimosaceae comprise most species (33) with most individuals (5.616). The Caesalpiniaceae are represented by 25 species with 1.831 individuals, and the Papilionaceae only by 20 species with 2.145 individuals. Of the Mimosaceae 19 species belong to height category A, 11 to a, and 3 to aa. Of the Caesalpiniaceae 20 are of height category A, 5 of a. And of the Papilionaceae all are of height category A. Of the Mimosaceae only one species is anemochorous, 4 are autochorous, 5 auto and zoochorous at the same time, and 23 are zoochorous (a high number!). Of the Caesalpiniaceae 4 are anemochorous, 8 autochorous, 5 auto and zoochorous at the same time, and 8 zoochorous. Of the Papilionaceae 12 are anemochorous (a very high number!), 3 autochorous, 2 auto and zoochorous at the same time, and only 3 zoochorous. Of the Mimosaceae 23 species have indehiscent pods and 10 dehiscent ones. Of the Caesalpiniaceae 12 species have indehiscent pods and 13 dehiscent ones. And of the Papilionaceae 15 species have indehiscent and 5 species have dehiscent pods. We thus observe the following gradients from the Mimosaceae towards the Papilionaceae: The number of species as well as the number of individuals declines. The number of high trees pertaining to height category A increases. While within the Mimosaceae all 3 height categories (A-a-aa) are still present, the Papilionaceae are composed only of trees from height category A, the highest one. The number of species which are anemochorous markedly increases from one to 12, while the number of purely zoochorous species declines from 23 to only 3. Likewise, the number of species with indehiscent pods declines. However, within the number of indehiscent species those with anemochory (a derived form) are included, this has to be taken into account when the relatively high number of indehiscent pods in the Papilionaceae is considered. The tendency to develop high trees with anemochorous (or autochorous) dispersal mechanisms becomes obvious. However, several species in each family do actually not belong to the rain forest. They are characteristic of more open landscapes (savanna, deciduous and semideciduous forest). These types of comunities are known for a higher percentage of anemochorous species. It is furthermore surprising that most species are represented by few individuals and that the Mimosaceae, the Caesalpiniaceae and the Papilionaceae show one species with an elevated number of individuals each. The species which are using diplochory (auto and zoochory at the same time) may still maintain a more ancient form of dispersal mechanism, which is zoochory, besides the more modern type of autochory (e.g. in the genus *Ormosia* with mimetic seeds). A further evidence of the more primitive character of the Mimosaceae is the usual polyspermy of their fruits, while monospermous fruits such as drupes and samaras occur in the Caesalpiniaceae and even more frequently in the Papilionaceae. The Papilionaceous seed is a specialization of the Mimosoid-Caesalpinioidone, according to Corner (1951).

Linaceae

Hebepetalum humiriifolium = Kerosenillo

Hab.: RF	Height: A
Fruit: Drupe	Ind.: 17
Disp.: Zoochorous	Fr. time: 12

The dry indehiscent fruit contains few seeds. The drupe has an ovoid shape measuring $0,7 \times 0,6$ cm, and is slightly 5-lobular. The surface is black with light-coloured longitudinal stripes. The 5-locular fruit is crowned by 5 persistent styles. Each of the 5 stones contains 1–2 seeds. The calyx is persistent. The drupe is dispersed by animals.

The species is of little importance in the forest studied. Judging from the low number of representatives, animal dispersal is disadvantageous in high trees. The tree is very common in Surinam, but occurs scattered in the rain forest.

Malpighiaceae

The fruits of the Malpighiaceae are generally either dehiscent or indehiscent schizocarps (winged or without wings), more seldom nutlets or drupes.

1. *Byrsonima aerugo* = Manteco de agua

Hab.: RF	Height: A
Fruit: Drupe	Ind.: 54
Disp.: Zoochorous	Fr. time: 12

2. *Mascagnia nerviosa* = Palo de colmena

Hab.: RF	Height: a
Fruit: Samara	Ind.: 42
Disp.: Anemochorous	Fr. time:

3. Manteco montañero

Hab.: RF	Height: A
Fruit: Drupe	Ind.: 7
Disp.: Zoochorous	Fr. time:

1. *Byrsonima aerugo:* The fleshy drupe is more or less globular with a diameter of 0,8–1 cm, of green to reddish colour, and has 3 persistent styles. The fruit is edible and tastes like a plum. The sepals are likewise persistent. The encodarp is bony. Each of the 3 locules contains a seed. Dispersal takes place by animals (birds). In Surinam, the species occurs sporadically in the rain forest, but may also be present in the high savanna as well as in the young secondary forest.

2. *Mascagnia nerviosa:* The light indehiscent samara has a papery wing and contains a single seed. Dispersal of the samara takes place by wind.

3. Manteco montañero: The fleshy indehiscent drupe contains few seeds and is dispersed by animals (birds).

The 3 species of Malpighiaceae studied have very few representatives in the forest studied. The 2 high trees are dispersed by animals, while the lower tree (height category a) is dispersed by wind. Neither wind dispersal in a lower tree nor animal dispersal in higher trees is very advantageous. This fact may explain the low rate of distribution.

Melastomaceae

The Melastomaceae are characterized by capsules or berries with numerous seeds, unless reduction of seeds takes place in the drupaceous type. The seeds generally lack an endosperm, but may develop relatively large cotyledons. Pulpy placentas occur.

1. *Miconia amacurensis* = Saquiyak rojo

Hab.: RF	Height: a
Fruit: Berry	Ind.: 6
Disp.: Zoochorous	Fr. time: 5

2. *Miconia* aff. *kappleri* = Palo de agua

Hab.: RF	Height: A

Fruit: Berry	Ind.: 3
Disp.: Zoochorous	Fr. time: 3/12

3. *Mouriria huberi* = Guarataro

Hab.: RF	Height: A
Fruit: Berry (drupe)	Ind.: 138
Disp.: Zoochorous	Fr. time: 2

4. *Mouriria sideroxylon* = Guaratarillo

Hab.: RF	Height: A
Fruit: Berry	Ind.: 3
Disp.: Zoochorous	Fr. time: 10

5. Saquiyak

Hab.: RF	Height: aa
Fruit: Berry	Ind.: 54
Disp.: Zoochorous	Fr. time:

1. *Miconia amacurensis:* The genus *Miconia* consisting in shrubs or small trees produces flowers in panicles and comprises species with edible fruits in the form of berries. The fruit of *M. amacurencis* is a many-seeded fleshy berry which is dispersed by animals (birds).

2. *Miconia* aff. *kappleri:* The globular fleshy berry measures 0,3 × 0,35 up to 1 cm in diameter and contains numerous seeds. It is first dark-green to red-orange, but later adopts a black colour. The calyx remains persistent. The berry is dispersed by animals (birds). In Surinam, the species is mostly represented by small trees, but may reach 30 m in height in certain specimens. Although rare, in Surinam, the tree mainly occurs in the secondary forest, in hilly areas of the rain forest and even as a shrub in the savanna.

3. *Mouriria huberi* (Fig. 91): The globular berry surrounded by a calyx cup ("cupule") is about 3,2 cm long and measures 3 cm in diameter. It generally contains a brown seed, 1,5 cm long and 1 cm broad, of ovoid shape and with a constriction. The seed surface is smooth and shiny, the texture hard. The transformation of an originally many-seeded berry into a single-seeded drupe is taking place here. Dispersal of the fruit takes place by animals.

The outer epidermis is very small-celled and covered by a thin cuticle. Below follow several (about 5) small-celled subepidermal layers. The fundamental tissue becomes collenchymatous towards the inside. The major part of the pericarp is composed of large sclerenchymatous groups scattered within the ground tissue. The sclereids are very large and have thick walls. The size of the stone cell groups increases towards the pericarp inside. The endocarp occupies a large part of the pericarp and is composed mainly of a cartilaginous collenchyma which replaces the usual endocarp sclerenchyma characteristic of most endocarp formations. The fruit could thus be called a drupe.

4. *Mouriria sideroxylon:* The globular berry surrounded by a persistent calyx cup is 1,5 cm long and measures 1,4 cm in diameter, exhibiting almost globular outlines. It has 3 locules each with a single seed. The pericarp measures 0,2 cm in thickness. The pericarp surface is covered with "knobs". The endocarp is woody. Here too, the transformation of an original berry fruit into a drupe is on the way, by reduction of the seed number and formation of an endocarp. Dispersal takes place by animals.

5. Saquiyak: The small fleshy berry contains many small seeds. The fruit is dispersed by animals.

All 5 species of Melastomaceae studied are characterized by the "berry" fruit type which is dispersed by animals. Some of the berries are, however, on the way to transform into drupes by reduction of the seed number from many to few or only one (*Mouriria huberi*) or by formation of an endocarp (*Mouriria sideroxylon*). The height of the trees is very different, all 3 height categories being represented (A-a-aa). All 5 species have the rain forest as their usual habitat. The number of individuals is, however, very low in all species studied, except in *Mouriria huberi* where it is a little higher. This species is a more derived one with drupes. The observations indicate that animal dispersal is not advantageous in high trees, in general (*Miconia* aff. *kappleri, Mouriria huberi, M. sideroxylon*), but in the lowest tree, Saquiyak, we do not observe a higher number of individuals either. All in all, the Melastomaceae show a very low representation in the forest studied.

Meliaceae

The ovary is mainly 4–5 locular. The fruits are usually capsules, and more seldom berries or drupes. The seeds of the capsules are often winged or furnished with an aril.

1. *Carapa guianensis* = Carapa
   Hab.: Lowland, semideciduous forest, RF, swampy forest — Height: A
   Ind.: 885
   Fr. time: 5
   Fruit: Capsule
   Disp.: Autochorous and zoochorous
2. *Carapa* sp. = Carapa lisa
   Hab.: RF — Height:
   Fruit: Capsule — Ind.: 29
   Disp.: Autochorous — Fr. time:
3. *Cedrela odorata* = Cedro rojo
   Hab.: RF — Height: A
   Fruit: Capsule — Ind.: 18
   Disp.: Anemochorous — Fr. time: 2/12
4. *Guarea schomburgkii* = Carapillo
   Hab.: RF — Height: a
   Fruit: Capsule — Ind.: 85
   Disp.: Autochorous and zoochorous — Fr. time:
5. *Guarea trichilioides* = Trompillo
   Hab.: Lowland — Height: A
   Fruit: Capsule — Ind.: 42
   Disp.: Autochorous and zoochorous — Fr. time:
6. *Trichilia propingua* = Bizcochuelo amarillo
   Hab.: RF — Height: a
   Fruit: Capsule — Ind.: 2
   Disp.: Autochorous and zoochorous — Fr. time: 2/3
7. *Trichilia schomburgkii* = Suipo
   Hab.: RF — Height: a
   Fruit: Capsule — Ind.: 1.376
   Disp.: Autochorous and zoochorous — Fr. time: 5–6
8. *Trichilia smithii* = Mijarro
   Hab.: RF — Height: a
   Fruit: Capsule — Ind.: 5
   Disp.: Autochorous and zoochorous — Fr. time: 2
9. *Trichilia* sp. = Bizcochuelo
   Hab.: RF — Height: a
   Fruit: Capsule — Ind.: 14
   Disp.: Autochorous and zoochorous — Fr. time:
10. *Trichilia* sp. = Bizcochuelo negro
    Hab.: RF — Height: a
    Fruit: Capsule — Ind.: 55
    Disp.: Anemochorous — Fr. time: 2/12

1. *Carapa guianensis* (Fig. 69): The globular woody capsule is up to 9–10 cm long and has a pointed top, opening septicidally with 4 valves and liberating about 16 seeds or more (2–3 seeds per locule). The surface is brown and corky, with 4 prominent longitudinal ribs. The vascular supply in the center of each locule is very prominent ending up in the tip of the opened valves. The seeds are about 5-7 cm long and 4,8 cm wide, 3-angular, with a woody shell and a prominent hilum. The seed coat is about 1/2 mm thick. The seed is nut-like and may be eaten by animals. The oily embryo contains up to 58–70% of the bitter-tasting *Carapa* oil which is commercially used for soap and candle manufacture. In popular medicine it is used against diseases of the skin and leprosy. Probably the species owes its high rate of distribution to the oily nutritious seeds. The seed has buoyancy due to its corky coat and the floating seed may also be dispersed by water. Besides in the rain forest, the species also grows on estuaries. In Surinam it mainly occurs in the marsh forest of the West. In Amazonian forests 10–11 individuals occur on one hectare (Black et al. 1950). *Carapa* is an old world genus and may have arrived in America from Africa by sea. The main equatorial current could carry the seeds from the Gulf of Guinea to the coast of Brazil in 3 or 4 months.

The pericarp consists of thin-walled parenchyma composed of roundish to ovalate cells, while the vascular bundles are scattered in this fundamental tissue.

2. *Carapa* sp. (Carapa lisa) (Figs. 71, 72): The

globular woody capsule with a warty surface has a length of 5 cm and a width of 6,6 cm. It has a partition wall in the center and contains 2 seeds. Dispersal is autochorous.

The pericarp is covered with a small cork layer composed of cells with horseshoe-shaped wall thickenings. The warts on the fruit surface originate by the favoured development of the phelloderm at short distances. The cork layer is interrupted by stratified lenticels (Fig. 71), in which layers of roundish "filling" cells alternate with layers of normal cork cells. The cork cells are suberized, while the roundish cells are thin-walled and not suberized. The fundamental tissue of the pericarp consists in cartilaginous collenchyma with irregularly thickened angular walls (Fig. 72). Especially abundant are scattered giant stone cells in the ground tissue of the fruit periphery. In contact with the vascular bundles are very large secretory spaces with a yellowish content (Fig. 72). These spaces are surrounded by a ring of short radial cell rows which originated from single cells repeatedly divided by periclinal walls, parallel to the cavity surface. Probably, these secretory cavities develop – at least partly – by tissue degeneration, as sometimes remnants of tracheids are included in the cavity center. The ring of cells bordering the spaces may possibly be interpreted as a limiting tissue protecting the neighbouring cells from degeneration (a separating sheath somehow resembling that of the secretory cavities in the fruit of *Anacardium occidentale*, according to Roth 1977). Possibly, this sheath originates from the epithelium of the secretory cavity.

3. *Cedrela odorata:* The elliptic woody capsule has a brown colour being covered by groups of lenticels. It reaches about 6 cm in length and 7,5 cm in width when open, dehiscing with 5 valves. It gives a star-like appearance when dehisced (Fig. 70). The surface is light-green and brown-black maculated. Pericarp thickness measures 0,2 cm. Dehiscence is septicidal and columnicidal at the same time so that the large 5-angular column remains in the center (Fig. 70). The numerous seeds (about 30–40) are winged "screw flyers", about 2,5–3 cm long and 1 cm broad, having a single papery wing (Fig. 70). Dispersal of the seeds takes place by wind. The fruit has a smell of garlic and secretes a whitish watery exudate when unripe.

The species occurs at low densities in mixed forest throughout much of the tropical lowlands of Latin America. Plantations are damaged in Latin America by a shoot borer, but are successful in Africa where the parasite is absent (Hartshorn 1972). The species is very common in Surinam, but occurs scattered in the rain forest.

4. *Guarea schomburgkii:* The dry dehiscent woody capsule is almost globular, measuring about 1,5–2 cm in diameter and has 4 indistinct longitudinal ribs. It opens with 4 valves and contains 4 brilliant seeds, 1 × 0,7 × 0,6 cm. Dispersal is autochorous, but may also take place by animals (birds), as the seeds are embedded in an aril-like flesh (= inner pericarp wall?). The species is rare in Surinam.

5. *Guarea trichilioides:* The dry dehiscent capsule is obovate, pear-shaped or trumpet-shaped and is composed of 4–5 locules each with a single seed. Dispersal is autochorous, but also takes place by animals (birds), as the seeds are furnished with an aril.

6. *Trichilia propingua:* The small approximately pear-shaped capsule has a delicate indumentum and measures about 3 × 1 cm. It is orange or salmon coloured, and 3-angular. The only 3-angular seed is white and surrounded by a violet-red aril of creamy consistency. The tree is cauliflorous. The seeds of the dehisced fruits are eaten by animals (birds). The species is found scattered in the riparian forest of Surinam.

7. *Trichilia schomburgkii:* The elongated capsule with a velvety surface through short hairs measuring 4,1 cm in length and 2 cm in width dehisces with 3 valves. The pericarp texture is shell-like. The fruit contains 3 seeds. These are 3-angular and measure 1,5 cm in length, 1,4 cm in width, and have an aril. Dispersal of the seeds takes place by animals (e.g. birds, monkeys).

8. *Trichilia smithii:* The woody 3-locular capsule is 1,5–2 cm long and 1,1 cm broad, of leathery consistency and dehisces with 3 valves leaving a column in the center. The outer surface is warty by big "knobs". The seeds are furnished with a

fleshy red aril. Dispersal takes place by animals (birds). The species also occurs in the dry forest, according to Frankie, Baker & Opler (1974).

9. *Trichilia* sp. (Bizcochuelo): The dry dehiscent capsule contains only few seeds (3?) which are probably furnished with an aril eaten by animals (e.g. birds).

10. *Trichilia* sp. (Biscochuelo negro): The ovoid capsule of brittle texture dehisces with 3 valves and has a velvety surface by short hairs. The fruit length measures 2,8 cm, the width 1,1 cm. The single seed is 2 cm long and 0,7 cm wide and develops a flat wing-like part. Dispersal takes probably place by wind. Possibly this is not a *Trichilia*, but a *Cedrela*?

The 10 species of Meliaceae studied are represented by 3 high trees of height category A and 6 trees of medium height (a). The fruits are capsules in all species and the dispersal mechanism is autochorous for that. Most of the species present in the forest studied show low rates of distribution. However, 2 species have a higher or high rate of frequency: *Carapa guianensis* with 885 individuals and *Trichilia schomburgkii* with 1.376 individuals. For having dehiscing capsules, the species studied show some kind of an autochorous dispersal mechanism. However, 7 species are also zoochorous, as their seeds are furnished with an aril. *Trichilia schomburgkii* with the highest number of individuals uses such a diplochorous dispersal type, if we can call it this way. Dehiscence of the capsules alone does not always provide a real way of dispersal, as the fruits remain on the tree. Arils that are exposed on the dehisced fruit which remains attached to the tree are mainly eaten by birds. Seeds that fall to the ground, on the other hand, may be taken by ground mammals. All the species of *Guarea* and *Trichilia* studied have probably all seeds furnished with an aril and are thus mainly zoochorous. *Carapa guianensis* has seeds with an oily embryo which may be dispersed by animals. *C. guianensis* is a tree characteristic of primitive forests, according to Pittier. This may be the main reason why it is represented by a high number of individuals. It furthermore has a wide range of tolerance concerning the environment (rain forest, lowland, semideciduous forest, estuaries) which explains the high number of individuals, at least partly. The fruits may even be carried by currents and possibly eaten by fish. Of *Trichilia schombrugkii* we have too few data to explain the high number of individuals, but perhaps it is also a primitive species, characteristic of the type of rain forest studied. Furthermore, the auto-zoochorous dispersal mechanism may be advantageous.

Two of the 10 species studied are anemochorous: *Cedrela odorata*, a high to very high tree which may reach a height up to 40 meters, and *Trichilia* sp. (Bizcochuelo negro). In the latter, anemochory may not be very well developed. Roosmalen indicates that *Cedrela odorata* is very common in Surinam, but occurs only scattered in the rain forest. This may be the reason for the low rate of distribution in the forest studied. The species is probably more apt to grow in open vegetation and may easily colonize new habitats for its anemochorous dispersal mechanism. In Mexico it abundantly occurs in the secondary forest. It is frequently cultivated as a shade-giving tree.

Moraceae

Of the 2 carpels developed only one is usually fertile. The fruits are drupes or nuts and frequently surrounded by a fleshy persistent perigone and receptacle. Fruits united in cymose inflorescences may develop infrutescences of very variable shape, in the genus *Ficus*, for example, in the form of urn-shaped syconia. The fleshy infrutescences are almost exclusively dispersed by frugivorous birds – often aided by fruit-bats and mammals, according to Ridley. The seeds are devoid of an aril.

1. *Brosium* sp. = Charo

Hab.: RF	Height: A
Fruit: Infrutescence	Ind.: 4
Disp.: Zoochorous	Fr. time:

2. *Brosimum?* = Charo negro
   Hab.: RF — Height:
   Fruit: Infrutescence — Ind.: 38
   Disp.: Zoochorous — Fr. time:
3. *Cecropia sciadophylla* = Yagrumo montañero
   Hab.: Lowland, RF — Height: a
   Fruit: Infrutescence — Ind.: 86
   Disp.: Zoochorous — Fr. time:
4. *Cecropia* sp. = Yagrumo blanco
   Hab.: RF — Height: a
   Fruit: Infrutescence — Ind.: 79
   Disp.: Zoochorous — Fr. time:
5. *Cecropia* sp. = Yagrumo morado
   Hab.: RF — Height:
   Fruit: Infrutescence — Ind.: 10
   Disp.: Zoochorous — Fr. time:
6. *Chlorophora tinctoria* = Mora amarilla
   Hab.: Deciduous forest — Height: A
   Ind.: 1
   Fruit: Infrutescence — Fr. time:
   Disp.: Zoochorous
7. *Clarisia racemosa* = Cajimán
   Hab.: RF — Height: A
   Fruit: Drupe — Ind.: 102
   Disp.: Zoochorous — Fr. time:
8. *Ficus* sp. = Higuerón
   Hab.: Lowland, RF — Height: A
   Fruit: Syconium — Ind.: 5
   Disp.: Zoochorous — Fr. time:
9. *Ficus* sp. = Higuito morado
   Hab.: RF — Height:
   Fruit: Syconium — Ind.: 13
   Disp.: Zoochorous — Fr. time:
10. *Ficus* sp. = Matapalo
    Hab.: RF — Height: A
    Fruit: Syconium — Ind.: 18
    Disp.: Zoochorous — Fr. time:
11. *Helicostylis tomentosa* = Charo macho
    Hab.: RF — Height:
    Fruit: Infrutescence — Ind.: 160
    Disp.: Zoochorous — Fr. time: 6/9
12. *Pourouma guianensis* = Chaparro de agua
    Hab.: RF — Height: a
    Fruit: Drupe — Ind.: 2
    Disp.: Zoochorous — Fr. time: 12

1. *Brosimum* sp. (Charo): The one-seeded fruits are enclosed within the fleshy receptacle of the inflorescence produced by coalescence of the flowers. The globular infrutescences which contain latex are dispersed by animals (birds, monkeys).

2. *Brosimum* sp.? (Charo negro): The one-seeded fruits are enclosed within the fleshy receptacle of the infrutescence. The globular infrutescences are dispersed by animals.

3. *Cecropia sciadophylla:* The catkins reach a lenght of up to 20 cm and a width of 1–1,5 cm. The one-seeded fruits are 0,25–0,3 cm long. The infrutescences are dispersed by animals, mainly birds and possibly bats too. In Surinam, the tree also occurs in the secondary forest (Eisenmann 1961, Leck 1972).

4. *Cecropia* sp. (Yagrumo blanco): The infrutescences in the form of catkins are dispersed by animals (birds, and bats).

5. *Cecropia* sp. (Yagrumo morado): The infrutescences in the form of catkins are dispersed by animals (birds, and bats).

6. *Chlorophora tinctoria:* The infrutescences are fleshy and of globular shape. They are resinous. The infrutescences are dispersed by animals (bats, according to Fleming & Heithaus 1981).

7. *Clarisia racemosa:* The fruit is of elliptic shape, 1,5 cm long and fleshy. The drupe is surrounded by an orange-red fleshy receptacle. Dispersal takes place by animals (birds, monkeys). In Surinam, the species is rare.

8. *Ficus* sp. (Higuerón): The urn-shaped fleshy syconium contains numerous small nutlets. The infrutescence is dispersed by animals, mainly birds.

9. *Ficus* sp. (Higuito morado): The fleshy syconium which contains numerous small nutlets is dispersed by animals, mainly birds.

10. *Ficus* sp. (Matapalo): The fleshy syconium which contains numerous small nutlets is dispersed by animals, mainly birds.

11. *Helicostylis tomentosa:* The globular infrutescence measures about 2,5–5 cm in diameter. The fruitlets are of elliptic shape and 0,6–0,7 cm long, being surrounded by 4 yellowish perigone leaves with a velvety indumentum. The seeds are 0,5 cm long. The infrutescences are dispersed by

animals. In Surinam, the species is very frequent and also occurs in the high savanna forest.

12. *Pourouma guianensis:* The fleshy drupe with a hard endocarp is united with an edible, more or less fleshy perigone. The ovoid fruit measures 1,5 × 1,1 cm up to 2 × 1,4 cm and contains a single seed. It adopts a red-brown somewhat hairy surface. Dispersal of the drupe takes place by animals. In Surinam, the species is common in the rain forest, but also occurs along the coast line.

Of the 12 species of Moraceae studied, only 3 belong to height category a, while 5 are of height category A. In 5 cases it is doubtful to which category they belong. All species are dispersed by animals and have fleshy fruits or infrutescences. Animal dispersal mainly takes place by birds which often are aided by fruit-bats and mammals. Although animal dispersal is usually more advantageous in lower trees, the 5 species belonging to height category A are also animal-dispersed. However, the Moraceae are very ill-represented in the forest studied, and only 2 species show a number of representatives higher than 100 (*Clarisia racemosa* with 102 and *Helicostylis tomentosa* with 160 individuals). Most species develop infrutescences and only 2 have drupes. The infrutescences of *Cecropia* correspond to fleshy catkins with 1-seeded dry nuts. *Helicostylis* has globular infrutescences with an apical slit-like opening resembling the syconia of *Ficus*. The infrutescences of *Ficus, Brosimum, Helicostylis* and *Cecropia* usually contain latex.

Myristicaceae

The fruits of the Myristicaceae frequently correspond to fleshy or woody capsules which open on the ventral and dorsal suture. The fruit composed of one carpel represents a follicle. The seeds are often furnished with a red and fleshy aril eaten by birds. The fruit type is relatively primitive. According to Ridley, the Myristicaceae are almost exclusively dispersed by frugivorous birds often aided by fruit-bats and mammals.

1. *Iryanthera lancifolia* = Cuajo grande
   - Hab.: RF — Height: A
   - Fruit: Capsule — Ind.: 43
   - Disp.: Autochorous and zoochorous — Fr. time: 6
2. *Virola sebifera* = Cuajo negro
   - Hab.: RF, swampy forest — Height: A
   - Ind.: 35
   - Fruit: Capsule — Fr. time:
   - Disp.: Autochorous and zoochorous
3. *Virola surinamensis* = Cuajo
   - Hab.: Flooded areas, RF — Height: A
   - Ind.: 36
   - Fruit: Capsule — Fr. time: 2/12
   - Disp.: Autochorous and zoochorous

1. *Iryanthera lancifolia:* The globular capsule has a diameter of about 2,8 to 4 cm (2,8 × 2,5–4 × 3,4 cm) with a pericarp thickness of 0,6 cm. The 0,2 cm thick endocarp is woody. The mesocarp is fleshy. The capsule which opens with 2 valves contains only a single seed. The ovoid seed is furnished with a red aril. Dispersal mainly takes place by birds (and monkeys). The species is very rare in Surinam.

2. *Virola sebifera* (Fig. 73): The ovoid capsule is about 2 cm long and 1,5 cm broad and opens at the ventral as well as at the dorsal suture dehiscing into 2 valves. Pericarp thickness measures 0,1 cm. The fruit surface is rough, the texture more or less fleshy. The only seed is 1,3 cm long and 1 cm broad, of ovoid shape, of light-brown colour, and has a brittle texture. The surface is smooth. The seed is surrounded by a laciniate purple aril deeply incised in a way resembling that of the nutmeg. Dispersal takes place by birds and mammalia. In Surinam, the species occurs very frequently.

The outer pericarp epidermis is small-celled and has thick outer walls. Beneath follow about 5 layers of small-celled parenchyma in which large secretory cells are dispersed. Towards the inside follow very large groups of stone cells (Fig. 74). The parenchyma cells become larger and adopt thickened walls with pits in a centripetal direction. Stripes of compressed parenchymatous tissue are interspersed between the normal parenchyma. The inner epidermis is very small-celled.

3. *Virola surinamensis:* The capsule of elliptic outlines resembles the true nutmeg and measures 2 × 1,7 cm and shows all-round longitudinal ribs. In a young state of the fruit, the surface has a yellow-brown velvety indumentum. The pericarp thickness is 0,2–0,3 cm. The seed measuring 1,5–1,7 × 1 cm is surrounded by a purple laciniate aril which is almost bipartite down to the base. It measures 0,1 cm in thickness. The dehisced capsule releases only one seed. The fruit is called in Venezuela "fruta de cuajo del Orinoco" and contains 47,5% oily substances (glycerine and myristic acid), according to the "Tropenflanzer" 1901 (p. 286-287). Large amounts of soluble tannins occur in capsule and seed. The tree is characteristic of the várzeas and the seed surrounded by a waxy substance is said to be eaten by fish. It is, however, mainly dispersed by birds and monkeys. In Surinam, the species is very frequent and mainly occurs in the swampy and marsh forest. In Amazonian forest 2–5 individuals occur on one hectare (Black et al. 1950).

The 3 species of Myristicaceae studied belong all to height category A. Nevertheless they are dispersed by animals. However, their rate of distribution is very low. *Virola surinamensis* may originally be dispersed by fish, probably an ancient form of dispersal. All 3 species are dispersed by birds, attracted by the colourful fleshy arils. The low rate of distribution of the Myristicaceae studied once again shows that animal dispersal in high trees is not very advantageous.

Myrtaceae

The fruit of the Myrtaceae is a loculicidal capsule, a berry or drupe with few to one seeds, often edible and crowned by the persistent calyx. Owing to similarities of the characters, difficulties often arise in the identification of the species.

1. *Calycorectes* sp. = Terciopelo
   Hab.: RF — Height: a
   Fruit: Drupe — Ind.: 140
   Disp.: Zoochorous — Fr. time: 2–3
2. *Eugenia anastomosans* = Guayabo montañero
   Hab.: RF — Height: a
   Fruit: Drupe — Ind.: 7
   Disp.: Zoochorous — Fr. time: 2/7
3. *Eugenia compta* = Guayabito blanco
   Hab.: RF — Height: A
   Fruit: Drupe — Ind.: 506
   Disp.: Zoochorous — Fr. time: 2/10/12
4. *Eugenia patrisii* = Pendanga negra
   Hab.: RF — Height: a
   Fruit: Drupe — Ind.: 24
   Disp.: Zoochorous — Fr. time:
5. *Marliera* sp. nov. = Guayabito amarillo
   Hab.: RF — Height: A
   Fruit: Drupe — Ind.: 3
   Disp.: Zoochorous — Fr. time: 11
6. *Myrcia amazonica* = Curtidor
   Hab.: Deciduous forest, RF — Height: a
   Ind.: 31
   Fruit: Drupe — Fr. time:
   Disp.: Zoochorous
7. *Myrcia paivae* = Carne asada
   Hab.: RF — Height: A
   Fruit: Drupe — Ind.: 95
   Disp.: Zoochorous — Fr. time: 2/7
8. *Myrcia splendens* = Guayabito negro
   Hab.: RF — Height: a
   Fruit: Drupe — Ind.: 243
   Disp.: Zoochorous — Fr. time:
9. *Myrciaria floribunda* = Guayabillo blanco
   Hab.: RF — Height: a
   Fruit: Drupe — Ind.: 70
   Disp.: Zoochorous — Fr. time:
10. Guayabillo negro
    Hab.: RF — Height: a
    Fruit: Drupe — Ind.:
    Disp.: Zoochorous — Fr. time:

11. Guayabillo rojo

Hab.: Deciduous forest, semideciduous forest	Height: A
	Ind.:
	Fr. time:
Fruit: Drupe	
Disp.: Zoochorous	

12. Guayabito piedrero

Hab.: RF	Height: a
Fruit: Drupe	Ind.: 78
Disp.: Zoochorous	Fr. time:

13. Guayabito zaba

Hab.: RF	Height: a
Fruit: Drupe	Ind.: 29
Disp.: Zoochorous	Fr. time:

14. Pendanga

Hab.: RF	Height: a
Fruit: Drupe	Ind.: 328
Disp.: Zoochorous	Fr. time:

1. *Calycorectes* sp. (Terciopelo): The ovoid drupe is 4 cm long and 2 cm broad. The pericarp, 0,2 cm in width, develops stone cells abundantly and, consequently, has a woody texture. Stamens and calyx are persistent. Due to a dense indumentum, the fruit surface is velvety, as indicated by the vernacular name (Figs. 75, 76). Secretory spaces are interspersed within the fundamental tissue of the pericarp. There is only one stone with a single seed.

The very small-celled outer pericarp epidermis is covered with an indumentum of long unicellular hairs which are occasionally twisted (Figs. 75, 76). The following subepidermal layers consist of small cells which are arranged in tangential rows due to the action of the dilatation growth. The fundamental tissue of the pericarp is represented by parenchyma in which groups of stone cells are dispersed abundantly. Larger secretory cavities surrounded by an epithelium occur here and there. Vascular bundles penetrate the parenchymatous ground tissue.

2. *Eugenia anastomosans:* The fleshy globular drupe is about 2,4–2,5 cm long and 1,5–2 cm broad, yellow to brilliant black, and has one locule with a single seed. Due to abundant stone cell formation the pericarp becomes somewhat woody when drying out. The endocarp inside is smooth. The fruit is crowned by 4 sepals and the stigma. The seed is 1,4 cm long and 1,5 cm broad, of globular shape, and has a smooth black outside. Two large cotyledons occupy most of the seed cavity. The fruit flesh is edible and dispersal of the fruit takes place by animals (birds). The species is rare in Surinam.

3. *Eugenia compta:* The small fleshy drupe, 1,8 cm in length and 1,1 cm in width, is of orange or red colour, of ovoid shape, being crowned by 4 sepals and the stigma. The fruit surface is uneven due to very small knobs. Secretory spaces are dispersed in the pericarp. The fruit encloses only one seed. This is 1,1 cm long and 0,7 cm broad and has a brown papery and smooth seed coat. The embryo shows 2 large cotyledons. The drupe is dispersed by animals (birds). In Surinam, the species is infrequent and occurs in the savanna. It there appears as a shrub (!) or seldom as a small tree.

4. *Eugenia patrisii:* The globular to pear-shaped drupe (berry?) measures 2,5 cm in diameter, is yellow-green to red, and has a juicy and edible fruit flesh. The calyx is persistent. The fruit contains one to few seeds. By the reduction of the seed number to one, transformation of the berry into a drupe is taking place. The fruit is dispersed by animals (e.g. birds). In Surinam, the species occurs frequently.

5. *Marliera* sp. nov.: The small fleshy indehiscent fruit contains only one seed. The drupe is dispersed by animals.

6. *Myrcia amazonica:* The small globular drupe measures 0,9 cm in length and 0,8 cm in width, is dark-green to black, and has a persistent calyx. The surface is devoid of hairs. The only seed is kidney-shaped. The fleshy drupe is dispersed by animals (e.g. birds). In Surinam, the species occurs as a shrub or small tree, especially in the high savanna forest, in the secondary forest, and in the shrub savanna, where it is frequent.

7. *Myrcia paivae:* The small ovoid drupe is 0,6 cm long and 0,5 cm broad, and has a hard endocarp. The fruit is crowned by the calyx and the stigma. The fruit surface is very hairy. There is only one locule with a single seed. The fleshy drupe is dispersed by animals (e.g. birds).

8. *Myrcia splendens:* The small fleshy elliptic to globular fruit is indehiscent and measures 1 × 0,5 cm. It is first yellow to red, but later becomes black and usually contains a single seed (sometimes 2). The drupe is dispersed by animals (birds). In Surinam, the species occurs very frequently, especially in the ridge forest and along creeks.

9. *Myrciaria floribunda:* The ovoid to globular drupe is small being crowned by the persistent calyx. It measures 0,7–1,5 cm in diameter. It is first dark-yellow, but later adopts a red colour. There is usually a single seed, but sometimes 2–3 of them. The seeds have a brilliant coat. The fleshy edible fruit is on the way to transform into a one-seeded drupe by reduction of the seed number. It is dispersed by animals. Roosmalen also observed larger deformed fruits which are dehiscent. In Surinam, the species is very common.

10. Guayabillo negro: The small fleshy indehiscent fruit corresponds to a drupe. It is one-seeded and dispersed by animals.

11. Guayabillo rojo: The small fleshy indehiscent fruit is one-seeded and corresponds to a drupe. It is dispersed by animals.

12. Guayabito piedrero: The fleshy indehiscent fruit is small and contains a single seed. The drupe is dispersed by animals.

13. Guayabito zaba: The small globular drupe measures about 0,9 cm in diameter and has a sclerenchymatous endocarp. It has a knobby surface and is crowned by the calyx and the stigma. The fleshy one-seeded fruit is dispersed by animals.

14. Pendanga: The almost globular drupe measures 2 cm in length and 1,8 cm in width. The texture is woody. The fruit is crowned by a papillose stigma. There are 1 (–2) locules and 1 (–2) seeds. The globular seed measures 1,5 cm in diameter, but is flattened on one side. It is black and has a smooth surface. The drupe is dispersed by animals.

Of the 14 species of Myrtaceae studied 10 belong to height category a and only 4 to height category A. Only some of the species are represented by high trees, while most of them are lower trees or even shrubs. All species resemble one another very much concerning their fruit formation. All species develop single-seeded drupes, usually of small size, which are dispersed by animals. The number of individuals is generally relatively low. Only 4 species are represented by over hundred individuals: *Calycorectes* sp. with 140, *Myrcia splendens* with 243, Pendanga with 328, and *Eugenia compta,* occasionally a high tree with a small fruit, is represented by 506 individuals. The fruits have a showy red or orange colour. The attractive colour and the sweetness of the fruit flesh attracts animals, mainly birds. There is no apparent reason visible, why certain species have a larger distribution rate than others. The case of the Myrtaceae shows once again that dispersal of fleshy fruits eaten by animals is more advantageous in the lower forest layers than in very high trees. *Eugenia compta* seems to make an exception, but apparently in Surinam it is a small tree or even a shrub, according to Roosmalen.

Nyctaginaceae

The gynaeceum of the Nyctaginaceae is usually composed of one carpel. The fruit is often a thin-walled nut entirely surrounded by the so-called "anthocarp" (see Roth 1977) which corresponds to the lower perigone part.

1. *Torrubia cuspidata* = Casabe
   Hab.: RF   Height: A
   Fruit: Nut   Ind.: 105
   Disp.: Zoochorous   Fr. time: 2/5
2. *Torrubia* sp. = Casabe blanco
   Hab.: RF   Height: a
   Fruit: Nut   Ind.: 350
   Disp.: Zoochorous   Fr. time:

1. *Torrubia cuspidata* (Figs. 77, 78): The indehiscent nut-like fruit is about 1,5 cm long and 0,6 cm broad. The anthocarp united with the degenerated pericarp is shell-like. There is only one locule with one seed. The anthocarp surface which is furnished with numerous longitudinal ribs (Fig. 77) is somewhat rough. The fruit shape is elongated. The elongated seed reaches 0,9 cm in length and 0,4 cm in width. It is brown and has a brittle texture. The seed outside shows numerous ribs. The fruit is edible and dispersed by animals.

The outer layers of the anthocarp are parenchymatous having wavy walls (Fig. 78). Towards the inside follows a ring of cells with a red-brown content (tannins?). This layer is in contact with a broad band of sclerenchyma. The inner epidermis is small-celled. The ribs of the anthocarp arise through the action of a "dorsal meristem" (Roth 1977), as Fig. 78 shows. Note the cell rows. The rate of distribution of this species is low.

2. *Torrubia* sp.: The dry indehiscent fruit has a light weight, is nut-like, and contains a single seed. It is dispersed by animals.

The 2 species of Nyctaginaceae studied resemble one another in the structure of their fruits. *T. cuspidata* belonging to height category A is represented by only 105 individuals, while *Torrubia* sp. belongs to a lower height category (a) and is represented by 350 individuals. This example shows once again that in the lower forest layers animal dispersal is more advantageous than in the higher ones.

Ochnaceae

1. *Ouratea guianensis* = Pilón morado

Hab.: RF	Height: A
Fruit: Drupelets	Ind.: 36
Disp.: Zoochorous	Fr. time:

2. *Ouratea sagotii* = Pilón amarillo

Hab.: RF	Height: a
Fruit: Drupelets	Ind.: 60
Disp.: Zoochorous	Fr. time:

1. *Ouratea guianensis:* The fruit is "apocarpous". The fruitlets are ovoid blue-black drupelets with a single seed. The receptacle is fleshy and of red colour. The aggregate fruit is edible being dispersed by animals, probably mainly by birds (red colour!).

The outer pericarp epidermis is palisade-shaped consisting of very long and thin-walled cells (Figs. 79,80). The outer cell walls are thickened and cutinized. The radial extension of the cells varies in a way that the fruit surface becomes somewhat undulated, as seen in t.s. The epidermis occupies about 1/3 of the entire pericarp transection and probably serves as a water reservoir. Beneath lies parenchyma composed of roundish cells (Fig. 80). In the following region which is composed of tangentially elongated and partially compressed parenchyma cells, the vascular bundles are embedded. They are surrounded by a strong sclerenchymatous cap on the phloem side. The endocarp consists in fiber-like lignified sclereids which are tangentially arranged, i.e. parallel to the inner fruit surface. The inner fruit epidermis is small-celled and sclerenchymatous being covered by a cuticle. In contact with the inner fruit epidermis is the seed coat composed of parenchyma and tracheidal cells.

The species is represented by a high tree in the forest studied, while Roosmalen indicates that it is a shrub or small tree between 5 and 10 meters in Surinam. It there frequently occurs in the rocky savanna and on rocky islands at the river side. The species is of little importance in the forest studied. However, the low number of individuals shows once again that dispersal by animals is less advantageous in high trees.

2. *Ouratea sagotii:* The dry indehiscent mericarps contain a single seed and correspond to drupelets. Several drupelets are assembled on the receptacle. The tree is not as tall as *O. guianensis*, but has the same dispersal mechanism. The number of individuals in *O. sagotii* is correspondingly (?) higher.

Olacaceae

Heisteria iquitensis = Corona de rey

Hab.: RF	Height: a
Fruit: Drupe	Ind.: 29
Disp.: Zoochorous	Fr. time: 5

The more or less globular single-seeded drupe is surrounded by the persistent calyx. It reaches about 1 cm in diameter and is dispersed by animals (birds). The species has little importance in the forest studied.

Opiliaceae

Agonandra brasiliensis = Aceituno

Hab.: RF	Height: A
Fruit: Drupe	Ind.: 7
Disp.: Zoochorous	Fr. time: 5

The fleshy drupe resembles an olive (vernacular name!) and contains a single seed. It is dispersed by animals. The tree of height category A with animal dispersal has a very low number of individuals in the forest studied.

Palmae

Usually 3 carpels are united in a three to unilocular ovary or – by reduction – 2 carpels become sterile and only one bears a seed. The fruits are berries or drupes with large seeds. They are almost exclusively dispersed by birds, often aided by fruit-bats or mammals, according to Ridley.

1. *Euterpe* sp. = Manaca

Hab.: Swampy forest, RF	Height: a
	Ind.: 36
Fruit: Drupe	Fr. time:
Disp.: Zoochorous	

2. *Jessenia* sp. = Seje

Hab.: RF	Height: A
Fruit: Drupe	Ind.: 149
Disp.: Zoochorous	Fr. time:

3. *Manicaria* (saccifera?) = Temiche

Hab.: RF	Height: a
Fruit: Drupe	Ind.:
Disp.: Zoochorous	Fr. time:

4. *Mauritia* (flexuosa?) = Moriche

Hab.: Savanna	Height: a
Fruit: Drupe	Ind.:
Disp.: Zoochorous	Fr. time:

5. *Maximiliana* sp. = Cucurito

Hab.: RF	Height: A
Fruit: Drupe?	Ind.: 57
Disp.: Zoochorous?	Fr. time:

6. *Sabal mauritiaeformis* = Carata

Hab.: Deciduous forest, RF	Height: a
	Ind.: 64
Fruit: Drupe	Fr. time:
Disp.: Zoochorous	

7. Seje macho

Hab.: RF	Height: A
Fruit: Drupe	Ind.: 21
Disp.: Zoochorous	Fr. time:

1. *Euterpe* sp.: The fleshy indehiscent fruit contains a single seed. The fruit flesh of the mesocarp is scarce. The fruit represents a drupe and is dispersed by animals.

2. *Jessenia* sp.: The fleshy indehiscent drupe contains a single seed. The fruit is sweet and the seed oily. Dispersal takes place by animals (birds).

3. *Manicaria* sp.: The fleshy indehiscent mericarp is of a larger size and contains a single seed. The double fruit illustrated in Fig. 81 is bilobed and has 2 locules, while one is aborted. In other cases, the fruit is tri-lobular with 3 locules. The fruit is covered with woody cork-warts, characteristic of the Phytelephanteae. It is dispersed by animals.

4. *Mauritia:* The drupe measures about 4 cm in diameter and is entirely covered with scales ar-

ranged in 2 contra-rotating helical lines (Fig. 81). The pulp of the fruit is used to prepare refreshments which are digestive and laxative at the same time. The "armoured" fruit with scales is characteristic of the Lepidocaryeae. The drupes are dispersed by animals, mainly birds.

5. *Maximiliana* sp.: According to the observation of Rollet, the fruit is dry and – at least partly? – dehiscent. Possibly it corresponds to a drupe with partly dehiscing exo and mesocarp?

6. *Sabal mauritiaeformis:* The fleshy indehiscent fruit contains a single seed. The inflorescence is large and the fruit when ripe has a blackish colour. Dispersal of the drupe takes place by animals.

7. Seje macho: The fleshy indehiscent fruit contains a single seed. The drupe is dispersed by animals.

Of the 7 species of Palmae studied, 4 belong to height category a, while 3 are of height category A. Probably all species studied are dispersed by animals, mainly birds. However, the representation of the Palmae in the forest studied is very low. At least 3 species belong to other habitats. Occasionally, *Mauritia* and *Manicaria* fruits are water-dispersed. The fruits of *Manicaria saccifera* are very buoyant and are drifted down rivers, but perish in sea water. *Mauritia flexuosa* occasionally grows singly at the riverside, the seeds having been placed there by the current.

Polygonaceae

The fruit composed of 3 carpels usually shows the form of a 3-angular nut. It is often surrounded by persistent tepals which increase in size during fruit development.

1. *Coccoloba* sp. = Arahueque

Hab.: RF	Height: a
Fruit: Nut	Ind.: 239
Disp.: Zoochorous	Fr. time:

2. *Coccoloba* sp. = Uvero blanco

Hab.: RF	Height: a
Fruit: Nut	Ind.: 114
Disp.: Zoochorous	Fr. time:

3. *Coccoloba* sp. = Uvero (negro)

Hab.: RF	Height: a
Fruit: Nut	Ind.: 31
Disp.: Zoochorous	Fr. time:

4. *Triplaris surinamensis* = Santa Maria

Hab.: RF	Height: A
Fruit: Samara	Ind.:
Disp.: Anemochorous	Fr. time:

1. *Coccoloba* sp. (Arahueque): The globular fruit together with its accessory parts (anthocarp) reaches a length of 0,9 cm and a width of 0,5 cm. The pericarp thickness measures 0,025 cm. The fruit surface (tepals) is smooth and shiny. There is only one locule with one seed and the fruit proper is a nut. Three tepals surround the fruit replacing a fleshy pericarp. The only seed is 0,6 cm long and 0,4 cm broad, of blackish colour and of brittle texture. The surface is smooth, but covered with very small knobs. On its apical end the seed shows 6 lobes. The very large endosperm is mealy and white. The tepals are united surrounding the fruit proper entirely and become fleshy and edible at maturity. Dispersal of the fruit takes place by animals (birds, monkeys).

2. *Coccoloba* sp. (Uvero blanco): The dry indehiscent nutlet contains a single seed and is surrounded by the persistent fleshy tepals. The fruit is dispersed by animals (e.g. birds, monkeys).

3. *Coccoloba* sp. (Uvero negro): The dry indehiscent nutlet contains a single seed and is surrounded by the persistent fleshy tepals. The fruit is dispersed by animals (e.g. birds, monkeys).

4. *Triplaris surinamensis:* The 3-angular ovoid fruit measures about 4,5 cm in length and 1 cm in width including the persistent tepals. The nutlet itself measures 1,3 cm in length and 0,7 cm in width. It contains only one seed. The pericarp texture is brittle and the fruit surface is smooth and shiny. The 3 big wings are modified outer tepals, while the 3 inner tepals are aborted. The 3 wings completely surround the nut with their united bases which are hairy. The white or rosy wings show reticulate innervation with a central raquis. Dispersal of the samara takes place by wind. The soft-wood tree is of rapid growth and grows in

other environments too. It is also cultivated. In Surinam it is very common and occurs in the swampy forest as well as in the secondary forest, and in swamps at the riverside.

Three of the species of Polygonaceae studied belong to the genus *Coccoloba* and have "berry-like" fruits with modified fleshy tepals. The fruits are dispersed by animals. All 3 species are represented by lower trees belonging to height category a. The highest tree, *Triplaris surinamensis*, has a samara with tepals transformed into wings.

This distribution of the berry-like fruits in the lower trees and wind-dispersed samaras in high trees shows once again that animal dispersal is more advantageous in the lower forest layers, while higher trees adapt to wind dispersal which is more convenient in the high forest levels.

Quiinaceae

Touroulia guianensis = Palmito

Hab.: RF	Height: a
Fruit: Berry	Ind.: 16
Disp.: Zoochorous	Fr. time:

The only species studied, *Touroulia guianensis*, forms small globular berries (2 × 2,5 cm) with a pericarp width of 0,5 cm. The ovary is plurilocular, habitually 7-locular, and the fruit flesh is edible. The fruit surface is distinguished by numerous small ribs at about 1 mm distance. Each locule contains an elongated seed (1,5 × 0,6 × 0,3 cm) with a tight indumentum, which is a family characteristic. The 5 sepals are persistent.

The inner structure of this berry is very interesting. Most prominent are the ribs being composed of different tissue types (Fig. 83). Each rib consists in a parenchymatous central part (fig. 82) and a sclerenchymatous peripheral part. Large thick-walled sclereids with pit canals are more or less radially arranged around the parenchymatous zone, being elongated in a radial direction (Fig. 84). Towards the outside, the sclerenchymatous part is covered by about 3 supepidermal layers of very small cells, while the epidermis itself is even more small-celled. In between the ribs lie large oil-pockets or oil-bodies (Fig. 82), consisting of thin-walled parenchymatous cells, densely filled with large oil drops. The cells are elongated parallel to the fruit surface, being arranged in 2 rows. Possibly, they develop from a single subepidermal cell which first divides anticlinally, while of the 2 daughter cells each produces many periclinal walls by repeated subdivisions. Up to 15 and more periclinal walls may be formed in this way. These oil-pockets certainly represent very nutricious tissues and possibly correspond to elaiosomes. Beneath the ribs lies parenchyma and a layer of large stone cells. Large-celled parenchyma in which groups of stone cells are regularly embedded follows towards the inside. The stone cell groups are roundish in outline and are arranged in one line, being separated by small "bridges" of parenchyma.

The berries are edible and very nutricious, being dispersed by animals, possibly even by ants for the presence of elaiosomes. Nevertheless, the rate of distribution of this species is very low in the forest studied with only 16 individuals. In Surinam, the species occurs in the secondary forest and in the creek forest too.

Rhamnaceae

Colubrina glandulosa = Cartancillo

Hab.: Deciduous forest, RF	Height:
	Ind.: 4
Fruit: Capsule	Fr. time: 1
Disp.: Autochorous	

The dry dehiscent capsule contains few light seeds. The dispersal mechanism is autochorous. The species is of very little importance in the forest studied.

Rhizophoraceae

Cassipourea guianensis = Mamoncillo blanco
Hab.: RF Height: a
Fruit: Capsule Ind.: 418
Disp.: Autochorous and zoochorous Fr. time: 2/5

The dry dehiscent capsule is elongated or elliptic in outline, measuring up to 1 × 0,4 cm, of dark-green colour, and with a smooth surface. It is 3-locular and contains 3 seeds, about 0,7 cm long, with a 0,3 cm long aril. The autochorous dispersal method is assisted by animals which eat the aril. The species shows a relatively good rate of distribution. In Surinam, the shrub or low tree is common, occurring in the secondary forest, in the riparian forest, in the rain forest, and on slopes.

Rosaceae

Characteristic is the ovary composed of many carpels of undetermined number. The fruits occur in the form of follicles opening at the ventral suture or are modified into nutlets or drupes. Receptacle and/or floral members (sepals, petals, stamens) occasionally take part in the fruit formation. The appendicular tissue develops in the form of a concave cup or a convex receptacle uniting the follicles in an aggregate fruit.

1. *Couepia glandulosa* = Merecure terán
Hab.: RF Height: A
Fruit: Drupe Ind.: 40
Disp.: Zoochorous Fr. time: 1/8/10
2. *Hirtella americana* = Terciopelo blanco
Hab.: RF Height:
Fruit: Drupe Ind.: 28
Disp.: Zoochorous Fr. time: 6/9
3. *Hirtella davisii* = Ceniza negra
Hab.: RF Height: A
Fruit: Drupe Ind.: 101
Disp.: Zoochorous Fr. time: 5
4. *Hirtella racemosa* = Picapica morada
Hab.: RF Height: a (A)
Fruit: Drupe Ind.: 29
Disp.: Zoochorous Fr. time:
5. *Licania alba* = Hierro
Hab.: RF Height: A
Fruit: Drupe Ind.: 679
Disp.: Zoochorous Fr. time: 2/5
6. *Licania apetala* = Mamoncillo rebalsero
Hab.: RF Height: a
Fruit: Drupe Ind.:
Disp.: Zoochorous Fr. time: 5
7. *Licania densiflora* = Hierrito
Hab.: RF Height: A
Fruit: Drupe Ind.: 3.258
Disp.: Zoochorous Fr. time: 1–2/10
8. *Licania* aff. *discolor* = Hierrillo
Hab.: RF Height: A
Fruit: Drupe Ind.: 343
Disp.: Zoochorous Fr. time:
9. *Licania hypoleuca* = Hierrito blanco
Hab.: RF Height: A
Fruit: Drupe Ind.:
Disp.: Zoochorous Fr. time:
10. *Licania parvifructa* = Hierrillo blanco
Hab.: RF Height: A
Fruit: Drupe Ind.: 24
Disp.: Zoochorous Fr. time: 7–8/10
11. *Licania rufescens* = Hierrillo negro
Hab.: RF Height: A
Fruit: Drupe Ind.: 85
Disp.: Zoochorous Fr. time: 2
12. *Licania* sp. = Pilón nazareno
Hab.: RF Height: A
Fruit: Drupe Ind.: 24
Disp.: Zoochorous Fr. time: 2
13. *Parinari excelsa* = Merecurillo
Hab.: RF Height: A
Fruit: Drupe Ind.: 133
Disp.: Zoochorous Fr. time: 2
14. *Parinari rodolphii* = Merecure montañero
Hab.: RF Height: A

Fruit: Drupe Ind.: 53
Disp.: Zoochorous Fr. time: 2/5/8/11

15. *Prunus sphaerocarpa* = Menta berti
Hab.: RF Height: A
Fruit: Drupe Ind.:
Disp.: Zoochorous Fr. time: 3

1. *Couepia glandulosa:* The ovoid to globular fleshy drupe reaches 3,6 cm in length and 2,4 cm in width with a pericarp thickness of 0,5 cm. The pericarp texture is woody due to numerous stone cells embedded in parenchyma. The fruit surface is rough through very small knobs and the colour is blackish. There is usually one locule with a single seed. However, occasionally a second aborted seed may be recognized. Progressing reduction of the seed number is apparent here. The fruit flesh is thin and the seed coat is brittle. Dispersal of the fleshy drupe takes place by animals. The very rare species is represented as a small tree up to 12 m in height in Surinam, occurring at the river side and at savanna margins.

2. *Hirtella americana:* The indehiscent drupaceous fruit contains only a single seed. The flowers develop in racemes. The drupe is dispersed by animals (e.g. birds).

3. *Hirtella davisii:* The flattened "stone" (endocarp) measures 2,5 cm in length and 1,4 cm in width and is only 0,1 cm thick. The fleshy outer part of the fruit was gone already in the material studied. The endocarp surface is longitudinally ribbed (Fig. 85), the ribs being small and closely together. The hard endocarp is composed of sclereids and covered inside by a dense indumentum. There is only one locule with a single seed. Dispersal of the drupe takes place by animals (e.g. birds).

4. *Hirtella racemosa:* The indehiscent drupaceous fruit measures 1,3 × 0,7 × 0,5 cm. It has a ribbed surface and a thin fleshy layer. The calyx is persistent. The drupe which contains a single seed is dispersed by animals (birds). In Surinam, the species is represented by a shrub or small tree, but occasionally it develops into a high tree. It is common in Surinam mainly at the river side.

5. *Licania alba* (Figs. 85, 86, 87, 89): The ovoid drupe has a more or less woody texture and measures 5,6 cm in length and 4,7 cm in width. The pericarp is 0,5 cm thick. The fibrous endocarp encloses a single seed. When dried out, the fruit surface has a reticulate appearance (Fig. 85). As in *Hirtella davisii*, the endocarp inside is hairy. The seed is 3,9 cm long and 2,6 cm broad, of ovoid shape and brown. Its texture is woody, its surface is "brain-like" through depressions generated by the imprints of the prominent vascularization in the endocarp. The seed coat is very thin. The embryo shows 2 large cotyledons with very smelly oil cells.

The outer pericarp epidermis is small-celled and densely covered with hairs. Most of the hairs are unicellular and have helicoidal wall thickenings. Occasionally, 2 helicoidal wall thickenings crossing each other strengthen the hair. Intermingled with the unicellular hairs are cone-shaped pluricellular hairs, probably of glandular nature. Beneath the epidermis follows a small-celled parenchyma in which individual stone cells are interspersed. Towards the inside the size of the parenchyma cells increases and the stone cells become united in groups augmenting in size towards the inside. The endocarp is fibrous, the fibers crossing one another.

Dispersal of the drupe takes place by animals. The rate of distribution of this species is high.

6. *Licania apetala:* The pear-shaped drupe measures 2,5 × 1,7 cm, is dark-green, and has a smooth "bold" surface. The fruit flesh layer is very thin. The only stone has a thin fibrous endocarp of woody texture which is hairy on its inside. There is only one seed. The drupe is dispersed by animals, and, for example, eaten by fish. In Surinam, the species is common and also occurs in the savanna forest. According to Prance (1973) *L. apetala* is an ochlospecies of wide distribution and of a polymorphic phenotype. The species probably originated by isolation into populations which did not develop sterility barriers during the time of isolations. None the less, the number of individuals in the forest studied is minimal.

7. *Licania densiflora* (Figs. 88, 89, 90): The small drupe has the shape of a little pear ("Urbirne"), is 2–3 cm long and 1,4 cm broad, and of woody texture. The reddish fruit surface is velvety

through short hairs. The pericarp measures 0,2 cm in thickness and contains many sclereids. The endocarp composed of stone cells has a very hairy inside with long "blond" hairs and encloses a single seed. The seed is 1,4 cm long and 0,6 cm broad. It is elongated and shows 4 angles. The brown seed coat has a somewhat rough surface. The seed is hard. Dispersal of the drupe takes place by animals. The species shows an enormous rate of distribution with 3.258 individuals. The tree seems to produce a large quantity of fruit. In Surinam, the tree reaches 30 m in height, but infrequently occurs in hilly regions of the rain forest.

8. *Licania* aff. *discolor:* The indehiscent pear-shaped drupaceous fruit contains a single stone with a single seed. The fruit surface is velvety and of a rusty brown colour. Dispersal of the drupe takes place by animals. In Surinam, the tree reaches 35 m in height, some specimens even 50 m, but is rare.

9. *Licania hypoleuca:* The drupe is ovoid to pear-shaped. It measures 1,5 × 1 cm. Its surface is green to pink or violet. The fruit flesh is thin. The only stone consists in a thin fibrous and hard endocarp which is rough and hairy on its inside, and contains a single seed. The drupe is edible and dispersed by animals. The tree reaches 35 meters in Surinam where it is very frequent in the savanna forest and in the rain forest, especially in hilly regions, at the river side and along creeks.

10. *Licania parvifructa:* The small unilocular drupe is about 1,8 cm long and 1,2 cm broad, having an ovoid shape. The fruit surface is smooth. The woody endocarp, rich in sclereids, is covered with hairs on its inside and surrounds a single seed. The rate of distribution of this species is very low. Dispersal takes place by animals, which eat the edible outer fruit part.

11. *Licania rufescens:* The unilocular drupe contains a single seed. It is dispersed by animals.

12. *Licania* sp. (Pilón nazareno): The small ovoid drupe, 2,1 cm long and 0,8 cm broad, has an endocarp composed of stone cells and encloses a single seed. No hairs are found on the endocarp inside. The fruit surface is shiny and knobby. The seed is 1,2 cm long and 0,6 cm broad, has an elongated shape and is black. Its surface is smooth. The drupe is dispersed by animals, but the rate of distribution is very low.

13. *Parinari excelsa:* The grey ovoid plum-like drupe with edible flesh is about 3–6 cm long and 2–3 cm broad, generally measuring 5 × 1,6 × 1,4 cm. The grey-brown surface is rough because of numerous warts. The white lenticels contrast with the brown colour of the surface. The pericarp thickness is about 0,7 cm. The fruit contains 1–2 stones, each with a single seed. The endocarp composed of sclereids measures 0,4 cm in thickness and develops plenty of hairs on its inside, adopting a woolly aspect. The endocarp, hard and of fibrous texture, is somewhat ribbed on its outside. The thin fruit flesh layer is edible. The seeds are large. In areas where the tree is dominant, both seedlings and saplings are scarce, according to Richards (1952). The drupe is dispersed by animals, in Africa by bats, according to Osmaton (1965). It is a wide-spread and variable species in tropical Africa, also eaten and dispersed by elephants, according to Hall & Swaine (1981). In Surinam, the tree reaches 40 meters in height, but is very rare.

14. *Parinari rodolphii:* The ovoid drupe has a woody texture and a brown colour. Its length is 5,5 cm, its width 3 cm. Pericarp thickness reaches 1 cm. The fruit surface is slightly rough. The drupe encloses 2 locules, each with a single seed. In some cases, there are 3 carpels and 3 locules. Apparently, reduction of the locules is advancing here.

The outer pericarp epidermis has thickened outer walls. Beneath follows a cartilaginous collenchyma of several layers. The ground tissue of the pericarp is represented by parenchyma with somewhat thickened cell walls. Very large stone cell groups are embedded in this tissue. Parenchyma cells often show an arrangement in the form of tangential rows due to the dilatation growth of the fruit. The pattern of sclerenchymatous groups becomes denser towards the fruit inside. Not seldom, sclerenchyma groups are radially extended, especially in the middle of the pericarp, following radial tensions. In the vicinity of the endocarp stone cell groups become very densely packed.

The endocarp is fibrous. Dispersal of the drupe takes place by animals.

15. *Prunus sphaerocarpa:* The fleshy drupaceous fruit is relatively large and contains a single seed. It has an ovoid shape. The drupe is dispersed by animals. Almost all species of *Prunus* are bird-dispersed, according to Ridley (1930).

The 15 species of Rosaceae studied form a very homogeneous group. They are very much alike in all important features studied. Almost all species are represent by high trees of height category A. All of them form flowers and fruits in dense racemes. In all cases the fruit type is about the same: a drupe with a hard endocarp and a parenchymatous mesocarp in which plenty of stone cells are dispersed. There is usually only one stone (locule) with a single seed, except in 2 examples where 2 locules each with one seed may be found. In all cases, dispersal of the drupes takes place by animals which eat the more or less fleshy mesocarp, while the hard endocarp remains intact until rotting begins which may probably be a slow process. Larger animals may, however, swallow the entire fruits and the endocarp of digested drupes may decompose sooner. The fruits are all relatively small measuring a few centimeters. Even the shape of the fruits is similar in all species varying between ovoid or occasionally pear-shaped. There is no real explanation through the type of dispersal mechanism why most of the species show a very low rate of distribution, while one shows a high rate (*Licania alba* with 679 individuals) and another one an extremely high rate (*Licania densiflora* with 3.258 individuals). One possible explanation is that *Licania densiflora* could be the most ancient established species in the forest. Other explanations could imply that it is a very aggressive and resistant species or that it is less afflicted by predators. The other 13 species with very low number of individuals once again give evidence that dispersal by animals in very high trees is usually less advantageous than other dispersal mechanisms. Prance (1973) could show that the few savanna species of *Licania* are all members of the more advanced sections of the genus, indicating that they are derived from what are basically lowland forest species. And the predominance of lowland forest species shows that the divergence within the group has mostly taken place within the lowland forest habitat. Prance explains this by dry periods with isolation of species into separate populations, and subsequent separation of species (refuge theory).

Rubiaceae

The usually 2-carpellary fruits are often crowned by the persistent calyx. The fruit forms are very different, appearing in the form of a capsule, a berry or a drupe.

1. *Alseis labatioides* = Carutillo blanco
   Hab.: RF — Height: A
   Fruit: Capsule — Ind.:
   Disp.: Autochorous — Fr. time:
2. *Amaioua guianensis* = Cacho de venado
   Hab.: RF — Height: a
   Fruit: Berry — Ind.: 77
   Disp.: Zoochorous — Fr. time: 6
3. *Chimarrhis microcarpa* var. *microcarpa* = Carutillo
   Hab.: Lowland — Height: A
   Fruit: Capsule — Ind.: 315
   Disp.: Autochorous — Fr. time: 6
4. *Coutarea hexandra* = Guatacare amarillo
   Hab.: RF — Height: A
   Fruit: Capsule — Ind.:
   Disp.: Anemochorous — Fr. time:
5. *Duroia* sp. = Conserva
   Hab.: RF — Height: aa
   Fruit: Berry — Ind.: 15
   Disp.: Zoochorous — Fr. time:
6. *Genipa americana* = Caruto montañero
   Hab.: Savanna, deciduous forest, RF — Height: A
   Ind.: 3
   Fr. time: 1–2/5
   Fruit: Berry
   Disp.: Zoochorous

7. *Guettarda* aff. *acreana* = Punteral negro
   Hab.: RF Height: aa
   Fruit: Drupe Ind.:
   Disp.: Zoochorous Fr. time: 10
8. *Rudgea hostmanniana* = Comida de danto
   Hab.: Riverside Height: aa
   Fruit: Drupe Ind.: 82
   Disp.: Zoochorous Fr. time:

1. *Alseis labatioides:* The dry dehiscent capsule contains numerous small seeds. The dispersal mechanism is autochorous.

2. *Amaioua guianensis:* The ovoid fleshy berry is crowned by a disc and has a leathery fruit wall. The numerous small seeds are embedded in a gelatinous fruit flesh. The fruit measures $1{,}3 \times 0{,}7$ cm up to $2 \times 1{,}5$ cm and has a dark violet-brown colour. Dispersal takes place by animals (birds). In Surinam, the species occurs in the savanna forest too.

3. *Chimarrhis microcarpa* var. *microcarpa:* The dry dehiscent capsule contains numerous small seeds. Dispersal is autochorous.

4. *Coutarea hexandra:* The woody ovoid but flattened capsule measures $3\text{–}8 \times 2{,}1 \times 0{,}4$ cm, is bi-locular, and opens with 2 valves beginning from the apex downwards. Each valve has 2 ribs at each margin. The calyx is persistent. The numerous seeds are winged, measuring $1{,}8 \times 0{,}8 \times 0{,}05$ cm. The seeds are dispersed by wind. In Surinam, the species is represented by a small tree up to 5 meters in height and mainly occurs in the coastal area, particularly on ridges.

5. *Duroia* sp. (Conserva): The fleshy berry contains numerous small seeds. The fruit shell is thick and woody. By union of the parietal placentas, the fruit becomes seemingly 5-6 locular. The seeds are embedded in a gelatinous edible pulp. The calyx is persistent. Dispersal of the fruit takes place by animals, e.g. monkeys.

6. *Genipa americana:* The more or less globular fleshy berry reaches 5–10 cm in diameter and is of gray-green to brownish-yellowish colour. The 0,2 cm thick fruit shell is hard, but smooth on its outside. The edible pulp has an aromatic flavour. The calyx is persistent. The numerous seeds measuring $0{,}8 \times 0{,}5 \times 0{,}2$ cm are embedded in a slimy mass. The edible fruits are dispersed by animals, e.g. monkeys and birds. Ichthyochory of fruits has been observed too. The species is common in Surinam in the swampy and marsh forest and in the ridge forest as well.

A liquid may be obtained from the fruits which turns dark-blue oxydizing at the air and which may be used as a substitute of ink. It contains tannins. The indians use it to paint their faces and bodies with it. Furthermore, the so-called "Maná" or sugar of caruto is obtained from the fruit which is somewhat laxative. The tree is bare of leaves during the dry season.

7. *Guettarda* aff. *acreana:* The fleshy ovoid drupe measures $0{,}75\text{–}0{,}8 \times 0{,}6\text{–}0{,}7$ cm, has a smooth surface and is 4-locular. It contains only few seeds. The fruit is dispersed by animals. The species is rare in Surinam where it occurs as a small tree up to 6 meters in height, mainly occupying slopes on granite and laterite.

8. *Rudgea hostmanniana:* The fleshy ovoid drupe measures $1 \times 0{,}8$ cm, is green to orange-red, and crowned by the persistent sepals. The stone is irregularly ribbed on its outside and contains a single seed. The drupe is dispersed by animals, e.g. tapirs and birds. The species is common in Surinam, mainly occurring in swampy and riparian forests.

Of the 8 species of Rubiaceae studied 4 belong to height category A, one to height category a, and 3 to height category aa. Of the 4 species belonging to height category A, 3 are dispersed in an autochorous way by dehiscing capsules, and only one, *Genipa americana*, is dispersed by animals. The lower and low trees of height categories a and aa are all dispersed by animals. This fact shows once more that autochorous or anemochorous dispersal is more advantageous in high trees, while in lower trees animal dispersal is more effective. More precisely: animal dispersal is kept in low trees, while in high trees an autochorous or anemochorous mechanism has developed by and by. All species studied have only few representatives in the forest studied except *Chimarrhis microcarpa* var. *microcarpa* with 315 individuals and an autochorous dispersal mechanism. Only one of the autochorous species with capsules develops winged seeds which are dispersed by wind.

Rutaceae

The fruits of Rutaceae are very variable and manyfold in their structure, as drupes, berries, winged fruits, and capsules occur. Speticidal capsules may develop into schizocarps separating into several dehiscing mericarps.

1. *Cusparia trifoliata* = Quina
   Hab.: RF — Height: aa
   Fruit: Capsule — Ind.: 33
   Disp.: Autochorous — Fr. time:
2. *Erythrochiton brasiliensis* = Cola de pava
   Hab.: RF — Height: aa
   Fruit: Capsule — Ind.: 203
   Disp.: Autochorous — Fr. time:
3. *Fagara macrocarpa* = Mapurite blanco
   Hab.: RF — Height: A
   Fruit: Capsule — Ind.: 4
   Disp.: Autochorous — Fr. time:
4. *Fagara* aff. *martinicensis* = Bocsuo amarillo
   Hab.: RF — Height: A
   Fruit: Capsule — Ind.: 37
   Disp.: Autochorous — Fr. time:
5. *Fagara quinduensis* = Bocsuo negro
   Hab.: RF — Height: a
   Fruit: Capsule — Ind.: 25
   Disp.: Autochorous — Fr. time:
6. Cozoiba redonda
   Hab.: RF — Height:
   Fruit: Capsule — Ind.:
   Disp.: Autochorous — Fr. time: 6
7. Erizo
   Hab.: RF — Height: A
   Fruit: Capsule — Ind.: 31
   Disp.: Autochorous — Fr. time:
8. Erizo morado
   Hab.: RF — Height: A
   Fruit: Capsule — Ind.: 4
   Disp.: Autochorous — Fr. time:
9. Mapurite negro
   Hab.: RF — Height: A
   Fruit: Capsule — Ind.: 5
   Disp.: Autochorous — Fr. time:

1. *Cusparia trifoliata:* The dry dehiscent mericarps contain only a single seed. The dispersal mechanism is autochorous.

2. *Erythrochiton brasiliensis:* The capsule (schizocarp) is composed of maximal 5 dehiscent mericarps, measuring up to $1{,}7 \times 1{,}2 \times 1{,}2$ cm, which are of red colour. The sepals generally persist. Each mericarp contains 2 seeds. The dispersal mechanism is autochorous. The species occurs on places in Surinam.

3. *Fagara macrocarpa:* The dry dehiscent mericarps contain only a single seed. The dispersal mechanism is autochorous.

4. *Fagara* aff. *martinicensis:* The dry dehiscent mericarps contain only a single seed. The dispersal mechanism is autochorous.

5. *Fagara quinduensis:* The dry dehiscent mericarps contain only a single seed. The dispersal mechanism is autochorous.

6. Cozoiba redonda: The dry dehiscent mericarps contain only a single seed. The dispersal mechanism is autochorous.

7. Erizo: The dry dehiscent mericarps contain only a single seed. The dispersal mechanism is autochorous.

8. Erizo morado: The dry dehiscent mericarps contain only a single seed. The dispersal mechanism is autochorous.

9. Mapurite (negro): The dry dehiscent mericarps contain only a single seed. The dispersal mechanism is autochorous.

The 9 species of Rutaceae studied belong to different height categories: 5 are of the highest category (A), one is of height category a, and 2 are of the lowest height category (aa), while in one species it is doubtful to which height category it belongs. All species have the same dispersal mechanism of dehiscing capsules with a very low number of seeds. The number of individuals of all species is very low and only *Erythrochiton* has 203 representatives in the forest. As only 2 species are low trees of height category aa, it becomes evident once more that autochory with dehiscing capsules is more common in high trees.

Sabiaceae

Meliosma herbertii = Carruache

Hab.: RF	Height: a
Fruit: Drupe	Ind.: 20
Disp.: Zoochorous	Fr. time:

The indehiscent fleshy drupe contains a single seed. The fruit is dispersed by animals. The number of individuals of this species in the forest studied is very low.

Sapindaceae

The ovary of the Sapindaceae is usually composed of 3 carpels, each locule being furnished with 1 (seldom 2 or more) ovules. The seeds often develop a large arillus rich in sugars, but have no endosperm.

1. *Cupania hirsuta* = Chaparillo

Hab.: RF	Height: aa
Fruit: Capsule	Ind.: 46
Disp.: Autochorous (and zoochorous?)	Fr. time:

2. *Dilodendron bipinnatum* = Helecho

Hab.: RF	Height: a
Fruit: Capsule	Ind.:
Disp.: Autochorous	Fr. time:

3. *Melicocca bijuga* = Mamón

Hab.: RF, Cultivated	Height: A
Fruit: Drupe	Ind.: 1
Disp.: Zoochorous	Fr. time:

4. *Sapindus saponaria* = Para-Para

Hab.: Deciduous forest, RF	Height: A
	Ind.: 1
Fruit: Drupe	Fr. time:
Disp.: Zoochorous	

5. *Talisia reticulata* = Cotoperiz montañero

Hab.: RF	Height: A
Fruit: Drupe	Ind.: 282
Disp.: Zoochorous	Fr. time: 11

6. *Toulicia guianensis* = Carapo blanco

Hab.: RF	Height: a
Fruit: Samara	Ind.: 795
Disp.: Anemochorous	Fr. time:

1. *Cupania hirsuta:* The dry dehiscent capsule contains only few seeds. It measures about 1,5 × 2 × 2 cm and has a brown indumentum. The calyx persists. Dispersal is autochorous. However, the seeds are possibly arillate and dispersed by animals. In Surinam, the species is represented as a tree of about 10 meters, while some specimens even reach 20 meters in height. It is very frequent in the rain forest, especially on slopes, and also occurs in the secondary forest.

2. *Dilodendron bipinnatum:* The dry dehiscent capsule contains only few seeds. The dispersal mechanism is autochorous.

3. *Melicocca bijuga:* The globular greenish drupe reaches about 2–4 cm in diameter and contains a single large seed. The yellow to salmon-coloured gelatinous pulp surrounding the seed has a sour-sweet pleasant taste and is edible. The mesocarp very much resembles that of grapes. The toasted seeds, although adstringent, are of agreable taste being used by the Indians to bake bread. The tree is also cultivated and used for ornamental reasons. Dispersal of the fruit takes place by animals.

4. *Sapindus saponaria:* The globular originally 3-locular drupe measures about 1–2 cm in diameter and has a transparent and brilliant exocarp when ripe. It has an orange to reddish colour. Usually only 1–2 locules develop so that the ripe drupe contains 1–2 seeds. The mesocarp is viscous and has a bitter taste. The seed, about 10–15 mm in diameter has a black and brilliant surface. The hard impervious shell is 1 mm thick. The embryo is surrounded by a bone-hard oily perisperm. The seed is poisonous. In a pulverized state the seeds are used as an insecticide. All parts of the tree contain a principle of saponification. The oil extracted from the cotyledons is used in medicine as

well as in the soap production. The fruit contains saponin and in contact with water produces foam abundantly. It may therefore be used to wash laundry and to kill fish by asphyxia, hence the vulgar name "soap berry". The fruits attract birds, but the toxic seeds with their hard shells remain intact in this way. The syndrome: edible fruits and protected, eventually toxic seeds, fits very well into the scheme designed in the papers of Janzen. Besides, the seed has buoyancy and drift seeds reach the Azores (Guppy 1917). According to Ridley (1930), seeds have even reached the Australo-Papuan region. The tree is characteristic of open woods in arid climate, and abundantly occurs in the secondary vegetation. It is resistant to drought and easily adapts to different soil types. The low occurrence of the tree in the forest studied is explained by these facts. The species is represented in Surinam as a small tree reaching up to 9 meters in height which infrequently occurs in the riparian and ridge forest. It also grows wild and is cultivated.

5. *Talisia reticulata:* The indehiscent fleshy fruit corresponds to a drupe which only contains a single juicy seed (sarcotesta). Dispersal takes place by animals (birds).

6. *Toulicia guianensis:* The 3-locular fruit is reversedly egg-shaped and measures about 4 × 3 × 1,5 cm. It is green and has a delicate indumentum on its surface. The fruit corresponds to a capsule in the form of a schizocarpium which splits into 3 indehiscent mericarps each with one seed. The mericarps are winged samaras with a wing along the midrib of the carpel. The papery wings show reticulate vascularization. The dark brown roundish seeds are 0,6 cm long and 0,4 cm broad and have a wrinkled surface. Dispersal of the samaras takes place by wind. The number of individuals of this species in the forest is high (795) and dispersal seems therefore to be advantageous. The species is rare in Surinam.

The 6 species of Sapindaceae studied do not play an important role in the forest studied. Three of the species belong to height category A, two to height category a, and only one to aa. The 3 highest trees form drupes which are dispersed by animals. Of the 2 species of medium height (a) one is anemochorous having winged mericarps, while the other one develops autochorous capsules. The smallest tree (height category aa) likewise forms autochorous capsules. *Toulicia guianensis* shows a relatively high occurrence with 795 individuals. Its anemochorous dispersal mechanism seems to be advantageous, although the tree is not very high. The autochorous dispersal mechanism of *Cupania hirsuta*, on the other hand, seems not to be very favourable judging from the low number of individuals. The 3 zoochorous species of the highest height category show very low or relatively low numbers of individuals. It thus seems that zoochory is disadvantageous for high trees in a dense forest. *Sapindus saponaria* belongs to another habitat, hence the low number of representatives.

Sapotaceae

The fruits of the Sapotaceae are berries or – by reduction of locules and seeds – transform into drupes of globular to ovoid shape. They are often fleshy and edible and develop large 5–50 mm long seeds which have a shiny surface and a large hilum, resembling in this way the seed of a horse-chestnut. The seeds are more or less albuminous. The inner part of the pericarp is penetrated by laticifers. The outer seed coat contains stone cells, while the inside is rich in tanniferous cells. The seeds have a light brown to black colour and are hard. An aril is absent. The large hilar zone is characteristically pale or white. The Sapotaceae are almost exclusively dispersed by frugivorous birds, often aided by fruit-bats and mammals. The fruit flesh may be sweet, while the seeds often taste bitter. Chicle and balatá are obtained from the laticifers.

1. *Chrysophyllum auratum* = Caimito morado

Hab.: RF	Height: A
Fruit: Drupe	Ind.: 458

Disp.: Zoochorous Fr. time: 2/12

2. *Chrysophyllum* sp. = Caimito amarillo
   Hab.: RF Height: A
   Fruit: Berry Ind.: 266
   Disp.: Zoochorous Fr. time:
3. *Chrysophyllum* sp. = Chupón
   Hab.: RF Height: A
   Fruit: Drupe Ind.: 28
   Disp.: Zoochorous Fr. time: 2
4. *Ecclinusa guianensis* = Chicle
   Hab.: RF Height: A
   Fruit: Berry Ind.: 1.658
   Disp.: Zoochorous Fr. time: 6/12
5. *Manilkara bidentata* = Purguo blanco
   Hab.: RF Height: A
   Fruit: Drupe Ind.: 387
   Disp.: Zoochorous Fr. time: 1–2/7
6. *Manilkara* sp. = Pendare
   Hab.: RF Height: A
   Fruit: Drupe Ind.: 129
   Disp.: Zoochorous Fr. time:
7. *Micropholis melinoniana* = Capurillo
   Hab.: RF Height: A
   Fruit: Drupe Ind.: 47
   Disp.: Zoochorous Fr. time: 5/8/12
8. *Oxythece* aff. *robusta* = Chuponcillo
   Hab.: RF Height: a
   Fruit: Berry Ind.: 164
   Disp.: Zoochorous Fr. time: 2/5
9. *Pouteria* aff. *anibaefolia* = Caimito blanco
   Hab.: RF Height: a
   Fruit: Berry Ind.: 94
   Disp.: Zoochorous Fr. time:
10. *Pouteria egregia* = Purguillo amarillo
   Hab.: RF Height: A
   Fruit: Drupe Ind.: 274
   Disp.: Zoochorous Fr. time: 8
11. *Pouteria eugeniifolia* = Purguillo felix
   Hab.: RF Height:
   Fruit: Drupe Ind.: 67
   Disp.: Zoochorous Fr. time:
12. *Pouteria* cf. *trilocularis* = Rosado
   Hab.: RF Height: a
   Fruit: Drupe Ind.: 722
   Disp.: Zoochorous Fr. time: 1
13. *Pouteria venosa* = Bámpara
   Hab.: RF Height: A
   Fruit: Berry Ind.: 103
   Disp.: Zoochorous Fr. time: 3/7
14. *Pouteria* sp. = Caimito negro
   Hab.: RF Height: a
   Fruit: Berry Ind.: 614
   Disp.: Zoochorous Fr. time: 8
15. *Pouteria* sp.? = Capure
   Hab.: Deciduous forest, RF Height: A
   Ind.: 276
   Fruit: Berry Fr. time: 8
   Disp.: Zoochorous
16. *Pradosia prealta* = Purguillo morado
   Hab.: RF Height: A
   Fruit: Drupe Ind.: 135
   Disp.: Zoochorous Fr. time: 8
17. Bámpara amarilla
   Hab.: RF Height: A
   Fruit: Drupe Ind.:
   Disp.: Zoochorous Fr. time:
18. Chicle rosado
   Hab.: RF Height: A
   Fruit: Berry Ind.: 301
   Disp.: Zoochorous Fr. time:
19. Mantequillo
   Hab.: RF Height: a
   Fruit: Berry Ind.: 4
   Disp.: Zoochorous Fr. time:
20. Pendarito
   Hab.: RF Height: A
   Fruit: Drupe Ind.: 28
   Disp.: Zoochorous Fr. time: 10
21. Purguillo
   Hab.: RF Height: A
   Fruit: Drupe Ind.: 795
   Disp.: Zoochorous Fr. time: 5/8–9

1. *Chrysophyllum auratum:* The ovoid drupe is 1,4–2,5 cm long and 0,8–1,5 cm broad, and contains 1 (–2) seeds. The seed is about 1,1 cm long and 0,8 cm broad, ovoid, light brown, of hard texture and has a smooth surface. The seed coat is vascularized and the hilum prominent. Dispersal of the drupe takes place by animals, mainly birds. In Surinam, the tree reaches 15–20 meters in height and is very common in the rain forest, in the riparian forest and on slopes, especially on sandy soil.

2. *Chrysophyllum* sp. (Caimito amarillo): The fleshy berry contains only few seeds. Dispersal takes place by animals, mainly birds.

3. *Chrysophyllum* sp. (Chupón): The fleshy edible drupe is 5,5 cm long and 2,4–3 cm broad with a pericarp thickness of 0,4–0,5 cm. The texture is woody and fibrous. The shape of the fruit is elongated and the surface almost smooth, but has light dots, possibly corresponding to lenticels. The unilocular fruit contains a single ovoid seed. The seed is about 2,9–3,5 cm long and 1,7–2,2 cm broad, of brown colour, and of hard and brittle texture. The surface is smooth and shiny. The seed coat is 0,1 cm thick and the elongated raphe (hilum) is very prominent. The embryo is very big.

The outermost layer of the pericarp consists in a periderm composed of cork cells with horseshoe-shaped wall thickenings and of a few layers of phelloderm without thickenings. The ground tissue is more or less collenchymatous. In the fruit periphery, large sclereids are scattered. In the middle region of the pericarp, irregular bundles of sclerenchyma and vascular bundles surrounded by a large sclerenchymatous sheath are abundant. Laticifers in the form of long chains of elongated cells (as seen in longitudinal section) are scattered all over the section, being most abundant in the middle region of the pericarp. The content of the laticifers stains black-blue with JKJ and deeply orange with Sudan III. Towards the pericarp inside, the collenchyma becomes compressed. The small endocarp consists in sclereids of irregular shape. Dispersal of the drupe takes place by animals, mainly birds.

4. *Ecclinusa guianensis:* The globular berry measures about 1,5–6 cm in diameter, is yellow to orange with a brown or gray indumentum, and has an edible fruit flesh. The 1–5 small elliptic seeds measure 2,6 × 1,3 × 1,1 cm. The hilar zone is 0,7 cm broad. Dispersal of the berry takes place by animals. The species is frequent in Surinam in the rain forest, the high savanna forest, in the rocky savanna, and in the marsh and riparian forest.

5. *Manilkara bidentata:* The fleshy ovoid drupe is 3 cm long and 2,5 cm broad, its surface is smooth and its colour is yellow-orange to light brown with black dots (lenticels?). The endocarp is coriaceous. There are 9 locules altogether, 8 of which are aborted and only one develops with a single seed. The seed is 2,2 cm long and 1,4 cm broad, light brown, of hard texture and with a smooth surface. The hilar region is prominent. The fruit flesh is sweet, sticky and edible and the fruit dispersed by animals (birds). The species is common in the rain forest and marsh forest in Surinam. The milky sap of the bark supplies the balatá.

6. *Manilkara* sp. (Pendare): The globular to ovoid drupe is 3,5 cm long and 2,1 cm broad and has a hard and brittle endocarp composed of stone cells. The number of locules is about 6, but 5 abort and only one develops with a single seed. The fruit surface is smooth and shows dots. The seed is 2 cm long and 1,4 cm broad, of ovoid shape and light brown colour and has a tough texture. Its surface is smooth. The hilar region is prominent. The drupe is dispersed by animals (birds?).

7. *Micropholis melinoniana:* The ovoid drupe is about 1,8 cm long and 1 cm broad with a pericarp thickness of 0,1 cm. The endocarp is woody and encloses a single seed. The fruit surface is longitudinally wrinkled. The seed, about 1,1 cm long and 0,8 cm broad, is brown, of ovoid shape and has a thin and brittle shell. The seed surface is smooth and the hilum prominent. The fruit is dispersed by animals.

8. *Oxytece* aff. *robusta:* The globular berry is 3 cm long and 2,7 cm broad and has a leathery to woody texture. The fruit surface is smooth. The pericarp measures about 0,3 cm in thickness. The fleshy fruit is 4-locular and possibly contains 4 seeds. The fruit is dispersed by animals.

9. *Pouteria* aff. *anibaefolia:* The fleshy berry or drupe, about 4,5 cm long and 3,7 cm broad, is composed of 4 carpels. Three of the locules are well developed, while one is aborted. The number of seeds is probably three. The endocarp is composed of stone cells. The seeds are 2,4 cm long and 1,1 cm broad, light-brown, of hard texture and with a smooth and shiny surface. The hilum is very large and prominent. The fruit is dispersed by animals (e.g. birds).

10. *Pouteria egregia:* The fleshy drupe measures 2,5 × 3 cm and is covered by a velvety indu-

mentum. The 4 sepals are persistent. The fruit generally contains a single seed, but sometimes 2. The drupe is dispersed by animals (birds). In Surinam, the tree mainly occurs on bauxit, but is rare.

11. *Pouteria* aff. *eugeniifolia:* The fleshy drupe contains a single seed and is dispersed by animals (e.g. birds).

12. *Pouteria* cf. *trilocularis:* The fleshy ovoid drupe is 3 cm long and 2,2 cm broad. It is of brown colour and has a hard endocarp. The endocarp inside is red-brown and shiny. The single seed is 2 cm long and 1,6 cm broad, of brown colour, and has a hard texture. Its surface is smooth and shiny. The hilar region is very prominent. Dispersal of the drupe takes place by animals (e.g. birds).

13. *Pouteria venosa:* The almost globular berry measures about 2,7 × 2,5 cm, but may reach a diamter of up to 6 cm. It is green and contains few seeds or only one. The calyx is persistent. The fleshy fruit is dispersed by animals (birds). In Surinam, the tree occurs infrequently and is found in the West.

14. *Pouteria* sp. (Caimito negro): The fleshy berry contains few seeds and is dispersed by animals (e.g. birds).

15. *Pouteria?* (Capure): The fleshy berry contains only few seeds and is dispersed by animals (e.g. birds).

16. *Pradosia prealta:* The fleshy drupe contains a single seed. The seed is about 1,6 cm long and 1,3 cm broad, of light brown colour and ovoid shape and has a hard texture. The surface is smooth and the testa vascularized. The hilum is very prominent.

17. Bampara amarilla: The drupe is 2,5 cm long and 3,5 cm broad being composed of 4 carpels. It has a woody and brittle texture. The style is persistent and the fruit sessile (without peduncle). The pericarp measures 0,7 cm in thickness. It is rich in stone cells and tannins, hence the red brown colour. The fruit surface is covered with short hairs. The fruit shape is roundish, but flattened on top. There are 3 aborted seeds and only one is well developed. The fruit somewhat resembles that of *Ecclinusa sanguinolenta*, according to the description of Roosmalen. Dispersal of the drupe takes place by animals.

18. Chicle rosado: The fleshy berry contains only few seeds and is dispersed by animals.

19. Mantequillo (Figs. 92, 93): The fleshy berry contains only few (2) seeds. The seeds are about 3,5 cm long and 3,4 cm broad, having an almost globular shape, but are flattened on one side. The convex dorsal side is smooth, while the ventral side is almost white and sligthly rough. A "collar" surrounds the hilum. The seed very much resembles that of the horse-chestnut. Dispersal takes place by animals.

20. Pendarito: The fleshy drupe contains only a single seed and is dispersed by animals.

21. Purguillo: The ovoid to globular fruit is about 2,5 cm long and 1,8 cm broad. The pericarp measures 0,15 cm in thickness. The texture is woody. The fruit contains one, sometimes 2, seeds. The seed is 1,9 cm long and 1,4 cm broad. It has an ovoid shape, is light brown, and its texture is hard. The surface is smooth. The hilum is very prominent. The drupe is dispersed by animals.

Fifteen of the 21 species of Sapotaceae studied belong to height category A, while only 5 belong to height category a. All species grow in the humid tropical forest except *Pouteria* sp. (Capure) which also occurs in the deciduous forest. The fruits and seeds of all species studied resemble one another very much. And all are dispersed in the same way, namely by animals, although most of the species are represented by high trees. Most of the fruits are ovoid to globular few-seeded berries or single-seeded drupes with a relatively hard pericarp that includes stone cells, but is probably edible. The seeds are comparatively large and in most cases the seed number is reduced to 2 or only one. Very conspicuous is the prominent hilum and the smooth and shiny testa. Irregularities in the number of locules and seeds indicate the progressive reduction in seed number and the continous transformation of a several-seeded berry into a single-seeded drupe. One species shows a high rate of distribution with 1.658 individuals (*Ecclinusa guianensis*). Others show a relatively high rate between 458 and 722 individuals (*Chrysophyllum auratum*, Caimito negro, and *Pouteria* cf. *trilocu-*

laris). Five species have very few individuals between 4 and 67. The remaining species show a number of individuals between near 100 up to about 300. *Ecclinusa guianensis* with most individuals has a fruit of yellow to orange colour with edible fruit flesh. It seems to be a resistant species which also grows in the savanna forest. But there is no real convincing explanation why some of the species have high numbers of individuals, while others have very low ones. Besides, dispersal by animals should be less effective in very high trees. But birds and bats seem to like these fruits. The Sapotaceae contain latex mainly in the bark and in the fruits and most species produce resins and gums, such as chicle and balatá. The fleshy berries are sticky by the latex and, therefore, appropiate for dispersal by birds.

Simaroubaceae

1. *Simaba multiflora* = Congrillo
   Hab.: RF — Height: A
   Fruit: Drupe — Ind.: 231
   Disp.: Zoochorous — Fr. time: 3
2. *Simarouba amara* = Cedro blanco
   Hab.: RF — Height: A
   Fruit: Drupe — Ind.: 98
   Disp.: Zoochorous — Fr. time: 1

1. *Simaba multiflora:* The globular fruit is about 1,5 cm long and measures 1,2 cm in width. It contains a single seed, 0,9 cm long and 0,7 cm broad, with a papery seed coat. Dispersal of the drupaceous fruit takes place by animals (e.g. birds).

2. *Simarouba amara:* According to Alvarado (Glos. 277) the name is correctly written "Simarruba" and may be deduced from the word "simaruppa" of the arauco, an Indian dialect. The fruit is drupaceous and dispersed by animals. The tree is deciduous. Parts of the tree are used as a diuretic.

Both species of Simaroubaceae studied are of the highest height category (A) and their fruits are drupaceous and dispersed by animals. Considering the number of individuals present in the forest studied, this type of dispersal seems not to be very advantageous. The family is of little importance in the forest studied.

Sterculiaceae

The carpels – usually five – are syncarpous or apocarpous. The fruit is often a dry or fleshy capsule opening with 5 valves. Dehiscence takes place either loculicidally or separation into mericarps occurs. A sarcotesta or an "imitation aril" may occur in *Sterculia* (v.d. Pijl 1966, Ridley 1930).

1. *Guazuma ulmifolia* = Guácimo
   Hab.: RF — Height: A
   Fruit: Capsule — Ind.: 95
   Disp.: Autochorous and Zoochorous — Fr. time: 5
2. *Herrania lemniscata* = Cacao montañero
   Hab.: RF — Height: aa
   Fruit: Capsule — Ind.:
   Disp.: Autochorous and zoochorous — Fr. time:
3. *Sterculia pruriens* = Majagua
   Hab.: RF — Height: A
   Fruit: Capsule — Ind.: 908
   Disp.: Autochorous — Fr. time: 5
4. *Sterculia rugosa* = Guaraunera
   Hab.: RF — Height: A
   Fruit: Capsule — Ind.:
   Disp.: Autochorous — Fr. time:

1. *Guazuma ulmifolia:* The small 2–4 cm long woody and almost globular to ovoid capsule is covered with small wart-like excrescences up to 0,25 cm long (Fig. 95) and opens loculicidally with 5 valves. Dehiscence is retarded. The numerous

ovoid seeds, 0,2–0,4 cm long, are embedded in a slimy edible fruit flesh probably produced by the endocarp in the same way as in *Theobroma cacao* (Roth 1977). Dispersal is zoochorous (e.g. by birds). The mucilaginous and adstringent fruit is edible and preferred by cattle.

The brown-black pericarp surface shows very conspicuous warty excrescences (Figs. 96, 97). These are covered with a very small-celled epidermis in which stomata occur. The ground tissue of the protuberances is parenchymatous being interrupted by scattered large slime cavities. Each protuberance is penetrated in its center by a vascular bundle which ends dead in the knob-like top of the protuberance. The endocarp consists in its outer part in crossing fiber bundles, some running parallel to the endocarp surface and some perpendicular to it. The pulpy part of the endocarp is possibly of the same structure as in *Theobroma cacao* and contains slime cavities and adstringent tannic substances. The species is uncommon in Surinam, mainly occurring on slopes.

2. *Herrania lemniscata:* The large capsule somewhat resembles that of the cacao (Fig. 94) and measures about 10,5 cm in length and 5 cm in diameter. It has 5 conspicuous longitudinal ribs which ramify laterally anastomosing with ramifications of neighbouring ribs and producing a reticulate pattern in this way (Fig. 94). The intercostal regions are covered with long stiff brown hairs. The pericarp has a woody consistence. The 5-locular capsule opens loculicidally and contains several seeds per locule. The seeds are embedded in a sweet tasting fruit flesh possibly produced by the endocarp in the same way as in *Theobroma cacao* (Roth 1977). The triangular seeds are black-brown and have a wrinkled surface. The fruit very much resembles that of *H. kanukuensis* R.E. Schultes, according to the description of Roosmalen. Dispersal takes place by animals, mainly birds.

3. *Sterculia pruriens:* The apocarpic fruit separates into mericarps. Each follicular mericarp measures up to 8,5 cm in length and 7,5 cm in width. The pericarp thickness is 0,5 cm. Each follicle of woody consistency opens along the ventral suture. The only seed is embedded in stiff blond hairs which protrude from the endocarp. The fruit type is relatively primitive.

The endocarp is composed of fiber-like sclereids elongated tangentially parallel to the inner fruit surface. Above extend tangential rows of parenchymatous cells (Fig. 95) following the growth in width of the locule. The cells elongate first tangentially and then divide by anticlinal walls. The parenchyma cells of the ground tissue of the pericarp are partially tanniferous and brown. Scattered in the parenchyma are large slime cells, characteristic of the family, and occurring, for example, in *Theobroma cacao* (Roth 1977). The vascular bundles are fortified by enormous caps of sclerenchyma.

The species is very common in Surinam, occurring in the marsh forest, on slopes, in the creek forest, and in the savanna forest, too.

4. *Sterculia rugosa:* The large fruit is apocarpic and measures about 7 cm in length and 5 cm in width. The large follicular mericarps are of woody consistency and heart-shaped when open (Fig. 94). The fruit surface is smooth, but very hairy because of stellate hairs, characteristic of the genus and of the entire family. The fruit interior is abundantly furnished with delicate hairs which easily penetrate the skin. Each follicle opens at the ventral suture and generally contains only one seed. Endo- and mesocarp are fibrous. The fruit type is relatively primitive. The dispersal mechanism is autochorous.

Of the 4 species of Sterculiaceae studied 3 belong to height category A and only one is a small tree of height category aa. All four species develop capsules and have consequently an autochorous dispersal mechanism. However, 2 species show a zoochorous dispersal besides. One of these, *Herrania lemniscata*, is a very low tree appropriate for animal dispersal. The other zoochorous species, *Guazuma ulmifolia*, represented by a high tree has a low rate of distribution with only 95 individuals. *Sterculia pruriens* with only autochorous dispersal has a rate of distribution about 10 times as much as that of *Guazuma ulmifolia* with autochorous and zoochorous dispersal methods combined. Judging from the number of individuals the autochorous dispersal seems to be superior over the combination of auto and zoochory. It is possible that the zoo-autochorous dispersal is more primitive than the autochorous mechanism, particularly because one auto-zoochorous species is represented by a very low tree (*Herrania lemniscata*). It is furthermore suggested in this book that the early angiospermous plants were represented by low trees and that high trees are more advanced.

Tiliaceae

The fruits of the Tiliaceae are often represented by larger capsules with numerous seeds. The fruit surface is tightly covered with prickles in the genus *Apeiba*. Winged seeds occur in *Luehea*. In *Apeiba*, the fleshy septs may form a pulpa in which the seeds are embedded.

1. *Apeiba echinata* = Onotillo

Hab.: RF	Height: A
Fruit: Capsule	Ind.: 142
Disp.: Autochorous and zoochorous?	Fr. time: 5

2. *Apeiba tibourbou* = Cabeza de mono

Hab.: RF	Height: A
Fruit: Capsule	Ind.: 19
Disp.: Autochorous and zoochorous	Fr. time: 2

3. *Apeiba* sp. = Cabeza de negro

Hab.: Deciduous forest	Height: a
	Ind.: 35
Fruit: Capsule	Fr. time: 8
Disp.: Autochorous and zoochorous?	

4. *Luehea candida* = Guácimo cimarron

Hab.: RF	Height: A
Fruit: Capsule	Ind.:
Disp.: Anemochorous	Fr. time: 2/5–6

1. *Apeiba echinata* (Fig. 100): The 10-locular disc-shaped to globular capsule is flattened on top and at the base in axial direction, measuring 6–8 cm in diameter and 2,5–3 cm in length, and contains numerous small seeds, about 0,3 cm long. The woody capsule is of black colour and opens in the apical region. The fruit surface is densely covered with short spines, about 0,2–0,3 cm in length. The fruit very much resembles a sea-urchin from which fact the species name is derived. The seeds are set free through an opening on top of the fruit (Fig. 100). Therefore, the capsule is not completely indehiscent, but shows apical dehiscence in the form of a small hole. In the mature state, the seeds are probably embedded in a pulp which is eaten by animals. The fruit has a fibrous endocarp, the fiber bundles running in different directions (Fig. 101). In Surinam, the tree occurs very frequently in the secondary forest and in the savanna forest too. There it bears fruit from June to August. In Amazonian forests 0–1 individuals occur on one hectare (Black et al. 1950).

2. *Apeiba tibourbou* (Fig. 98): The globular capsule measures 7–8 up to 10 cm in diameter and has a depression in the center. It is densely covered with long bristles, about 2,5–3,5 cm long, and contains numerous seeds, about 0,3 cm long, separated in 10–20 locules. The gold-brown bristles are furnished with hairs, 0,2 cm long. Tbe pericarp thickness measures 0,15 cm. The fruit wall has a woody texture. The globular fruit with its ash-blond hairs very much resembles a head with a wig. From this aspect the vernacular name "Head of a monkey" is derived. The endocarp is very fibrous and strongly penetrated by vascular bundles. The seeds are embedded in clusters in a sweet pulp. The seeds are rich in oil which is used commercially. The dehiscence of the capsule is delayed. The dispersal is autochorous and zoochorous at the same time. In Surinam the species is common in hilly regions of the rain forest.

3. *Apeiba* sp. (Cabeza de negro) (Fig. 98, 101): The globular capsule, 4 cm in length and 8 cm in diameter, is densely covered with bristles, about 1 cm long, which are furnished in turn with unicellular hairs. The pericarp has a woody texture and is 0,2 cm thick. The 0,3 cm long seeds are numerous. The seed coat is vascularized. The seeds escape through an opening on top of the fruit. The vernacular name indicates the resemblance of this brown fruit with the head of a negro. The seeds are possibly embedded in a sweet pulp which is eaten by animals.

The outer fruit epidermis is very thick-walled and furnished with unicellular hairs. Beneath follow 2 layers of stone cells. The main bulk of the pericarp consists in fibrous elements which cross one another in all directions (Fig. 101).

4. *Luehea candida* (Fig. 99): The woody capsule is ellipsoid to globular measuring 2,5 cm in length and 2,5 cm in diameter, and opens with 5–10 valves, i.e. loculicidally and septicidally. Pericarp thickness is 0,05 cm. The fruit surface is spiny, the length of the spines reaching 0,3 cm. Dispersal of the few winged seeds takes place by wind.

The 4 species of Tiliaceae studied are of little importance in the forest studied. Three species are of height category A, while one is a lower tree of height category a. *Luehea candida* of height category A shows dispersal of winged seeds by wind. This example makes clear again that anemochory mainly occurs in high trees. The species of *Apeiba* develop capsules with delayed dehiscence in which seeds may remain enclosed for a long time ensuring dispersal over a long period of time. However, besides an incomplete apical dehiscence, another dispersal mechanism may take place here, i.e. zoochory, as the seeds may be embedded in a sweet pulp. *Apeiba echinata* shows the highest number of individuals (142). *Apeiba tibourbou* is a species which also occurs in arid regions and in secondary forest, hence the low number of individuals.

Verbenaceae

The Verbenaceae frequently produce drupaceous fruits.

1. *Citharexylum macrophyllum* = Totumillo blanco

Hab.: RF	Height: A
Fruit: Drupe	Ind.: 23
Disp.: Zoochorous	Fr. time:

2. *Vitex stahelii* = Totumillo

Hab.: RF	Height: A
Fruit: Drupe	Ind.: 51
Disp.: Zoochorous	Fr. time: 5–6

3. Cenicero

Hab.: Deciduous forest	Height: A
	Ind.:
Fruit: ?	Fr. time:
Disp.: ?	

1. *Citharexylum macrophyllum:* The ovoid drupe measures 1 × 0,7 cm, and has a smooth red surface. The calyx is cup-shaped and 0,4–0,5 cm long. The fleshy fruit contains a single seed. The drupe is dispersed by animals (e.g. birds). In Surinam, the species is common and occurs in the marsh forest too.

2. *Vitex stahelii:* The 2 cm long and 1,4 cm broad ovoid drupe of the size of an olive has a tough texture and a hard endocarp. The fruit is completely surrounded by the persistent 5-dentate calyx. The mesocarp is edible and the fleshy drupe dispersed by animals. The species is uncommon in Surinam.

3. Cenicero: The fruit is unknown.

The 3 species of Verbenaceae studied are represented by high trees of height category A. Their fruits are, nevertheless, dispersed by animals. However, their number of individuals in the forest studied is very low. This fact proves once more that dispersal by wind or autochorous dispersal of fruits and seeds in high trees is more advantageous than dispersal by animals.

Violaceae

The unilocular ovary is composed of 3 carpels. The capsules dehisce with 3 valves, while berries are rare. The placentation is parietal, each valve bearing many to one seed. Dehiscence is loculicidal. The seeds are often furnished with an aril, but sometimes are winged, e.g. in some woody lianas.

1. *Paypayrola longifolia* = Gaspadillo negro

Hab.: RF	Height: a
Fruit: Capsule	Ind.: 91
Disp.: Autochorous	Fr. time:

2. *Rinorea riana* = Gaspadillo marrón

Hab.: RF	Height: aa
Fruit: Capsule	Ind.:
Disp.: Autochorous	Fr. time:

1. *Paypayrola longifolia:* The small 3-angular capsule opens with 3 valves. Fruit length is about 2 cm, width 1,9 cm. The green fruit only contains few seeds. The dispersal mechanism is autochorous. The species very frequently occurs in British Guiana.

The outer fruit epidermis is thin-walled. The outer part of the mesocarp consists in parenchyma composed of roundish thin-walled cells (Fig. 102). Vascular bundles are dispersed in the parenchymatous ground tissue. The dilatation growth becomes very obvious in the outermost layers of the mesocarp by very clear tangential cell rows. The cells here divide only anticlinically to the surface. Towards the inside the tissue becomes somewhat looser and the cells enlarge. The cells in contact with the endocarp become partly compressed. The endocarp is proportionally prominent, consisting of elongated fiber-like sclereids which are arranged in 3 different layers. The outermost layer is composed of a single row of sclereids which pass in a longitudinal direction. Towards the inside follow several layers (about 15) of sclereids which are mostly arranged in a tangential direction crossing those of the outermost layer. The innermost endocarp layer corresponding to the inner epidermis of the pericarp shows a single row of sclereids elongated longitudinally, parallel to those of the outermost endocarp layer. The small tree is rare in Surinam.

2. *Rinorea riana:* The woody explosive capsule is elliptic to globular, but slightly 3-angular, and measures 2 x 1,3 cm. The surface is covered by a fine brownish indumentum, equally as that of the seeds. Each of the 3 valves encloses a single seed before dehiscence. The seeds measure about 0,6–0,7 cm in diameter. The dispersal mechanism is autochorous. The species very frequently occurs in British and French Guiana, but is rare in Surinam.

The Violaceae studied are of very little importance in the floristic composition of the forest studied. Both species have an autochorous dispersal mechanism. The number of individuals is very low. However, the higher tree shows a higher number of individuals.

Vochysiaceae

The ovary of the Vochysiaceae is usually composed of 3 carpels. It may be 1–3 locular and contains numerous seeds. Loculicidal capsules are often developed, but seldom are indehiscent fruits with only one seed.

1. *Erisma uncinatum* = Mureillo

Hab.: RF	Height: A
Fruit: Samara	Ind.: 487
Disp.: Anemochorous	Fr. time: 2/9

2. *Qualea dinizii* = Guarapo

Hab.: RF	Height: A
Fruit: Capsule	Ind.: 22
Disp.: Anemochorous	Fr. time: 2/5–6/8

3. *Vochysia lehmannii* = Canelito

Hab.: RF	Height: A
Fruit: Capsule	Ind.: 135
Disp.: Anemochorous	Fr. time: 3/9

4. *Vochysia tetraphylla* = Lorjena

Hab.: RF	Height: A
Fruit: Capsule	Ind.: 1
Disp.: Anemochorous	Fr. time: 2

1. *Erisma uncinatum:* The winged fruit is of leathery consistency. The calyx increasing in size during fruit development is persistent, while the fruit remains indehiscent. The fruit is egg-shaped and reaches 1,5–2 cm in length. The 4 sepals develop

unequally, one surpassing the 3 other ones in size. The grown-up sepals are of woody consistency, green and blue to violett in colour, and are distinguished by conspicuous veins. The largest sepal reaches up to 7 cm in length. The number of seeds is reduced to one. The samara is dispersed by wind. In Surinam the species sparcely occurs in the rain forest.

2. *Qualea dinizii:* The loculicidal capsule opens with 3 valves and contains 3 winged seeds (Fig. 103). The woody capsule is egg-shaped to elongated measuring 3,7–5 cm in length and 2,2–2,5 cm in width, and has a rough gray surface. Dispersal of the winged seeds takes place by wind. In Surinam the species is common and occurs in the savanna forest too.

The outer epidermis of the pericarp is small-celled. Beneath lies parenchyma in which small stone cells are scattered. The fundamental tissue consists in a cartilaginous collenchyma (Fig. 107). Fiber-like sclereids arranged in radial bundles run through the mesocarp (Fig. 108). Slime cells of giant size with stratified walls accompany the larger vascular bundles (Fig. 106). Towards the inside reappear sclereids either solitary or in groups. The endocarp is composed of sclereids elongated parallel to the fruit long axis (Fig. 108).

3. *Vochysia lehmannii:* The elongated loculicidal capsule is triangular and contains 3 winged seeds. Dispersal of the seeds takes place by wind.

The outer epidermis of the pericarp is small-celled. Several layers of small-celled parenchyma follow. A small region of angular collenchyma lies beneath. The major part of the mesocarp is occupied by a large-celled parenchyma in which very large rhexigeneous lacunas occur. The vascular bundles are concentric with xylem surrounded by phloem. In contact with the vascular bundles are large slime cavities which probably pertain to the phloem (Fig. 104). The tissue in contact with the endocarp is thin-walled and the cells are elongated parallel to the surface. The encocarp is composed of prosenchymatous thicker-walled cells likewise elongated parallel to the surface (Fig. 105). A conspicuous cuticle borders the seed cavity.

4. *Vochysia tetraphylla:* The green loculicidal capsule measures 3,5 × 1,3 × 1 cm. Each of the 3 locules contains a single seed. Three ribs are prominent on the fruit which opens with 3 valves. The seeds are winged. Dispersal of the seeds takes place by wind. In Surinam, the species very frequently occurs in the marsh forest at the riverside and along creeks.

The Vochysiaceae occupy a subordinate position in the rain forest studied concerning the number of existing individuals (645 altogether). *Erisma uncinatum* shows the highest rate of distribution with 487 individuals, followed by *Vochysia lehmannii* with 135 individuals. *Qualea dinizii* contributes 22 individuals and *Vochysia tetraphylla* only one. The original fruit type of the Vochysiaceae is represented by a capsule opening with 3 valves and releasing winged seeds. Fruit indehiscence found in *Erisma uncinatum* is probably of secondary nature. Consequently, the seeds do not develop wings, but the calyx substitutes the wings of the seeds. Possibly, this type of dispersal is more advantageous. Including the calyx into consideration, the fruit of *Erisma uncinatum* is the largest. However, it has a low specific weight, the ripe gynoeceum being small. Possibly, the fruit may be carried away further by wind (over longer distances) than the flat winged seeds of the other species of Vochysiaceae studied. However, other facts may be the reason of the respective numbers of individuals. At any rate, dispersal by wind is common to all 4 species which are all represented by high trees. Noteworthy is the sticky resinous content of the large secretory cavities in the pericarp of *Vochysia lehmannii, Qualea dinizii* and *Erisma uncinatum.* In *V. lehmannii* and *Qualea dinizii*, the secretory cavities are in contact with the vascular bundles. The pericarp of *Qualea dinizii* is hard and woody due to its richness in sclerenchymatous elements, partly originating from the xylem and partly from the well developed endocarp. The brittle pericarp of *V. lehmannii*, on the other hand, has smaller vascular bundles and a small endocarp. Large rhexigeneous lacunas in the mesocarp render the pericarp even more fragile. In the Vochysiaceae studied it becomes very obvious that wind dispersal is advantageous in high trees.

Unidentified species without family

"Cachicamo": 93 individuals, height Category A
"Casabe amarillo": 4 individuals, drupe with a single seed
"Congrio amarillo": 41 individuals
"Jebe negro": 1 individual, fruit a samara, height category a
"Mandingo montañero": 7 individuals, height category A
"Palo colorado": 40 individuals
"Picatón negro": 13 individuals, height category A
"Pilón": 19 individuals, height category A
"Pilón sabanero": 3 individuals, savanna, RF
"Sabañon": 17 individuals, height category A
"Tasajo": 5 individuals, height category A, deciduous forest, RF

These unidentified species with very few representatives in the forest studied will be neglected in the further studies *.

Number of species and individuals studied and their height categories

For a survey over the families studied with their number of species and individuals see Table 1. In this Table all the families studied are integrated indicating the number of species and individuals with the height category to which they belong. In total 58 families were studied with 375 species and 66.977 individuals. Of the 375 species 233 belong to height category A, 107 to height category a, and only 19 to height category aa. Of the 66.977 individuals 45.543 belong to height category A, 18.814 to height category a, and 2.262 to height category aa. Families very well represented concerning their number of species and individuals are:

Lecythidaceae	with 11 species and 8.824 individuals
Euphorbiaceae	with 24 species and 6.873 individuals
Sapotaceae	with 21 species and 6.550 individuals
Mimosaceae	with 33 species and 5.616 individuals
Burseraceae	with 12 species and 5.449 individuals
Rosaceae	with 15 species and 4.797 individuals
Apocynaceae	with 9 species and 3.047 individuals
Guttiferae	with 14 species and 2.513 individuals
Meliaceae	with 10 species and 2.511 individuals
Papilionaceae	with 20 species and 2.145 individuals
Caesalpiniaceae	with 25 species and 1.831 individuals
Myrtaceae	with 14 species and 1.554 individuals
Sapindaceae	with 6 species and 1.125 individuals
Boraginaceae	with 7 species and 1.114 individuals
Lauraceae	with 14 species and 968 individuals

About 68% of the total individuals belong to height category A, 28,09% to height category a, and only 3,37% to height category aa. Some families are distinguished by very high trees. Most species and individuals of the Apocynaceae, Boraginaceae, Lecythidaceae, Papilionaceae, and Rosaceae belong to height category A. In a further chapter we have to study which are the dispersal mechanisms of the different families and whether there are relations between the height of the trees and their dispersal methods.

Number of species and individuals studied and their dispersal methods

As Table 2 indicates, it is distinguished between anemochory, autochory, auto-zoochory, and zoochory. As autochorous are designated all plant species which have dehiscent fruits. As auto-zoochorous are considered those species which have dehiscent fruits, but are additionally dispersed by animals be it through fleshy arils surrounding the seeds or by fleshy fruit parts.

In total, 50 species with 8.088 individuals are anemochorous, 66 species with 13.845 individuals are autochorous, 57 species with 20.829 individuals are auto-zoochorous, and 197 species with 24.115 individuals are zoochorous.

13,33% of all species are thus anemochorous
17,50% of all species are thus autochorous
15,20% of all species are thus auto-zoochorous, and
52,53% of all species are thus zoochorous.

However, when we consider the number of individuals percentages change in the following way:

12,07% of all individuals are anemochorous
20,67% of all individuals are autochorous
31,09% of all individuals are auto-zoochorous, and
36,00% of all individuals are zoochorous.

* The tables 1–4 should have been placed here, but by mistake of the printery they were inserted on pages 39–48. The reader is referred to these pages.

The percentage values of anemochorous species and individuals are more or less maintained. The autochorous dispersal method has somewhat gained in percentages, when the number of individuals is concerned. Likewise, the auto-zoochorous method has gained almost by the double, while the zoochorous method has lost quite a bit in percentages when individuals are considered. Possibly, we may draw the conclusion from these observations that autochory and auto-zoochory with increasing numbers of individuals are advancing methods, while zoochory with diminishing numbers of individuals is on the descending line, the more so as it is considered the more original form of dispersal. Anemochory and autochory, on the other hand, are possibly derived dispersal methods. At least in certain families a certain kind of autochory seems to be derived. A share of 12% anemochorous individuals is a relatively high proportion as compared with other rain forests. Likewise the percentage of autochorous individuals is high. As autochorous were considered those species which have dehiscent fruits. Whether autochory is effective in all examples presented here cannot be examined in these studies. The auto-zoochorous species have dehiscent fruits, but dispersal rather takes place by animals. We thus could unite the auto-zoochorous species with the exclusively zoochorous ones to comprise all species actively dispersed by animals and would come to a percentage of 67. This seems to be a realistic percentage, the more so as within the autochorous species there may be found some more which are rather dispersed actively by animals. But this part will be discussed later.

In the following, the families studied with a larger number of individuals and their dispersal mechanisms are considered.

Within the Annonaceae we observe that zoochory prevails over autochory. The zoochorous species with most individuals is represented by the lowest tree of height category aa. Of the Apocynaceae 6 species are anemochorous including 3.022 individuals, while the 3 zoochorous species only contribute 25 individuals. All 6 species of the Bignoniaceae are anemochorous contributing 899 individuals. Within the Bombacaceae, one species contributes most individuals. This is autochorous. Most species of the Boraginaceae are zoochorous including 1.098 individuals. All 12 species of the Burseraceae are auto-zoochorous with 5.449 individuals. Within the Celastraceae, the autochorous species has most individuals. In the Combretaceae, the 2 anemochorous species have most individuals. The zoochorous species of the Dichapetalaceae with drupaceous fruits has a very elevated number of individuals. Of the Elaeocarpaceae the 3 auto-zoochorous species contribute most individuals, while the 3 autochorous species relatively few. Most species with most individuals of the Euphorbiaceae are autochorous, while the zoochorous species and individuals are few. In the Flacourtiaceae, on the other hand, zoochory prevails in the number of species and individuals. In the Guttiferae, the number of zoochorous species prevails (10), while the plants with autochorous dispersal methods are abundant with 2.233 individuals. This is mainly due to one species with 1.893 individuals. In the Lauraceae, all 14 species are zoochorous, but the number of individuals is relatively low. In the Lecythidaceae, 2 species with 669 individuals altogether are anemochorous, while the 8 auto-zoochorous species contribute 8.002 individuals.

In the Mimosaceae, there is only one anemochorous species with 185 individuals; there are 4 autochorous species with 118 individuals, 5 auto-zoochorous species with 3.328 individuals, and 20 species with 1.985 individuals. The zoochorous species thus prevail. In the Caesalpiniaceae, the anemochorous species increased to 4 with 631 individuals (as compared with the Mimosaceae), the autochorous species increased to 8 with 455 individuals, the auto-zoochorous species remained 5 with 236 individuals, and the 8 zoochorous species comprise only 509 individuals.

In the Papilionaceae, the numbers of species and individuals changed accordingly, as compared with the Caesalpiniaceae and Mimosaceae. The anemochorous species increased to 12 with 559 individuals, the 3 autochorous species contribute 1.450 individuals, only 2 auto-zoochorous species with 106 individuals are present, and only 3 zoochorous species produce not more than 30 individuals.

The 5 species of Melastomaceae are all zoochorous. Within the Meliaceae there are 2 anemochorous species, one autochorous species, and 7 auto-zoochorous species which contribute most individuals (2.409). All 12 species of the Moraceae are zoochorous, but the number of individuals is low (518).
The 3 species of Myristicaceae are auto-zoochorous with only 114 individuals. The 14 species of Myrtaceae are all zoochorous including 1.554 individuals.

The Nyctaginaceae, Ochnaceae, Olacaceae, Opiliaceae, Palmae, and Quiinaceae are zoochorous exclusively, but their number of species and individuals is very low. The 3 important species of Polygonaceae are likewise zoochorous, but with a low number of representatives. The 15 species of Rosaceae are all zoochorous contributing 4.740 individuals. Two species of the Rubiaceae are autochorous, while 5 are zoochorous. However, the number of individuals is very low. The 9 species of the Rutaceae studied are all autochorous, but contribute very few individuals (342 altogether). Three species of the Sapindaceae are zoochorous, 2 are autochorous, and only one species is anemochorous. However, this species contributes most individuals (795).

The 21 species of Sapotaceae are all zoochorous with 6.550 individuals. The 2 species of Simaroubaceae are zoochorous with relatively few individuals. Two species of the Steruliaceae are autochorous, while the other 2 species are auto-zoochorous. However, the autochorous species provide 10 times more individuals. Three species of the Tiliaceae are auto-zoochorous. But the number of individuals is very low. The 2 species of Verbenaceae are zoochorous with only 74 individuals. The 2 autochorous species of Violaceae contribute only 91 individuals. All 4 species of Vochysiaceae are anemochorous with 645 individuals.

Families with a high number of anemochorous individuals (over 500) are thus: the Apocynaceae (3.022), the Bignoniaceae (899), Lecythidaceae (669), Caesalpiniaceae (631), Papilionaceae (559), Sapindaceae (795), Vochysiaceae (645).

Families with a high number of autochorous individuals are: Bombacaceae (767), Euphorbiaceae (6.391), Guttiferae (2.233), Papilionaceae (1.450), Sterculiaceae (908).

Families with a high number of auto-zoochorous individuals are: the Burseraceae (5.449), Lecythidaceae (8.002), Mimosaceae (3.328), Meliaceae (2.409).

Families with a high number of zoochorous individuals are: Annonaceae (609), Boraginaceae (1.098), Dichapetalaceae (1.527), Lauraceae (968), Mimosaceae (1.985), Caesalpiniaceae (509), Moraceae (518), Myrtaceae (1.554), Rosaceae (4.740), Sapotaceae (6.550).

Anemochory is represented in 19 families. Autochory is represented in 18 families. Auto-zoochory is represented in 16 families, and zoochory is represented in 40 families. Zoochory is thus most common among plants of the tropical humid forest studied.

When we compare these results with the data of other authors, we recognize some coincidence as well as some differences. Hall & Swaine (1981) found winged or plumed propagules (anemochorous) in about 20-30% of tall trees and lianas, and in only 5% of understory trees of a forest in Ghana. The same authors indicate that about 10-15% of the species have explosive fruits (autochorous) regardless to which stratum they belong. About 70% of tall trees and climbers have fleshy fruits. However, the proportion of species with fleshy fruits is higher among understory trees (80%). For a forest in Panamá it is estimated that about 80% of the canopy and understory trees are dispersed by animals (Foster 1973, cited by Karr 1976).

However, in the present studies anemochory and autochory show relatively high species percentages as compared with other humid tropical forests studied, while the zoochorous and auto-zoochorous species united do not reach the percentage of 80, as indicated above by Foster, but only 67,73%.

Plant families and their dispersal methods related to height categories

When checking all the families on our list (see Tables 1 and 2), it becomes quite obvious that certain families are distinguished by their special dispersal methods, e.g. the Bignoniaceae by wind-dispersed seeds, the Sapotaceae with fleshy fruits dispersed by birds and bats. Most of the Anacardiaceae studied are dispersed by animals, except one species which is wind-dispersed. This is the highest tree. In total, 5 species of the Anacardiaceae were studied, 4 of which belong to height category A and one to a. Three species have drupaceous fruits and one develops nuts. These 4 species are dispersed by animals. In the drupe and the nut, as well as in the anemochorous samara the number of seeds is reduced. It is probably not by accident that the high trees have fruits with a reduced seed number. These fruits are derived forms and the height of the trees is possibly also a secondarily developed character. Of the Annonaceae it is well known that most species have edible and animal-dispersed fruits. However, 2 species studied are autochorous. One species belongs to height category aa, and this species is zoochorous. Two species belong to height category A and 4 to a. Five of these species are zoochorous. The animal-dispersed species of layer aa reaches the highest number of individuals, possibly because this is the layer most adequate for zoochory. Most species of the Apocynaceae studied are anemochorous – these are all represented by high trees. However, 3 species are zoochorous and one of them belongs to height category a, the other 2 are of height category A. Their number of individuals is very low. Of the 6 anemochorous species one reaches 1.760 individuals. The 3 species of Araliaceae studied are animal-dispersed and probably all 3 are of height category a. Their number of individuals is very low. Of the 6 species of Bignoniaceae 5 belong to height category A and only one to a. All 6 species are anemochorous. All 3 species of Bombacaceae studied are high trees of height category A, and 2 of them are anemochorous, while one is autochorous. All of them have capsular fruits, but in one species the number of seeds is reduced to one. All 7 species of Boraginaceae are high trees of height category A, but only one is anemochorous, while 5 are zoochorous. Four of these species develop drupes with a reduced number of seeds, in other words, derived fruit forms. Of the Burseraceae 12 species were studied. All of them are auto-zoochorous having drupaceous capsules which dehisce. Six species belong to height category A and 6 to a. Again it is worth mentioning that the drupaceous fruits have a reduced seed number and represent derived fruit forms. Of the 3 Capparidaceae studied one species belongs to height category A, another to a, and the third to aa. All 3 are zoochorous. The highest tree bears drupes, while the smaller trees develop berries. All 3 species of Celastraceae belong to height category A, one is anemochorous, another autochorous, and the third zoochorous. *Cochlospermum* sp. is worth mentioning in so far, as this tree comes from the semideciduous and tropophilous forest and has an anemochorous dispersal mechanism, characteristic of drier and more open habitats. All 3 species of Combretaceae are of height category A. Two of them are anemochorous with single-seeded samaras, while one is zoochorous having a single-seeded drupe. All 3 fruit forms are, consequently, derived. All 3 species of Ebenaceae studied are high trees of height category A and zoochorous at the same time. All the 6 species of Elaeocarpaceae studied belong to the genus *Sloanea*. Four are of height category A, while 2 belong to height category a. All 6 species form capsules and are, consequently, autochorous. However, 3 species are also dispersed by animals, as the seeds are furnished with an aril. Of the 24 species of Euphorbiaceae studied 12 belong to height category A, 9 to a, and only one to aa. 21 species have capsules and, therefore, use an autochorous dispersal mechanism, but 3 develop drupes and are zoochorous. Two of them belong to height category A. Here again we meet with a derived fruit form in high trees. Of the 8 species of Flacourtiaceae studied 3 are of height category A, 4 of a, and one of aa. Five are zoochorous, 3 of which belong to height category a, one to A and another one to aa. One species of height category a is autochorous, while 2 species of height category A are auto-zoochorous. Of the 14 species of Guttiferae studied 7 are of height category A and 6 of a. Three species of height category a are autochorous, having capsules. One species is auto-zoochorous having arils on the seeds. This species is of height category A. The remaining 10 species are zoochorous. Four of these have drupes and belong to height category A, while 6 have berries. Three of these are of height category a, while 2 belong to height category A. *Hernandia guianensis* of height category A has a single-seeded nut, a derived fruit form. The zoochorous drupe of *Saccoglottis cydonioides* is a derived fruit form. The tree belongs to height category A. Of the 14 species of Lauraceae studied 6 belong to height category A and 5 to a. All species develop drupes, i.e. derived fruit forms. Of the 11 species of Lecythidaceae studied all are very high trees of height category A except one of height category aa. Two species are anemochorous, while all the other species are autochorous having pyxidia, except the zoochorous *Gustavia augusta*. At the same time they are probably also zoochorous. Of the 33 species of Mimosaceae studied 19 belong to height category A, 11 to a, and only 3 to aa. 23 species have indehiscent pods and only 10 species have dehiscent fruits. Only one species is anemochorous (of height category A), 4 are autochorous, 3 of which are of height category A and one of a. Of the 5 auto-zoochorous species 3 belong to height category A, one to a, and another one to aa. Of the 23 zoochorous species 2 are of height category aa, 9 of a, and 12 of A. Anemochorous and autochorous species are thus preferably high trees, while very low trees of height category aa are usually zoochorous. Of the 25 species of Caesalpiniaceae studied 20 are of height category A, 5 species belong to height category a. In total, 8 species are autochorous, 4 are anemochorous, 8 zoochorous, and 5 auto-zoochorous. Of the species belonging to height category a, 2 are autochorous, one is anemochorous, and 2 are zoochorous. Most of the anemochorous, autochorous, and auto-zoochorous species are, consequently, high trees. All 20 species of the Papilionaceae studied are of height category A. The Papilionaceae is considered the most derived family. Three of the 20 species are autochorous, as many as 12 are anemochorous, 3 only are zoochorous, and 2 auto-zoochorous. It is quite evident that within the Leguminosae studied evolution has led towards anemochory of high trees. Within the Malpighiaceae (5 species) we observe formation of derived fruit forms (drupes) with a reduction of the seed number in the high trees (height category A), while the samara belongs to a tree of height category a. Of the 5 species of Melastomaceae, 3 belong to height category A, one to a, and another one to aa. All 5 species form berries which are zoochorous. However, *Miconia kappleri* is usually a small tree. The fruit of *Mouriria huberi* is on the way to transform from a berry into a single-seeded drupe. Likewise in *M. sideroxylon* transformation of a many-seeded berry into a drupe with 3 stones is on the way. *M. huberi* and *M. sideroxylon*, both of height category A, have thus derived fruits. The 10 species of Meliaceae are represented by 3 high trees of height category A, and 6 trees of height category a. The fruits are capsular in all species and the dispersal mechanism is autochorous for that. However, 7 species are besides zoochorous, their seeds being furnished with an aril. Of the 12 species of Moraceae studied only 3 belong to height category a, while 5 are of height category A. All species are nevertheless zoochorous having fleshy fruits or infrutescences. The 3 species of Myristicaceae studied belong to height category A and have capsules. The autochorous dispersal mechanism is, however, aided by zoochory, as the seeds are furnished with an aril. Of the 14 species of Myrtaceae 10 belong to height category a, and only 4 to height category A. All are zoochorous and have drupes. Their rate of distribution is, however, low. The 2 species of Nyctaginaceae, one belonging to height category A, the other to a, have single-seeded nuts which are considered derived fruit forms. Another example of seed reduction in high trees is the drupe of *Agonandra brasiliensis*, Opiliaceae. The 7 species of Palmae studied form all drupes. Four species belong to height category a, while 3 are of height category A. The 7 species are all zoochorous and usually do not attain a very great height. Within the Polygonaceae we observe a clear distinction between 3 zoochorous species of lesser height, and one anemochorous species with a single-seeded samara of height category A. The 15 species of Rosaceae are probably all of height category A and develop derived fruits in the form of – usually – single-seeded drupes. Of the 8 species of Rubiaceae studied 4 belong to height category A, one to a, and 3 to aa. Of the 4 species of height category A 3 are autochorous, while one is zoochorous. The lower and low trees of height category a and aa are all dispersed by animals. Five of the 9 species of Rutaceae studied, are of height category A, one belongs to height category a, and 2 are of the lowest height category aa. All species have dehiscing capsules with a very low number of seeds. And all species have a very low number of individuals. Of the 6 species of Sapindaceae studied 3 belong to height category A, 2 to a, and one to aa. The 3 highest trees form drupes dispersed by animals. Of the 2 species of height category a, one is anemochorous, the other autochorous. And the smallest tree of height category aa is likewise autochorous. 15 of the 21 species of Sapotaceae studied belong to height category A, while 5 are of height category a. All species are zoochorous, having few-seeded berries or single-seeded drupes. Irregularities in the number of locules and seeds indicate the progressive reduction in seed number and the continuous transformation of a several-seeded berry into a single-seeded drupe. The 2 species of Simaroubaceae belonging to height category A form likewise drupes. Of the 4 species of Sterculiaceae 3 belong to height category A, while one is of aa. All 4 species are autochorous having capsules, while 2 are zoochorous besides. Of the 4 species of Tiliaceae 3 are of height category A, while one is of a. Three species are auto-zoochorous, while one of height category A is anemochorous. The 3 species of Verbenaceae are all of height category A and 2 of them have derived fruits in the form of drupes. The 2 species of Violaceae are autochorous having capsules. One species is of height category a, the other one of aa. And the 4 species of Vochysiaceae studied, finally, are all of height category A and are anemochorous at the same time, although only one species develops a samara, while the other 3 species have capsules with winged seeds. This is probably the more original fruit type, while the samara is more derived.

We may thus summarize the following results. Anemochorous (with anemochorous fruits or seeds) are mostly high to very high trees so that anemochory is almost restricted to the uppermost forest layer (Bignoniaceae, Apocynaceae, Vochysiaceae). This phenomenon explains itself by the fact that wind and air turbulence mainly act in the uppermost story where wind-dispersal is, consequently, most advantageous. Otherwise anemochory is more common in drier and more open habitats. Another result is that autochory by means of dehiscing fruits is also much more common in high trees of the upper forest layers (Euphorbiaceae, the genus *Sloanea*, Caesalpiniaceae). Autochory seems to be advantageous when the seeds fall from a great height, in this way they may be removed further from the parent tree. Explosive mechanisms are certainly most effectful and the best example of this type is *Hura crepitans*. Dehiscing capsules of high trees are usually devoid of an aril, possibly by reduction. The auto-zoochorous species we could possibly consider more primitive, as the capsules are fleshy or the seeds have an aril, while dry capsules with seeds devoid of an aril would be more advanced. We furthermore observe that high trees often develop derived fruit forms such as single-seeded drupes (Rosaceae, Sapotaceae, Lauraceae, Boraginaceae, Burseraceae, Malpighiaceae, Melastomaceae), nuts or samaras (anemochorous high trees), all single-seeded or with a reduced number of seeds. Within the Leguminosae we observe a certain development starting with a majority of species with indehiscent fruits in the Mimosaceae, over autochorous species with dehiscent fruits in the Caesalpiniaceae, and ending up with anemochory in the Papilionaceae. Auto-zoochorous and zoochorous species are mostly trees of the lower forest layers, except the above mentioned examples with derived fruit forms. There is thus a certain layering of the forest becoming obvious concerning fruit types and their dispersal mechanisms. This layering depends on the fauna of the forest and its prefered stories, on the one hand, and on environmental influences, on the other.

Important characteristics of certain families, genera or species and their number of individuals

Although *Bravaisia integerrima*, Acanthaceae, has a dispersal mechanism advantageous for high trees and in spite of the fact that it is a resistant species, supporting drought, and has a profound radical system, adapting to different soils and climates, it has only 2 representatives in the forest studied.

Within the Anacardiaceae there is one anemochorous species represented by a high tree. The other 4 species are zoochorous, 3 of them being high trees and from there the low number of individuals may be explained. *Spondias mombin*, however, has a much higher rate of distribution in more open regions and in the secondary forest. The fruit types of all species studied are derived for the reduced number of locules and seeds. The lowest zoochorous tree has the highest number of representatives.

Within the Annonaceae the zoochorous species represented by the lowest tree shows the highest number of individuals, whereas the 2 highest zoochorous tree species have least individuals. The 2 autochorous species of height category a have higher numbers of individuals than the zoochorous species of the same height category a.

The 3 zoochorous species of Apocynaceae (2 of height category A and one of a) have a very low rate of distribution. The other 6 species, all of height category A, are anemochorous and partly have high numbers of individuals (1.760, 650, 550) which fact proves again that wind-dispersal is advantageous in high trees.

The 3 zoochorous species of Araliaceae, all probably of height category a, have very low numbers of individuals.

Five of the 6 species of Bignoniaceae studied belong to height category A and one to a. All species are anemochorous, as most species of the Bignoniaceae. Abundance and ecological importance of the Bignoniaceae is greatest in dry areas, but only abundance of individuals, while species diversity is higher in wet than in dry forest (Gentry 1974, 1976).

Two of the species of Bombacaceae studied are anemochorous, while one is autochorous. All 3 are high trees (height category A). The autochorous species has most individuals (767).

All 7 species of Boraginaceae studied are high trees of height category A, but only one is anemochorous. In 4 species the fruits are drupes with a reduced number of seeds. These species have the highest numbers of individuals. The species with the berry fruit has a low number of representatives. Possibly, the ancestors of *Cordia* species had fleshy zoochorous fruits. Animal dispersal is common in shrubs and understory trees, while the higher trees are dispersed by wind. Species represented by shrubs use birds as their dispersers, species of medium height are dispersed by mammalia. Sticky *Cordia* fruits with slimy contents are epi-ornithochorous (Opler et al. 1975). The drupaceous fruits of the high *Cordia* species are considered derived on account of their seed reduction.

The 12 species of Burseraceae studied have all zoochorous drupaceous fruits. At the same time their dispersal mechanism is autochorous, as the fruits are dehiscent. An obvious reduction of seeds is apparent in all species. Six species are of height category A and 6 of a. The highest number of individuals has a species of height category A. Whether the fruits were originally capsular being on the way to transform into indehiscent drupes or whether they were berry-like transforming into few-seeded capsules or schizocarps so that zoochory becomes converted into autochory, is left open.

Two of the species of Capparidaceae studied have berry fruits, one of which belongs to height category a and the other to aa, whereas the highest tree of height category A has a derived fruit in the form of a drupe.

Caryocar nuciferum with its "giant-drupe" is of height category A. The fruit form is considered derived so that progressive reduction of stones and seeds may be observed.

The 3 species of Celastraceae studied are all of height category A. Although having 3 different fruit types, the fruits are derived to some degree in all 3 species. One species has a few-seeded berry showing reduction of seeds, the capsular fruit of the second species is likewise few-seeded, and the third species develops a samara with a single seed. The species with the capsular fruits shows the highest number of individuals.

Cochlospermum with anemochorous seeds is a tree characteristic of more open and drier habitats in the semideciduous and tropophilous forest.

Of the 3 species of Combretaceae studied one is zoochorous having single-seeded drupaceous fruits, the other 2 develop anemochorous samaras. One of the anemochorous species, *Terminalia amazonia*, is represented by the highest number of individuals. It has the smaller samaras, and in Mexico the tree reaches 70 m in height. The zoochorous species fruits in August, while the 2 anemochorous species fruit in February/May and May respectively. All species belong to height category A.

Tapura guianensis, Dichapetalaceae, of height category A has a derived fruit in the form of a drupe. Its number of individuals is high (1.527).

The 3 species of *Diospyros*, Ebenaceae, have zoochorous berries although they are of height category A. However, *D. guianensis* has poisonous fruits of mimetic character. Even so, this species has most individuals as compared with the other 2 species (176). It fruits in February, while the other 2 species fruit in June and December respectively.

Four of the 6 species of *Sloanea*, Elaeocarpaceae, are of height category A, while 2 are of a. They develop dehiscing capsules with a reduced number of seeds. An autochorous dispersal mechanism seems to be advantageous for the high trees. With certainty, 3 of the species studied have arillate seeds which are dispersed by animals. It is supported here that few-seeded capsules in high trees correspond to derived fruit forms.

Of the 24 species of Euphorbiaceae studied only 3 develop drupes and are zoochorous, all the other species have capsules and are autochorous. Twelve belong to height category A, 9 to a, and only one to aa. According to Van der Pijl, zoochory (saurochory) is an original feature in Euphorbiaceae. This would signify that dehiscing capsules in the Euphorbiaceae are derived fruit forms. This speculation coincides with my observation that high trees frequently have an autochorous dispersal mechanism by dry dehiscing capsules. *Hura crepitans* with a very complicated dehiscence mechanism is characteristic of more open and drier habitats. Its fruit certainly represents a derived form. The Euphorbiaceae would thus provide an example that dry dehiscing capsules in high trees are derived fruit forms adapted to the height of the tree, as an autochorous dispersal mechanism is more advantageous in the upper forest stories than dispersal by animals.

Of the 8 species of Flacourtiaceae studied 3 belong to height category A, 4 to a, and only one to aa. Five species are zoochorous, 2 are auto-zoochorous, and only one autochorous. The species of height category aa has zoochorous berries. The 2 auto-zoochorous species are of height category A. It is possible that the berry represents the more original fruit type, while capsules are more derived. This suggestion becomes the more probable as the capsular fruits are found in high trees and because the number of seeds is reduced to one in *Homalium racemosum*. But it remains doubtful whether the berry or the fleshy capsule possibly with arillate seeds is more original.

Seven of the 14 species of Guttiferae are of height category A, and 6 of height category a. Six species develop berries, 4 have drupes, and 4 capsules. The species with drupes are all of height category A. The drupe is considered a derived fruit form not infrequently found in high trees. *Caraipa richardiana* with capsular fruits has most individuals (1.893). Next follows an autochorous species with 264 individuals having capsular fruits. The 4 species with capsules are of height category a and possibly develop a more primitive fruit type, while drupes are considered derived.

The 14 species of Lauraceae studied have all zoochorous drupes. Six of them belong to height category A, while 5 are of height category a. The lower trees of height category a have more representatives in the forest which fact proves again that zoochory is more advantageous in lower trees. Some of the species have seeds protected against predators by bitter-tasting substances or by an astringent taste.

Ten of the 11 species of Lecythidaceae are represented by high trees of height category A, and only one is of height category aa. All species develop pyxidia. The 2 species of *Couratari* are anemochorous having winged seeds. *Gustavia augusta* of height category aa, on the other hand, is dispersed by animals. For the presence of dehiscing pyxidia 10 species have to be considered autochorous.

However, in the genus *Eschweilera* and in *Lecythis* the seeds are probably also dispersed by animals. Four species of *Eschweilera* have very high to high numbers of individuals (3.446, 2.374, 1.025, 803). Here again it becomes obvious that an autochorous dispersal mechanism seems to be favourable in high trees.

The Mimosaceae studied play a very important part in the composition of the forest concerning number of species and individuals (33 species with 5.616 individuals). However, most species are represented by few individuals. At least 7 species are characteristic of other habitats, hence the few representatives. The fruits of all species studied of *Inga* are considered indehiscent pods which fall to the ground where they are eaten by animals for the sweet pulp surrounding the seeds, no matter to which height category the trees belong, as dispersal takes place on the ground. The genus *Inga* is primitive and the ancient way of saurochory is still being observed within the genus, according to Van der Pijl. The number of individuals is, however, not very high in any studied species of *Inga.* Of the 33 studied species of Mimosaceae, 19 belong to height category A, 11 to a, and 3 to aa. Of these 33 species one is anemochorous, 4 are autochorous, 5 are auto-zoochorous, and 23 are zoochorous. Of the 3 lowest species of height category aa 2 are zoochorous and one is auto-zoochorous. Of the 4 autochorous species 3 are of height category A and one of a. The anemochorous species belong to height category A. It thus becomes obvious that high trees are preferably anemochorous or autochorous, while lower trees are zoochorous. Furthermore, zoochory prevails in the Mimosaceae. The species with most individuals is *Pentaclethra macroloba* with 2.534 individuals. Its fruits are explosive pods which throw the seeds as far as up to 10 meters. Autochory is combined with hydrochory and scatterhoarding by animals. Autochory is the more advantageous, as the tree may reach a height of up to 40 m. Germination of the seeds is very efficient, and seedlings of the species are abundant in the forest. The species is resistant growing on soils of low fertility reducing interspecific competition in this way. In prehistoric times the species had a much larger extension. These facts explain the frequency of individuals in the forest studied.

Of the 25 species of Caesalpiniaceae studied as many as 20 are of height category A, and only 5 of height category a. However, most species have a very low number of individuals. With 1.831 individuals altogether the Caesalpiniaceae are much less important than the Mimosaceae. Of the 20 species of height category A, 7 are zoochorous, 3 auto-zoorochous, 7 autochorous, and 3 anemochorous. Of the 5 species of height category a, 2 are zoochorous, 2 autochorous, and one anemochorous. This fact shows again that anemochory as well as autochory are more frequently found and more advantageous in high treees. The species which have most representatives of all Caesalpiniaceae studied are autochorous (with 434 individuals) and anemochorous (with 421 individuals) respectively. The wind-dispersed species shows the highest number of individuals in the A story, a further prove that anemochory is an adaptation of high story trees. Interesting differences in the fruiting time of 2 species of *Cassia* make obvious that anemochorous fruits are preferably dispersed during the dry season (January in this case), while zoochorous fruits are frequently dispersed during the wet season (August/November).

All species of Papilionaceae studied are of height category A. The Papilionaceae contribute 20 species with 2.145 individuals. Of the 20 species 3 are zoochorous, 2 auto-zoochorous, 3 autochorous, and 12 anemochorous. This fact shows again that anemochory is an adaptation of high trees to the environmental conditions. Furthermore, several of the anemochorous species are more characteristic of the semideciduous or deciduous forest or of the savanna. Most species contribute only few representatives. And only an autochorous species is represented by a high number of individuals (1.262).

Comparing the 3 families of Leguminosae with one another one observes that the number of species as well as the number of individuals declines from the Mimosaceae towards the Papilionaceae. The number of high trees, on the other hand, increases. While within the Mimosaceae still all 3 height categories are present, the Papilionaceae are composed only of trees of height category A. The number of anemochorous species markedly increases, while that of zoochorous species declines. Likewise the number of species with indehiscent pods declines. However, within this category also the anemochorous species with samaras are included. Species which are characteristic of more open and drier habitats show anemochory. Most species are represented by few individuals and the Mimosaceae as well as the Papilionaceae have only one species with a high or elevated number of individuals each. Zoochory thus seems to be more primitive within the Leguminosae, while autochory and anemochory seem to be derived dispersal mechanisms, especially adapted to high trees.

The 5 species of Melastomaceae studied are all zoochorous having berries. Three of them belong to height category A, while one belongs to a and another one to aa. The species with the highest number of individuals has a more derived fruit form (drupe).

Three of the 10 species of Meliaceae studied belong to height category A, while 6 belong to a. The fruits of all species are capsules. Most species have a low distribution rate in the forest except 2 autozoochorous species with high numbers of individuals. *Carapa guianensis* with 885 individuals is a tree characteristic of primitive forests. This may be the reason for the high number of individuals. It furthermore has a wide range of tolerance concerning the environment (rain forest, lowland, semidecidious forest, estuaries). The fruits may even be carried by currents and possibly be eaten by fish. Two species are anemochorous, one of which is *Cedrela odorata*, a high tree which may reach as much as 40 m. Its low number of individuals may be explained by the fact that the species prefers more open vegetation and easily colonizes new habitats. In Mexico it abundantly occurs in the secondary forest.

The 12 species of Moraceae studied are all zoochorous. Five of them belong to height category A, while 3 belong with security to a. In the other cases the height category is doubtful. However, concerning the number of individuals the Moraceae are very ill represented in the forest.

The 3 species of Myristicaceae studied are autozoochorous and of height category A. *Virola surinamensis* may originally be dispersed by fish, an ancient form of dispersal. All 3 species are dispersed by birds which are attracted by the colourful fleshy arils. The low number of individuals of the 3 species once again demonstrates that animal-dispersal in high trees is not advantageous.

Of the 14 species of Myrtaceae studied only 4 belong to height category A and 10 to height category a. All species develop single-seeded drupes dispersed by animals. The species with most individuals has showy red or orange fruits with a sweet flesh (*Eugenia compta*). In Surinam it is a shrub or small tree, especially apt for animal dispersal.

Of the 2 zoochorous species of Nyctaginaceae the one of height category a has a larger number of individuals than that of height category A. The same is true of the Ochnaceae.

Of the 7 zoochorous species of Palmae studied 3 belong to height category A, while 4 are of height category a. All species have very few representatives in the forest and partly belong to other habitats. The fruits of 2 species are occasionally water-dispersed.

The 3 species of *Coccoloba*, Polygonaceae, have berry-like fruits with modified fleshy tepals and are zoochorous. All 3 species are lower trees of height category a. The high tree of height category A, *Triplaris surinamensis*, has a samara and is wind-dispersed. Zoochory in lower trees and anemochory in high trees is advantageous, anemochory being an adaptation to the microclimate in the upper tree story.

The 15 species of Rosaceae studied form a very homogeneous group. They are almost all represented by high trees with zoochorous drupes of ovoid to pear-shaped form, measuring a few centimeters. The 7 species of *Licania* show partly very low numbers of individuals, except one species with 679 individuals and another one with 3.258 individuals. According to Prance (1973) savanna species of *Licania* are derived from lowland forest species. The predominance of lowland forest species proves that the divergence within the group has mostly taken place within the lowland forest habitat during dry periods so that isolation of species into separate populations took place. *Licania densiflora* with 3.258 individuals could be the most ancient species in the forest or it could be a very resistant species which reproduces very well. However, by the dispersal mechanism alone the differences in the number of individuals of the different species of *Licania* cannot be explained by any means.

Of the 8 species of Rubiaceae studied 4 belong to height category A, one to height category a, and 3 to aa. Of the 4 species of height category A, 3 are autochorous or anemochorous and only one is zoochorous. The lower and low trees are all zoochorous. Animal dispersal is kept in the lower trees where it is more efficient, while in higher trees an autochorous or anemochorous dispersal mechanism has developed by and by. The species with most individuals is autochorous.

Five of the 9 species of Rubiaceae studied belong to height category A, one to a, and 2 to aa. All species have dehiscing capsules with a low number of seeds and are thus autochorous. The number of individuals of all species is very low. As only 2 species are low trees of height category aa, it becomes evident once more that autochory is more advantageous in high trees.

Three of the 6 species of Sapindaceae studied belong to height category A, 2 to a, and one to aa. The highest trees have derived fruits in the form of drupes. Of the 2 species of medium height one is autochorous, the other anemochorous. The smallest tree is likewise autochorous. The anemochorous species has most individuals (795). The 3 zoochorous species of height category A have low to very low numbers of individuals showing that zoochory in high trees is disadvantageous.

The 21 species of Sapotaceae studied are very much alike in their fruit formation and way of dispersal. The fruits are few-seeded berries or single-seeded drupes dispersed by animals. The seeds are relatively large and reduced in number having a smooth shiny testa and a prominent hilum. Irregularities in the number of locules indicate a progressive reduction of the seed number and a continuous transformation of a several-seeded berry into a single-seeded drupe. Fifteen of the 21 species belong to height category A, while only 5 belong to height category a. Drupes (12 species) are more common in the high trees (A) and berries (9 species) in the lower trees (a). Drupes are combined with high trees (A) 10 times, berries with high trees 5 times, berries with lower trees (a) 4 times, drupes with lower trees one time. A diagram with the following pattern results:

DA DA DA DA DA DA DA DA DA DA
BA BA BA BA BA
Ba Ba Ba Ba
Da

In this diagram D = species with drupes
B = species with berries
A = height category A
a = height category a

It becomes obvious again that derived fruit forms such as the drupe more frequently occur in high than in lower trees. *Ecclinusa guianensis* has most individuals (1.658), other 2 species have 722 and 458 individuals respectively, *E. guianensis* has fruits of yellow to orange colour with edible flesh. It seems to be a resistant species which also grows in the savanna forest. The species is certainly indigenous of Guiana. But the high number of individuals cannot be explained by the dispersal mechanism and the resistance of the tree alone.

The 2 species of Simaroubaceae studied are both of height category A and develop drupes which correspond to a derived fruit form.

Three of the species of Sterculiaceae studied belong to height category A and one to aa. Two of the high trees (A) are autochorous, while one is auto-zoochorous. The species of height category aa is likewise auto-zoochorous. *Sterculia pruriens* with autochorous dispersal mechanism has a rate of distribution about 10 times as high as the auto-zoochorous *Guazuma ulmifolia* of the same height category. Judging from the number of individuals, autochorous dispersal seems to be more advantageous in high trees than an auto-zoochorous mechanism.

Two of the 3 species of *Apeiba*, Tiliaceae, belong to height category A and one to a. *Luehea candida* of height category A is anemochorous demonstrating again that anemochory is more advantageous in high trees. The species of *Apeiba* of height category a is more characteristic of the deciduous forest, hence the low number of individuals. All 3 species of *Apeiba* have capsules with an autochorous dispersal mechanism. However, the dehiscence is delayed so that the seeds remain enclosed within the capsule for a long time ensuring dispersal over a long period. But as the seeds are embedded in a sweet pulp, zoochory may also take place here.

The 3 species of Verbenaceae studied are all of height category A and probably have drupes, a derived fruit form more often found in high trees. However, the number of individuals remains very low in all 3 species.

The 2 species of Violaceae studied have an autochorous dispersal mechanism. However, the higher tree has a higher number of individuals.

The 4 species of Vochysiaceae studied belong all to height category A and are anemochorous. However, *Erisma uncinatum* develops samaras, whereas the other 3 species have capsules with winged seeds. The original fruit type of the Vochysiaceae is a capsule opening with 3 valves and releasing winged seeds. Fruit indehiscence as found in *E. uncinatum* is probably of secondary nature. The calyx here substitutes wings. Possibly this type of dispersal is more advantageous than the original one judging from the number of individuals. The specific weight of the samara is very low. In the Vochysiaceae it is again demonstrated that high trees are often adapted to wind dispersal.

As most important results of this chapter we may summarize: High trees frequently show a derived fruit form such as few-seeded or single-seeded drupes, nuts or samaras. Wind-dispersal is more frequently found in high trees. Anemochorous trees not infrequently have a higher number of individuals. Likewise, an autochorous dispersal mechanism, often found in high trees, seems to be advantageous judging from the higher numbers of individuals. In the medium and lower forest strata the original zoochory is kept, layers where it is more favourable, because animal life is here most intense. Species with their main distribution in other habitats usually show lower numbers of individuals. Even very resistant species adapted to adverse climatic conditions or with a very resistant root system do not reach the very high numbers of individuals expected, although there are exceptions. In the Leguminosae the number of species and individuals declines from the Mimosaceae towards the Papilionaceae, but the number of high trees with anemochory markedly increases.

Fruiting times and dispersal mechanisms

Anemochory

Table 3 indicates the fruiting times of anemochorous, autochorous, auto-zoochorous, and zoochorous species. The ciphers 1–12 signify the different months of the year. Each species fruiting in a month is marked by the number of individuals present in the forest studied. When the fruiting time of a species extends over several months, the number of individuals is repeated in the respective months following the same horizontal line, e.g. 421 in January, 421 in February, and 421 in May. This does not mean that all 421 individuals fruit in all 3 months. Species with no number of individuals are represented by a cross (+) only. As table 3 a shows, most anemochorous species fruit in February, but exclusively in February fruit 5 species, the other 7 species fruit in other months too. February coincides with the dry season, and species fruiting in January or March likewise prefer the dry season. Even December is usually a dry month. However, another fruiting peak is found in May, at the end of the dry season and the beginning of the heavy rain falls (approximately middle of May, but sometimes earlier). Possibly this time is less advantageous for the dispersal of anemochorous fruits and seeds, but more favourable for immediate germination of seeds. High winds arising at this time may carry the dispersal units over larger distances. Fruiting times in February and May overlap in 4 species. Other species even extend their fruiting time into June and August, taking advantage of the wet season too. Species which fruit during the dry season as well as during the wet season may have better chances of dispersal and germination (6 species). Possibly May is a very favourable time, because of the starting rains and the germination possibilities for the seeds. All in all, February and May are the preferred months of fruiting, but more species fruit in February. In Panamá, Karr (1976) observed that wind-dispersed plant species fruit during the late dry and early wet season (March-May), the time of the highest winds.

Autochory

Fruiting times change, when we consider the autochorous species (Table 3 b). It becomes apparent on Table 3 b that the fruiting times are more spread out over the year when autochory is considered. However, a certain preference of May/June, August and December is obvious. Some species fruit in January, August and December, others in May, August and December, or even in February, June and December or in March and May. These are the species with the highest numbers of individuals. They actually cover rainy periods as well as drier or dry months. They thus cover a broader spectrum with different seasons than the anemochorous species and this may be the reason of their success. Their numbers of individuals are quite high. Surprising is the fruiting peak in December which, however, partly extends into January and February. December which is usually a dry month seems not to be very favourable for germination of seeds, although it may be advantageous for dispersal. The seeds would probably have to wait until May for germination when the rain falls start.

Auto-zoochory

Although the fruiting times of the auto-zoochorous species are more or less spread out over the entire year (Table 3 c), a certain concentration of fruiting species in February, on the one hand, and in May/June, on the other, is obvious. Twenty auto-zoochorous species fruit in February. However, there are quite a few species which prefer the very dry season (February and March) as well as the very wet season (May/June) at the same time. Others fruit in February and again in August/September, towards the end of the wet season. These species show partly very high numbers of individuals.

Zoochory

Most species studied are zoochorous. As Table 3 d demonstrates, a striking concentration of fruiting species becomes evident in February (34 species). These species prefer the dry season which also extends over January and March. Although fruiting is spread out over the entire year, a second peak becomes obvious during the wet season (May/June) which extends over July and August. In Panamá a bird peak was observed in the high stratum during May, according to Karr (1976). May and June seem to coincide with the peak of biomass in Panamá. A third concentration of fruiting species is visible at the end of the year, mainly in December. Quite a few species fruit during the dry season, in January/February, and at the beginning of the wet season, in May/June, or at the end of the wet season. Others spread out fruiting only over the wet months (May to October/November). Those species which fruit during the wet season may be adapted to the more vigorous animal life during these months. Fruiting during the dry season, on the other hand, may satisfy animals when other food sources as well as water become scarce. Furthermore, many birds and other animals may start in February with their mating season or already with sexual reproduction, nesting or laying eggs. The third accumulation of fruiting species in December is partly due to species which extend their fruiting time from the wet season to the starting dry season, and partly by others which prefer the drier season (January-February-March) starting already in December. Preference of 2 different seasons is thus obvious here.

When we compare the fruiting of the 4 different dispersal types on Table 3 a-d we come to the following results: Anemochory is best distinguished by 2 clearly separated fruiting peaks in February and in May. More or less extended over the entire year seems to be the fruiting of autochorous species, however, with a certain accumulation towards the end of the year. A certain concentration of fruiting species in February and May becomes obvious in the auto-zoochorous species. However, fruiting is more spread out over the year than it is with the anemochorous species. A combination of all possibilities is realized in the zoochorous species. A fruiting peak in the dry season is conspicuous. Fruiting during the wet season is more spread out over several months starting with May, during the time of increasing rain falls, and thinning out towards September/October, the later part of the rainy season, while we observe a revival of fruiting in December. However, this third peak is explained by species which belong either to the dry-fruiting or to the wet-fruiting species, the first starting already in December, the latter finishing in this month. Nevertheless, fruiting of zoochorous species is well spread out over the entire year. One of the explanations of the fruiting time of the species studied is the influence of the climate: the alternation of dry season and wet season seems to be one driving factor. Particularly, anemochorous species seem to have adapted best to this climatic change, either prefering the dry season for dispersal or the beginning of the wet season for germination. Highest wind favourable for dispersal also occurs during this time. Furthermore, we may distinguish between 3 different fruiting types: one type prefering the dry season (many anemochorous species), another type prefering the wet season, and a third one prefering the dry as well as the wet season. One could assume that the dry fruiting types may be represented by dry dehiscing or anemochorous dispersal units, the wet fruiting types by fleshy drupes or berries, and the dry-wet fruiting types by intermediate types. This will be studied in the next chapter.

A second important factor which influences the fruiting times is certainly – besides the climate – the animal life (see also chapter "Seasonality of animals life in the forest"). This is most vigorous during the wet season which extends over the major part of the year. Certain zoochorous plant species are, however, adapted to the necessities of certain animal species, e.g. those fruiting during the dry season when food and water become scarce. The staggering of fruiting times over the year by closely related species (e.g. in the genus *Miconia*), (see p. 9/10, 35/36, 146) may strongly be influenced by the animal life. But finally, also animal life is mostly adapted to the seasons so that a complete circle is closed again.

Snow (1965) in his studies of Melastomaceae (*Miconia*) came to the result that fruiting species are clumped around June and July forming the major peak, while a smaller peak was observed towards the end of the year. In Trinidad where observations were made, the dry season lasts from January to mid-May, while a short dry spell may act in September or October. The latter corresponds to the main south-tropical dry season being a last trace of this periodicity in the northern hemisphere. Snow distinguishes accordingly a northern seasonal pattern with the dry season from January to mid-May and a southern seasonal pattern with the dry season in September-October. In British Guiana 2 wet and 2 dry seasons may be distinguished, according to Snow. Snow reasons further that species of *Miconia* which fruit in April-June follow the northern tropical pattern, while those fruiting at the end of the year belong to the southern tropical pattern. However, a large number of species are very widespread belonging to both regions. The suggestion of Snow will be discussed later. The geographical distribution of the species whether they come from the northern or the southern hemisphere may play its part. However, to a large degree the seasonal periodicity of fruiting seems to depend on the local climatic factors. As we have pointed out before, some species prefer the dry season for fruiting, others the wet season, and a third group fruits in the dry as well as in the wet season. The picture of all fruiting species whether anamochorous, autochorous, auto-zoochorous or zoochorous is too homogeneous as to allow any other interpretation besides the one of climatic factors and, to a certain degree, influences through animal life, particularly when specialists are concerned.

Fruiting times at different heights (A-a-aa) of the forest

Using the data of Rollet concerning fruiting times, so far as available, there seems to occur little difference in fruiting times between the 3 different strata studied (A-a-aa. See Table 4 a-c). Fruiting times are spread out over the year in all strata, except in April when no fruits were recorded. However, an accumalation of fruiting trees in February is very obvious. This fruiting time thus coincides with the drier climatic period. In March fruit maturation is declining, while the peak of fruit maturation in February is preceded by a smaller peak in December, when the dry period begins. However, this peak partly comprises other species too besides those fruiting in February. Another very obvious peak of fruit maturation occurs in May which declines towards June and July. This peak coincides with the main rainy season in the area. In August then fruit maturation revives, but to a larger degree with different species. We observe several species that have fruiting peaks in December/January and again in August (*Caraipa richardiana*) or in December, May and August (*Mabea piriri*), in June and December (*Ecclinusa guianensis*), in January/February and October (*Licania densiflora*) or in January/February and July (*Manilkara bidentata*) making obvious that there are species with 2 fruiting peaks. Even in the lowest stratum (aa) 2 peaks may be observed, e.g. in *Inga rubiginosa*, where one in February and another one in August occurs, or in *Duguetia megalophylla* fruiting in March and December. So far the survey over the fruiting situation in the whole area.

Beginning with the lowest stratum (aa) (Table 4 c), we have too few data to make significant statements. But possibly fruiting is more common from May to August when rain falls prevail and animal life is vigourous.

Considering the trees of the a-stratum, we observe fruit maturation mainly in February and in May. No fruit maturation was observed in April and September, and very little in October and November. Fruit maturation then revives in December. Considering the number of individuals bearing fruit during the year, the highest number occurs in May (with 5.233 individuals), followed by February (with 4.482 individuals), August (with 4.276 individuals) and December (with 4.187 individuals). However, we have to take into account here that we are counting the number of individuals of certain species twice or even 3 times when fruiting extends over several months. Otherwise August is relatively poor in the number of fruiting species. *Mabea piriri* with 1.706 individuals is included here as well as *Caraipa richardiana* with 1.893 individuals. Eighteen species altogether mature fruits in February, 17 species in May, 10 species in December, and 8 species in June, while 7 fruit in March. Quite a few species fruit in February and May (9). *Caraipa richardiana* fruits in August and December/January. *Mabea piriri*, on the other hand, has fruits in May, August and December.

In the highest story (A), fruiting times seem to be somewhat more stagerred, except for February in which month a marked concentration of fruiting trees is found. However, we have to take into account that the A-story comprises a larger number of species and that this may be the reason for the more staggered pattern of fruiting. Counting the number of species we find 49 species fruiting in February and 31 species in May. The fruiting peak in February is declining rapidly towards March when only 14 species fruit. The same phenomenon may be observed during the second fruiting peak in May. The number of fruiting species declines in June to 21, and in July to 12. In August the number of fruiting species then revives reaching 25. In each month, September and November, only 9 species were found, while in October 14 species were recorded. A revival of fruiting occurs again in December with 20 species, while in January the number of species declines to 10. Concerning the number of individuals, we find a maximum of fruiting individuals in February (21.152) which is followed by 13.589 individuals in May. The least number of individuals is found in July (1.371) when we leave out April in which month no observations of fruiting trees were made. We thus come to the following results: A maximum of species and individuals fruits in February when the dry season is in full action. In March when the dry season is still reigning, we observe a rapid decline in number of species and individuals. January is usually also a dry month and the fruiting of several important species extends over January-February-March. There are other important species with a high number of individuals which extend their fruiting time from December to February. All these species are actually adapted to fruiting during the dry period.

The second obvious fruiting peak is full in action when the rainy season is most vigorous with very heavy rainfalls, i.e. in May. Then fruiting declines rapidly until reaching a minimum in July. A revival starts in August which then declines towards the end of the year. It is obvious that several important species extend their fruiting time from May over June finding a revival in August, October or November, or even in December. All these species show a preference for the rainy season. The revival of fruiting in August largely occurs with other species than those starting in May. But these species also show preference for the wet season, sometimes extending their fruiting time over September or October, respectively. A third group, however, fruits during the dry season as well as during the wet season, e.g. in February and May/June or in January/February and May. We shall see later which dispersal type is characteristic of the 3 fruiting types: dry-fruiting, wet-fruiting, and dry-wet-fruiting. All in all, there is not much difference perceptible concerning fruiting times in the different strata (A-a-aa) of the forest.

Families, dispersal types and fruiting times

In the following, plant families are studied concerning their dispersal methods related to the respective fruiting times.

The species of Annonaceae are partly autochorous and partly zoochorous. Surprisingly, one autochorous and 2 zoochorous species have approximately the same fruiting times: 3/12, 3/12, 2/12, while one zoochorous species fruits only in December. The zoochorous*Xylopia frutescens* fruits in August, but *Xylopia nitida* var. *nervosia*, likewise zoochorous, in February and December. Both zoochorous species of *Xylopia* differ thus very much in their fruiting times. One autochorous species fruits in October.

Two anemochorous species of *Aspidosperma*, Apocynaceae, are fruiting in February or May, respectively, while a zoochorous species fruits in February and May.

Of 3 anemochorous species of Bignoniaceae, one fruits in February, the other 2 in May: *Tabebuia serratifolia* and *T. stenocalyx*, although from different habitats, fruit at the same time.

The autochorous species of Bombacaceae fruits in January, while one anemochorous species fruits in August, the other in April and May. The latter is more characteristic of open habitats.

Two of the zoochorous species of Boraginaceae (*Cordia bicolor* and *C. fallax*) fruit in June and July, but one of them also in December. Another zoochorous species of *Cordia* fruits in February, while *Lepidocordia punctata* bears fruit in September.

One auto-zoochorous species of Burseraceae, *Protium decandrum*, has a very wide range of fruiting time, namely, February-May and June-November. It is represented by 2.669 individuals. Another zoochorous species, *Protium* sp., fruits in February and August, a third (*Tetragastritis panamensis*) in February and May. A fourth zoochorous species fruits only in February, whereas a fifth (*Tetragastris?*) in November and December. The fruiting times of the Burseraceae are thus spread out over the entire year, although a certain preference for February is obvious.

Within the Combretaceae, the zoochorous species fruits in August, the 2 anemochorous species in February/May and May respectively (*Terminalia guayanensis* and *T. amazonia*).

Tapura guianensis of the Dichapetalaceae with 1.527 individuals has a wide range of fruiting, i.e. June, August, and October. It develops a zoochorous drupe.

The 3 zoochorous species of *Diospyros*, Ebenaceae, fruit at very different times, namely in February, June, December, respectively, although they belong to the same height category (A) and to the same habitat. In the Ebenaceae it thus becomes very obvious that the 3 species of *Diospyros* "avoid" one another concerning fruiting times, as they fruit partly during the dry and partly during the wet season.

The autochorous and auto-zoochorous species of *Sloaneae*, Elaeocarpacea, differ very much in their fruiting times: 2 species fruit in June, another species in August, a fourth in March, a fifth in February and May, and a sixth in August/September and in December. The species of *Sloanea* thus spread out their fruiting times over the year.

The 2 autochorous species of *Mabea*, Euphorbiaceae, fruit at very different times. One fruits from October to November having 0 individuals, while the other species fruits in May, August and December having 1.706 individuals. Of the autochorous species of *Sapium* one fruits in August and October, the other in July. The species with most individuals in the forest (3.679), the autochorous *Chaetocarpus schomburgkianus*, fruits in February, June and December. But another autochorous species fruiting in February, July and October has only 14 individuals (*Conceveiba guianensis*). In the 2 zoochorous species of *Drypetes* fruiting times are delayed or anticipated respectively. Both species start to fruit in March, the second fruiting month is May in one species and July in the other, while the third fruiting month is December in one species and October in the other. The third zoochorous species, *Hieronyma laxiflora*, fruits at a completely different time, namely in August and November. Other autochorous species fruit in May, July or October. Considering all the autochorous species fruiting times are spread out over the year and several species prefer 2 or 3 different months of the year for fruiting.

The zoochorous species of Flacourtiaceae fruit from May to July or from May to June or in January and November. An auto-zoochorous species fruits in February and August.

Of the autochorous species of *Tovomita*, Guttiferae, one fruits during the dry season in March, the other during the wet season in June. The autochorous species with the highest number of individuals (1.893), *Caraipa richardiana*, fruits in January, August and December, covering dry and wet months as well. The 3 zoochorous species fruit at very different times. One zoochorous species fruits in February, another in July, and the third in June and from November to December. None of them has a higher number of individuals.

In the Lauraceae it becomes obvious that species of the same genus usually differ in their fruiting times, e.g. the 2 species of *Aniba*, and the 2 species of *Nectandra*, or the 2 species of *Ocotea*. Snow (1966) states of the Lauraceae in Trinidad that they follow the Northern seasonal pattern with a main peak of fruiting from April to June and a minor peak in November. However, none of the species studied here follows this pattern. Three species fruit in June only. One of them is *Nectandra grandis*. The other species of *Nectandra* fruits in December. *Beilschmiedia curviramea* also fruits in December. *Aniba riparis* fruits in August and December, whereas *A. excelsa* fruits from February to March and in June. *Ocotea nicaraguensis*, finally, fruits in February and November. Considering this species name, it should be a species following the very Northern pattern without recognizing a second dry period at the end of the year.

The 2 anemochorous species of *Couratari*, Lecythidaceae, fruit in May and February/May, respectively, the latter adding an additional month, February. All the other auto-zoochorous species fruit at very different times. Of the 6 species of *Eschweilera* one fruits in February and May, in a second species the fruiting in May is extended to June, in a third species September is added to February and May, and in a fourth species May is left out, but the fruiting during the wet season is extended from August to September. Two further species fruit at quite different times. Extending, adding or shifting of fruiting times is very obvious here. The species with the highest number of individuals fruit in February and May (3.446 individuals), in March and June (2.374 individuals) and in February, August to September (1.025 individuals). *Gustavia augusta*, exclusively zoochorous, fruits in August, and *Lecythis davisii* in January.

Of the zoochorous species of Mimosaceae, a certain preference of the dry months February or March is obvious. To these dry months another month of the wet season may be added. Fruiting starts likewise with a dry month, Februaru or March, in *Parkia*, however, one species of *Parkia* adds December. Of the 3 species of *Pithecellobium* 2 fruit in February, but one of them is auto-zoochorous, the other zoochorous. The third species fruits in January and March, all dry months. While one species of *Stryphnodendron* fruits in August and October, fruiting is delayed towards September and December in the other. Of the 16 species of *Inga*, all zoochorous, one fruits in February, 3 fruit in March, another one in February and March, a sixth in February and August, 2 further species in May, and a ninth in February and May. One anemochorous and 2 auto-zoochorous species of Mimosaceae fruit in February. One of these auto-zoochorous species fruiting in February is *Pithecellobium* cf. *claviflorum*. Another auto-zoochorous species, *Pithecellobium pedicellare*, fruits in January and March, while *Pithecellobium* sp. likewise fruits in February.

The anemochorous *Cassia multijuga*, Caesalpiniaceae, fruits in January while the zoochorous *C.* sp. fruits in August and November. One autochorous species fruits in May and September. While one of the auto-zoochorous species of *Peltogyne* fruits in October, the other species has added February and June to October. The latter has more individuals. *Peltogyne porphyrocardia*, auto-zoochorous and of the decidious forest, fruits in February, June and October. *P.* sp. of the rain forest, and likewise auto-zoochorous, fruits only in October. The auto-zoochorous *Macrolobium acaciaefolium* fruits in February. The anemochorous *Sclerolobium paniculatum* fruits from January to February and in May. The zoochorous *Swartzia* sp. fruits in February and May.

The anemochorous species of *Pterocarpus*, Papilionaceae, prefer March and May respectively for fruiting. Otherwise, fruiting is very much spread out over the year no matter whether the species are auto-, auto-zoochorous or anemochorous. *Pterocarpus officinalis*, anemochorous, of the lowland and the swampy forest, fruits in March. *P. rohrii*, likewise anemochorous, fruits in May. Two other anemochorous species of different genera fruit in November. A further anemochorous species fruits in June. A zoochorous species fruits in November. And the autochorous species with most individuals (1.262) fruits in March and May.

The 2 species of *Miconia*, Melastomaceae, as well as the 2 species of *Mouriria*, fruit at very different times, although all 4 are zoochorous. *Miconia amacurensis* fruits in May, *M.* aff. *kappleri* in March and December. *Mouriria huberi* fruits in February, *M. sideroxylon* in October.

Of the 3 auto-zoochorous species of *Trichilia*, Meliaceae, one fruits in February, another has extended the fruiting time from February to March, and the third fruits from May to June. The 2 anemochorous species prefer February and December for fruiting, although pertaining to 2 different genera (*Trichilia* sp. and *Cedrela odorata*). The 2 auto-zoochorous species which fruit in May (and June) have the highest numbers of individuals.

Of the auto-zoochorous species of Myristicaceae, one fruits in June, the other in February and December.

Four of the zoochorous species of Myrtaceae with drupes start to fruit in February. Two species add July, another extends fruiting to March, and the fourth species has October and December as additional months. Only *Marliera* sp. fruits in November. All Myrtaceae are zoochorous.

Couepia glandulosa, Rosaceae, fruits in January, August and October. *Hirtella davisii* fruits in May, while *H. americana* has shifted to June and September. Of the zoochorous species of *Licania*, 2 fruit in February, one in February and May, and a fourth in May. *L. densiflora* with most individuals (3.258) fruits from January to February and in October. *L. parvifructa* fruits from July to August and in October, but has only 24 individuals. Four of the species of *Licania* start thus to fruit in the dry month February, one species adding January and October. Another species adds May. A fifth species of *Licania* fruits in May only, and a sixth species also starts in the wet season, but fruits from July to August adding October. Of the 2 species of *Parinari*, one fruits in February, the other adds May, August and November to February. *Prunus sphaerocarpa*, finally, fruits in March. All species of the Rosaceae are zoochorous. In the

Rosaceae as a whole, fruiting is very much extended over many months.

The species of Rubiaceae studied belong all to different genera. One autochorous and one zoochorous species fruit in June. A second zoochorous species fruits in October, and a third zoochorous species from January to February and in May. Two species of the Rubiaceae thus fruit at the same time, but one is autochorous, while the other is zoochorous.

Of the 2 zoochorous species of *Chrysophyllum*, Sapotaceae, one fruits in February, the other additionally in December. The zoochorous *Ecclinusa guianensis* which reaches the highest number of individuals of all Sapotaceae fruits in June and December. *Pouteria egregia*, *P.* sp. (Caimito negro) and *P.* sp. (Capure) fruit in August, while *P.* cf. *trilocularis* fruits in January. *Pradosia prealta* likewise fruits in August. "Pendarito" fruits in October, *Manilkara bidentata* fruits from January to February and in July, *Oxythece* in February and May, *Micropholis melinoniana* in May, August and December, and finally "Purguillo" in May and from August to September. Many species thus spread out their fruiting over the year. All Sapotaceae are zoochorous.

Guazuma ulmifolia and *Sterculia pruriens* of the Sterculiaceae fruit in May, one species being auto-zoochorous, the other autochorous.

Of the 3 auto-zoochorous species of *Apeiba*, Tiliaceae, one fruits in May, the second in February and the third in August. The anemochorous *Luehea candida* fruits in February and from May to June. February and May are thus the preferred months.

The 4 anemochorous species of Vochysiaceae prefer the dry season for fruiting. *Vochysia tetraphylla* fruits in February, *V. lehmannii* in March and September. *Erisma uncinatum* fruits in February and September, and *Qualea dinizii* fruits in February, from May to June, and in August. February is the preferred month. However, *V. lehmannii* shifted to March and added September, while *Qualea dinizii* added 3 months to February (May, June and August).

From these observations we may resume: Most obvious become again the tight relations of the anemochorous species with the climate. The preferred months are February of the dry season and May, when the dry season ends and the wet season starts with high winds. This is easily explained, because anemochory prevails in the high and highest trees which are most exposed to drought, wind, and isolation. The climate has thus the greatest influence on their exposed crowns. On the other hand, no biological factors such as animals and their periodicity exercise any influence over the anemochorous species. The dry months are most favourable for the dispersal of anemochorous fruits and seeds, while the starting wet season is more advantageous for immediate germination of the seeds. Anemochorous species growing in drier habitats likewise prefer the dry season for fruiting, and vice versa, species growing in dry habitats are frequently anemochorous.

However, within one and the same family we observe that fruiting times are spread out more or less over the year. This is partly due to the fact that different dispersal methods frequently prevail within a family. But also in families where one and the same dispersal mechanism predominates, e.g. zoochory or autochory, the different species often fruit at different times. This becomes most obvious in species belonging to the same genus (e.g. *Sloanea*). The above mentioned examples show that 2 or more species of the same genus fruit at very different times so as to avoid one another in fruiting. In many examples we observe a certain variation of an original basic "fruiting theme" through extension, adding or shifting of fruiting times. By delay or anticipation of the fruiting time, by adding of further fruiting months or by shifting to other fruiting months and by extension of the fruiting, the basic pattern may be changed and new dispersal possibilities may arise for the shifting species. As an example we may assume that a certain species fruits in February. By delay of fruiting a shifting to March may result – or by anticipation a third species may already start to fruit in January. All 3 months belong to the dry season. A fourth species may simply extend the fruiting time not only fruiting in February, but extending the fruiting to March, so that 2 months are covered with fruiting, February and

March. A fifth species may add a month, e.g. of the wet season, thus fruiting in February and August. In this way, the other 4 new species have further chances of dispersal, especially zoochorous species may find other dispersers in other or additional months. This type of fruiting pattern imitates a pattern of mutual avoidance. However, we have to emphasize here that not only zoochorous species avoid one another (see also Janzen) but also species with other dispersal mechanisms such as autochorous or anemochorous methods where selection of fruiting times through the influence of animals with much probability does not play any part.

A wide range of fruiting extended over several months of the year, and possibly covering some months of the wet as well as of the dry season may be advantageous under certain circumstances. Some species with a high number of individuals show a more "staggered" fruiting period extending over the year. But there are many species, on the other hand, with few individuals which have the same staggered fruiting periods. Different "niches" may be responsible for the distinct fruiting times of species of the same genus, for the adding of fruiting months and the shifting to other months by delay, anticipation or extension of the fruiting. Besides, expansion and contraction of the forest areas in earlier periods connected with the formation of refuges could be the reason for different fruiting times of species of the same genus or of the same family. This could apply to very homogeneous groups, such as the Sapotaceae, the Apocynaceae (*Aspidosperma*), the Burseraceae or the Elaeocarpaceae (*Sloanea*). However, this subject will be discussed more thoroughly in the chapter "Rufuge Theory" (p. 169, 170).

Geographical distribution, habitats and fruiting times

In the following, selected species are studied concerning their geographical distribution, their special habitats and fruiting times. Only those species are treated which also occur in other habitats besides the humid tropical forest or are also recorded in other geographical regions (e.g. Surinam, Mexico, Amazon). As far as known, fruiting times in the different regions are indicated and, if possible, compared with one another.

Of *Bravaisia integerrima*, Acanthaceae, (with 2 individuals) we do not know the fruiting time, but we are informed that the species also occurs in Mexico at the Golf side as well as at the Pasific side. It is not a typical species of the rain forest, but occurs in the semideciduous forest too. It grows on plain alluvial territories with clay which are inundated most of the year, and in certain parts of Mexico it forms almost pure local stands.

Of the Anacardiaceae, *Anacardium giganteum* also occurs in Surinam in the savanna forest and along creeks. *Spondias mombin* (with 36 individuals) is a wide-spread species, which also occurs in the dry deciduous forest as well as in the secondary forest. Although native of South America it occurs in Africa too and is widely planted in South America and Africa (Ghana). Likewise in Mexico the tree has a wide range of distribution in different types of forests, reaching from sea-level up to 1200 m. In Surinam, the tree is common in the ridge forest as well as in the marsh forest. In Venezuelan Guiana it fruits in August during the wet season.

Anaxagorea sp. (with 142 individuals), Annonaceae, is found in the dry forest too and fruits in October, i.e. towards the end of the wet season. *Unonopsis glaucopetala* (240 individuals) with 2 different fruiting peaks (3/12) frequently occurs in Surinam, particularly along creeks.

Of the Apocynaceae, *Aspidosperma megalocarpon* fruits in February. In Mexico the species occurs at the Golf side as well as at the Pacific side growing in the semideciduous forest too. It is represented by 650 individuals in the forest studied. *A. oblongum* occurs in Surinam as well as in French Guiana. *Himatanthus articulatus* (with only 59 individuals) is found in Surinam in the ridge forest and in the savanna forest. *Parahankornia amapa* (with 4 individuals) fruits in February and May. In Surinam the species is found in the savanna forest and in the swampy forest too.

Of the Araliaceae *Dendropanax arboreus* (with

one individual) fruits in June, during the wet season, in Venezuelan Guiana. In Mexico the species fruits almost throughout the year, but preferably from March to August. In Mexico the tree is also found in the deciduous forest and has an ample range of soil tolerance. It is found at the Golf side as well as at the Pacific side. *Didymopanax morototonii* (with 40 individuals) also fruits in June in Venezuelan Guiana. However, in Mexico fruits ripen from January to March, during the dry season. In Mexico the species has a restricted occurrence at the Golf side exclusively growing in humid climate. It is abundant in advanced secondary vegetation of high evergreen forest being associated with *Terminalia amazonica* on alluvial soils of very ancient origin.

Of the Bignoniaceae, *Jacaranda copaia* (with 49 individuals) occurs in Surinam in the swampy forest too. It fruits in February. *Tabebuia serratifolia* (with 18 individuals) is characteristic on ridges and along rivers.

Ceiba pentandra of the Bombacaceae is a tree of very ample distribution. It occurs in Surinam in the swampy forest, on sandy ridges, and along rivers. In Mexico it is found at the Golf side as well as at the Pacific side. It frequently occurs in disturbed areas of evergreen and semideciduous forests. To my knowledge it fruits only once a year in Venezuela, during the dry season.

Cordia alliodora of the Boraginaceae fruits during the dry season (Janzen 1976). It is abundant in the secondary vegetation and in the semideciduous forest. In Mexico the tree fruits from September to April. The species frequently occurs in the secondary forest at the Golf side as well as at the Pacific side. *Cordia bicolor* fruiting from June to July also occurs in Surinam.

Tetragastris panamensis of the Burseraceae fruits in February and May. The species is very common in the rain forest of Surinam.

Caryocar nuciferum of the Caryocaraceae occurs in Surinam too, particularly in hilly regions of the West, but is a rare species.

Buchenavia capitata of the Combretaceae is a rare species in the humid forest of Surinam. *Terminalia amazonia* fruiting in May frequently occurs in Surinam, being dispersed in the humid forest, in the ridge forest and savanna forest. In Mexico the fruits ripen from April to May. The species is restricted to the Golf side in Mexico. It is one of the dominant species of the high evergreen forest. It grows on different types of soil and has an altitudinal span from sea level to 900 m.

Tapura guianensis of the Dichapetalaceae fruits in June, August and October. In Surinam it is a small tree up to 13 m in height occuring frequently in the humid forest, in the high savanna forest and in the secondary forest as well.

Diospyros guianensis of the Ebenaceae is a shrub or small tree of up to 8 m (up to 15 m) height in Surinam, frequently occuring in the swampy forest along rivers and creeks. Its fruiting time is February in Venezuelan Guiana. *Diospyros melinonii* fruiting in June is very frequent in Surinam reaching a height of 18 m.

Sloanea grandiflora of the Elaeocarpaceae is frequent in Surinam along creeks and in the high savanna forest. It fruits in February and May. *Sloanea guianensis* is frequent in the humid forest as well as in the savanna forest in Surinam. It fruits in March. *Sloanea robusta* fruiting in August is a very rare species in Surinam, restricted to slopes.

Amanoa guianensis, Euphorbiaceae, fruiting in May occurs in Surinam in the swampy forest and in the savanna forest exhibiting transformations from a 5 m low shrub to a 21 m high tree and even appears as a liana. *Chaetocarpus schomburgkianus* (with 3.679 individuals) fruiting in February, June and December is frequent in the humid forest as well as in the high savanna forest of Surinam. *Conceveiba guianensis* (with 14 individuals) fruiting in February, July and October is found in Surinam in the humid and swampy forest, particularly at the river side. *Croton matourensis* fruiting in October is found in Surinam in the humid forest as well as in the savanna forest. *Drypetes variabilis* fruiting in March, July and October is frequent in the humid forest of Surinam. *Mabea piriri* (with 1.706 individuals) fruiting in May, August and December is frequently found in the humid and swampy forest of Surinam, particularly along creeks. *Mabea taquiri* fruiting from October to November frequently occurs in the humid forest at the river side and in the savanna of Surinam.

Margaritaria nobilis frequently occurs as a shrub, liana or small tree especially in the swampy forest of Surinam. *Pera schomburgkiana* fruiting in February and from May to June is very frequently found in the high savanna forest and near creeks in the savanna of Surinam. *Sapium aubletianum* fruiting in August and October is very frequent in the ridge forest, in the swampy forest , in hilly regions, and as a shrub even occurs in open areas of Surinam.

Banara guianensis of the Flacourtiaceae is frequent in Surinam at the river side, at forest margins, and in the secondary forest. *Homalium racemosum* is very frequent in Surinam mainly along rivers, and on granitic islands in valleys. *Laetia procera* fruiting in February and August is frequently found in the humid forest of Surinam, especially on sandy soil. *Zuelania guidonia* is restricted to the Golf side in Mexico occurring from sea level up to 700 m. It forms part of the upper or medium stratum of high evergreen or semi-deciduous forests. It only grows on limy soils with good drainage. It may be found abundantly in the secondary vegetation of high evergreen forests and semideciduous forests.

Calophyllum brasiliense of the Guttiferae fruiting in February is common in Surinam, particularly in the savanna forest, in the wet savanna forest and in the swampy forest. In Mexico the fruits ripen from October to December. The species occurs at the Golf side as well as at the Pacific side in high evergreen and semideciduous forests with good drainage. In certain areas it abundantly occurs as a codominant tree together with *Terminalia amazonia, Dialium guianense* and *Guatteria anomala. Symphonia globulifera* is common in swampy forests of Surinam where it develops stilt roots. *Vismia guianensis* is very common mainly in the secondary forest, in the humid as well as in the swampy forest of Surinam. *Vismia macrophylla* fruiting in June and from November to December is very frequent in the humid forest of Surinam.

Saccoglottis cydonioides of the Humiriaceae sparsely occurs in the humid forest of Surinam.

Aniba riparia of the Lauraceae fruiting in August and December is a very rare species in Surinam, while *Nectandra grandis* fruiting in June is a common tree of the humid forest in Surinam.

Couratari multiflora of the Lecythidaceae is anemochorous and fruits in May. In Surinam the species is common in the humid and swampy forest, e.g. at the river side. It flowers at the beginning of the great dry period after having lost the leaves. *Eschweilera chartacea* fruits in February and from May to June. It sparsely occurs in the humid forest of Surinam. *E. corrugata* fruiting in February, May and September is frequent in the humid and swampy forest of Surinam. *E. odora* (with 1.025 individuals) fruiting in February and from August to September is common in the humid forest of Surinam. *E. subglandulosa* (with 2.374 individuals) fruiting in March and June is common in the swampy forest of Surinam, particularly along creeks and rivers. *Gustavia augusta* fruiting in August in Caracas, is a very common species mainly in the swampy forest of Surinam. *Lecythis davisii* fruiting in January sparsely occurs in the humid forest of Surinam.

Enterolobium cyclocarpum of the Mimosaceae has an ample distribution at the Golf side as well as at the Pacific side in Mexico, occurring in disturbed areas of evergreen and semideciduous forests. *Inga alba* fruiting in February is commonly found in Surinam in the humid forest and in the high savanna forest. *Inga capitata* fruiting in March is common in Surinam in the humid forest, especially in hilly regions. *Inga heterephylla* is common in Surinam along creeks in the forest. *I. lateriflora* fruiting in March sparsely occurs in the humid forest and in the savanna forest of Surinam. *I. leiocalycina* likewise fruiting in March is an uncommon species in the swampy and humid forest of Surinam. *I. rubiginosa* fruiting in February and August frequently occurs in the humid forest, especially on slopes and on bauxit in Surinam. *I. splendens* fruiting in May is frequent in the humid forest of Surinam. *I. scabriuscula* fruiting from February to March is common in Surinam along creeks in the forest and in hilly regions. *Parkia oppositifolia* fruiting in February and December is rare in Surinam. *P. pendula* fruiting in March is very common in the humid forest of Surinam. *Pentaclethra macroloba* fruiting in February is common in Surinam in the swampy forest along rivers. In Venezuelan Guiana the species reaches 2.534 individuals in the forest studied. *Pithecellobium* cf. *claviflorum* fruiting in February sparsely occurs in Surinam. *P. pedicellare* fruiting in January and March is found dispersed in the humid forest of Surinam. *Stryphnodendron polystachyum* fruiting in September and December is found dispersed in the humid forest of Surinam. *S. purpereum* fruiting in August and October is rare occurring along the Surinam river.

Cassia multijuga of the Caesalpiniaceae fruiting in January is a very common small tree in Surinam, particularly on rocks and along rivers. *Dialium guianense* frequently occurs in the humid forest of Surinam. In Mexico the tree fruits from March to June. It is restricted to the Golf side and in the high evergreen forest it is codominant with *Terminalia amazonia, Calophyllum brasiliense* and *Guatteria anomala. Hymenaea courbaril* is very frequent in the humid and swampy forest of Surinam, particularly on rocks and on river banks. In Mexico the species occurs at the Golf side as well as at the Pacific side, where it is more abundant. Its ample altitudinal range reaches from sea level up to 900 m. The tree reaches its largest size in the high humid forests, but also grows in semideciduous and deciduous forests where it may form almost pure stands. *Macrolobium acaciaefolium* fruiting in February is common in swampy forests on river banks in Surinam.

Alexa imperatricis of the Papilionaceae fruiting in March and May is a rare species in Surinam being more numerous in places on slopes. In the forest studied of Venezuelan Guiana it reaches 1.262 individuals. *Diplotropis purpurea* fruiting in November is common in the humid forest of Surinam. *Dipteryx odorata* likewise fruiting in November occurs dispersed in the humid forest of Surinam. *Pterocarpus officinalis* fruiting in March is common in the swampy forest and along rivers in Surinam. *Pterocarpus rohrii* fruiting in May is common in the humid forest in Surinam.

Hebepetalum humiriifolium of the Linaceae fruits in December and is very common in Surinam, reaching only 17 individuals in the forest studied of Venezuelan Guiana.

Byrsonima aerugo of the Malpighiaceae fruits in December and occurs dispersed in the humid and high savanna forest of Surinam.

Miconia aff. *kappleri* of the Melastomaceae fruiting in March and December is not common in Surinam mainly growing in the secondary forest, in the humid forest in hilly regions, while some individuals occur as shrubs in the savanna.

Carapa guianensis (with 885 individuals) of the Meliaceae fruiting in May is very common in Surinam, especially in the swampy forest in the West of Surinam. The anemochorous *Cedrela odorata* fruiting in February and December is very common in Surinam. In Mexico the species abundantly grows in the secondary forest of different types in areas constantly protected by man. It grows best in regions with a high pluvial precipitation, but supports different types of soils. It is cultivated as an ornamental and shade tree. *Trichilia propingua* fruiting in February and March is very common in Surinam occurring dispersed in the humid forest on river banks.

Helicostylis tomentosa of the Moraceae fruiting in June and September is very frequent in the humid forest and the high savanna forest of Surinam. *Pourouma guianensis* fruiting in December is common in Surinam in the humid forest and at the young coastal plane.

Iryanthera lancifolia of the Myristicaceae fruiting in June is rare in Surinam. *Virola surinamensis* fruiting in February and December very frequently occurs in the swampy forest of Surinam.

Eugenia anastomosans of the Myrtaceae fruiting in February and July is very sparse in Surinam. *Eugenia compta* fruiting in February, October and December occurs in the savanna of Surinam, but is rare.

Cassipourea guianensis of the Rhizophoraceae fruiting in February and May is common in Surinam in the humid forest, on river banks and slopes and in the secondary forest.

Couepia glandulosa of the Rosaceae (Chrysobalanaceae) fruiting in January, August and October is very common in Surinam on river banks and at the margin of savannas. *Licania apetala* fruiting in May is common in Surinam in the humid forest and in the savanna forest. *L. densiflora* (with 3.258

individuals) fruiting from January to February and in October is an uncommon species in the humid forest mainly in hilly areas of Surinam. *Parinari excelsa* fruiting in February is very rare in Surinam in the humid forest, and mainly occurs in hilly areas.

Amaioua guianensis of the Rubiaceae fruiting in June is a common tree or shrub in the humid forest and the savanna forest of Surinam. *Genipa americana* fruiting from January to February and in May is a common species in the swampy forest and the ridge forest of Surinam. *Guettarda acreana* fruiting in October is rare occurring on slopes of granitic or lateritic ground in Surinam.

Sapindus saponaria of the Sapindaceae is an uncommon species in the ridge forest and along rivers at the coastal area of Surinam, also being planted and growing wild. In Mexico it fruits from January to September having an ample distribution in all humid and warm regions of Mexico, and is very abundant in the secondary vegetation. It grows in different types of forests and on different soils. Its altitudinal range reaches from sea level up to 700 m.

Chrysophyllum auratum of the Sapotaceae fruiting in February and December is very common in Surinam in the humid forest, along rivers, along creeks and on slopes, particularly on sandy soils. *Ecclinusa guiananesis* (with 1.658 individuals) fruiting in June and December is common in the humid forest, in the high savanna forest, in the swampy forest on river banks, and in the rocky savanna of Surinam. *Manilkara bidentata* fruiting from January to February and in July is common in the humid and swampy forest of Surinam. *Pouteria egregia* fruiting in August is a rare species in the humid forest of Surinam mainly growing on bauxit. *P. venosa* fruiting in March and July is an uncommon species in the humid forest of the West of Surinam.

Guazuma ulmifolia of the Sterculiaceae fruiting in May very sparsely occurs in the high forest on slopes in Surinam. The fruits mature almost the year round in Mexico, and particularly from September to April remaining on the tree for a long time. The species occurs abundantly in the secondary vegetation, especially where the dry season is well developed or in savannas. It develops indifferently on soils of vulcanic or sedimentary origin. In advanced secondary vegetation the species may form an important component giving the impression as if it was a primary element. *Sterculia pruriens* (with 908 individuals) fruiting in May is a very common species in the humid and swampy forest, on slopes, along creeks, and even in the savanna forest of Surinam.

Apeiba echinata of the Tiliaceae fruiting in May in Venezuelan Guiana is a very common component of the humid forest, the secondary forest and the savanna forest in Surinam where it fruits from June to August and flowers from October to December. *Apeiba tibourbou* fruiting in February in Venezuelan Guiana is a common tree in the humid forest of Surinam, particularly in hilly regions. In Mexico the fruits ripen from November to January remaining on the tree for some time. The species occurs at the Golf side as well as at the Pacific side and is an important component of the secondary vegetation of an age of 15–30 years.

Vitex stahelii of the Verbenaceae fruiting from May to June is an uncommon species in Surinam.

Erisma uncinatum (with 487 individuals) of the Vochysiaceae fruiting in February and September occurs scattered in the humid forest of Surinam. *Qualea dinizii* fruiting in February, from May to June, and in August is a common species in the humid forest and savanna forest of Surinam. *Vochysia tetraphylla* fruiting in February is a very common tree in the swampy forest of Surinam along creeks.

Mexico and the studied area in Venezuelan Guiana do not have many species in common. On the contrary, Surinam and Venezuelan Guiana have a great many species in common. However, only few species have an ample distribution in Venezuelan Guiana as well as in Surinam. Species which are very common in Surinam are frequently rare in the studied area of Venezuelan Guiana and vice versa.

In summary we could say that species which grow well in other habitats such as the semideciduous, the deciduous forest, secondary forest or savanna usually have a low number of individuals in the humid forest of Venezuelan Guiana. Although the species may be resistant having a good tolerance of different habitats and soils, resistance to drought and insolation, with an ample altitudinal range, and growing well in disturbed areas, they may be poorly represented in the forest studied in Venezuelan Guiana so that resistance, rapid growth or easy adaptation to different climatic conditions generally are not the important characteristics of a species to increase the number of individuals in the area in question. Examples of this type are: *Bravaisia integerrima, Anacardium giganteum, Spondias mombin, Himatenthus articulatus, Parahankornia amapa, Dendropanax arboreus, Tabebuia serratifolia, Ceiba pentandra, Cordia alliodora, Pera schomburgkiana, Zuelania guidonia, Symphonia globulifera, Cedrela odorata, Hymenaea courbaril, Couepia glandulosa, Genipa americana, Sapindus saponaria* and many others.

Snow (1965) observed that around the equatorial region in Middle and South America 2 different types of fruiting times occur, depending on the environmental changes. In Central America and in Northern Venezuela he recognizes a single "North-tropical dry season" from January to May alternating with the long wet season. Species belonging to this hemisphere show a single fruiting peak. On the contrary, 2 dry and 2 wet seasons occur in the equatorial climate, e.g. of Northern British Guiana. Towards the South, in Northeastern Brazil where the main "South-tropical dry season" reigns in September and October again a single fruiting peak occurs. The region where 2 fruiting peaks develop is so-to-say in an intermediate position being influenced by the Northern as well as by the Southern climate, according to Snow. Apparently, fruiting seasons are markedly influenced by climatic factors. North of the equator the dry season from January to May determines the fruiting peak of the North-tropical species, while South of the equator the main dry season in September-October exercises its influence over the fruiting of South-tropical species. In the intermediate area 2 dry seasons may be observed, a main dry season from January to May, and a minor dry season in September/October. Following the observations of Snow, species of this region may show 2 fruiting peaks, accordingly. In the material studied by me, many species have a single fruiting peak, others have 2, and some even 3 peaks. In accordance with Snow I suppose that fruiting of the species studied is largely influenced by climatic conditions. In Venezuela, there is a marked dry season from January to May, when heavy rainfalls start. The wet season may progressively diminish towards September or October, however, this very much depends on the year. In some years rains continue until the middle of December or even January, in others September is already a more or less dry month. One could speculate on the reflection whether indigenous species of the North-tropical region keep their North-tropical rhythm in the form of an endogeneous rhythm continuing with a single fruiting peak depending on the dry season from January to May. Checking the present material in comparison with fruiting data given by Pennington and Sarukhan for Mexico we note that species which also occur in Mexico usually show only a single fruiting peak. Examples are: *Aspidosperma megalocarpon*, in Venezuela fruiting in February, *Dendropanax arboreus* fruiting in June in Venezuela, in Mexico from March to August, *Didymopanax morototonii* fruiting in Venezuela in June, in Mexico from January to March, *Terminalia amazonia* fruiting in Venezuela in May, in Mexico from April to May, *Calophyllum brasiliense* fruiting in Venezuela in February, in Mexico from October to December, *Guazuma ulmifolia* fruiting in Venezuela in May, in Mexico from September to April (almost the year round), *Apeiba tibourbou* fruiting in Venezuela in February, in Mexico from November to January.

As may be seen, the fruiting times in Venezuelan Guiana and Mexico do not so well coincide. *Didymopanax morototonii*, for example, fruits during the wet season in Venezuelan Guiana, while it matures fruits during the dry season in Mexico. And more differences may be taken from the above given examples.

Genera or species which have their center of distribution in the Amazon should thus show a late dry season in September/October and a fruiting peak influenced by this periodicity. Some examples were selected in the following:

Caryocar nuciferum fruiting in Venezuelan Guiana 2/6,
Couratari pulchra 2/5,
C. multiflora 5,
Lecythis davisii 1,
Gustavia augusta 8,
Tapura guianensis 6/8/10,
Hirtella americana 6/9,
Hirtella davisii 5,
Licania apetala 5,
Licania alba 2/5,
Licania densiflora 1–2/10,
Licania parvifructa 7–8/10,
Licania rufescens 2,
Licania sp. 2.

As may be seen from these examples, some species ripen fruits during the South-tropical dry season (9/10). About one half of the species have 2 (or 3) fruiting peaks, the other half only one. It is thus suggested here that an "endogeneous rhythm" of fruiting if present at all is only kept partly, while probably most species have adapted to the local climatic conditions. In any case, fruiting times in tropical wet forests seem to depend more markedly on climatic circumstances than was admitted until now.

Another observation may be added here. Comparing the relative size of the trees as well as the size of their fruits and seeds we often observe the largest size in individuals in the high humid forest, while the sizes decrease in drier areas such as the deciduous forest, the savanna or other drier regions in Surinam or Mexico. This is explained by the fact that in the humid tropical forest conditions are optimal. The corresponding data of Surinam were taken from Roosmalen.

Animal life and forest layering

Stratification of tropical forests is not only visible through the different stories (3–4) produced by plants of different height, but also becomes obvious in the animal life. According to Foster (1973) 80% of the canopy and understory tree species of a tropical forest on Barro Colorado Island are dispersed by animals. In this connection it is of importance to know in which stratum of the forest certain animals develop their activities.

Charles-Dominique et al. (1981) could observe in a forest of French Guiana that nocturnal frugivorous mammalia species with a similar diet and a comparable body size spread out in different forest layers.

Ground animals

There are principally all kinds of ground animals which may eat fruits fallen to the ground. Rodents such as agoutis (*Dasyprocta punctata*) and pacas which have an active night life may be of great importance in the dispersal of fruits and seeds. They do not only eat the fallen fruits, but also scatterhoard them. Not all hidden fruits are later found again and seed dispersal is assured in this way. *Cuniculus paca*, for example, is very frequent in Guiana and its meat is said to be the most delicious of game. Even mice and rats may contribute to seed dispersal. Other ground animals useful for dispersal are the pecaries, likewise present in large numbers in the forest studied. A further ground animal which eats fruit is the tapir. Also deer may play a certain part in fruit and seed dispersal. McClure (1966) mentions deer, wild pig, and rodents as seed dispersers for a Malaysian tropical rain forest. Even turtles may be facultative seed dispersers. André (1964) mentions *Emyx tracaxa* which often eats fruits with a bitter taste so that the eggs become indedible. We have to emphasize here that the height category of the tree does not play any part when zoochory is happening on the ground. This may be the reason for the fact that some zoochorous species are very high trees, although zoochory is more characteristic of and more advantageous in the lower forest layers.

Tree animals

However, most animals live on the trees searching for their food between the twigs and leaves. A marsupial which is very wide-spread and often occurs so frequently that it can do great damage to the harvest of cultivated fruit trees (e.g. mango) is, for example, the so-called "rabipelado", *Didelphis marsupialis*, the animal with the naked tail which leads a nocturnal life on the trees. It is certainly also a seed disperser. Within the rodents, the squirrels are of great importance in the dissemination of seeds which they may scatterhoard too. *Guerlinguetus aestuans* is one of the most common squirrels in Venezuela. However, squirrels may also damage a great part of the harvest. McClure reports for winged fruits of *Shorea platyclados* that they are eaten unripe by squirrels permitting only very few to mature and fall. Likewise the porcupine, *Coendu prehensilis*, climbing up and down the trees, eats fruits and leaves, dispersing some seeds in this way. Within the Procyonidae, the cuchicuchi, *Potos flavus*, and the likewise nocturnal skunk, *Nasua solitaria*, eat besides other food also fruits and seeds. Although the sloth, *Bradypus*, one of the oldest mammals of the world, may not be a very important seed disperser, it eats some fruits besides its preferred leaves. Finally, we should not forget the reptiles which are fruit eaters, for example, the iguanas, *Iguana iguana*, living on the trees and eating buds, flowers and fruits. Being good swimmers they prefer the vicinity of rivers and lakes. Monkeys are certainly very active and mobile climbers as well as eager fruit eaters being important seed dispersers in this way. Howling monkeys are very wide-spreed animals in hot and mountainous tropical forests (*Alouatta ursina*). Typical of Guiana is *Ateles paniscus*, the coaita monkey, and *Cebus capucinus* likewise occurs there. McClure mentions for Malaysia the siamang and gibbons as important fruit eaters and dispersers, e.g. of *Ficus* fruits, of *Artocarpus lanceifolius, Myristica gigantea* and even of *Erythroxylon cuneatum* fruits with a high content of cocaine. Of the flying animals, bats may frequently search for their food in the canopy of the forest. Most important fruit eaters and dispersers are, however, the many kinds of birds with their enormous phalanx of different species of varying habit. Although not all bird species are fruit and seed eaters, some of the insectivorous birds also eat fruits sometimes or at certain stages of their lives (e.g. woodpeckers, flycatchers), while others are omnivorous. The range of tree animals is thus very large, but the exact demarcation of their territories in the different strata or their association with a definite stratum in the forest is sometimes difficult. Many birds, for example, rather live in the lower story between shrubs and low trees. The clever mammalia, and particularly the egg-searchers climb up high in the trees to find the bird's nests. The monkeys do not have any difficulties to go up to the crowns of the trees. Likewise any climbing animal may go high up in the trees to find special fruits or buds and young leaves, e.g. in the crowns of 30 m high fig trees. Certain birds, such as *Crax nigra*, put up their nests in the crowns of the highest trees. However, these nests are also more exposed to wind and weather and to birds of prey. A parrot, *Ara chloroptera*, puts its nests in such trees as *Dimorphandra excelsa* which reaches 50 m in height. But these may be exceptions. As a rule, however, we have to consider the lower and middle strata as the areas where most of the animal life occurs. Consequently, the many red, blue and black berries attracting birds mainly appear on shrubs and smaller trees of the forest. Karr (1976) emphasizes that the avian community composition changes with the stratum. Within the microhabitats he distinguishes a ground stratum, a low, medium and high foliage stratum, a bark stratum, and open spaces within and above the vegetation. Seasonal instabilities occur in the high stratum. The bird species of this area are either rare or appear in flocks to exploit temporal and spatial instabilities in resource distribution. Furthermore, the high stratum in forest habitats is only slightly richer in the tropics than in temperate regions, according to Karr, while the more buffered low stratum is richer in the tropics.

Of the frugivorous and seed eating birds which may occur in Venezuelan Guiana a short survey is given in the following. Of the Cracidae *Crax nigra* (Paují de copete) which is typical of Guiana likes berries. *Pipile cumanensis*, a "pava", is frequently mentioned by André in his description of Guiana. Parrots and parakeets are an important component of the Guianan fauna. There are many species of "aras", "loros" and "Guacamayas". *Ara macao* and *Ara chloroptera*, the latter nesting in very high trees such as *Dimorphrandra excelsa*, are frequently mentioned in the forest of Guiana. Other Psittacidae are *Deroptyus accipitrinus, Amazona amazonica, Amazona ochrocephala, Pionites melanocephala, Amazona festiva, Aratinga pertinax, Touit batavica, Aratinga solstitialis, Forpus passerinus.* Of the Ramphastidae, the toucan is an important fruit eater, e.g. *Ramphastros haematorhynchus* or *Onychorhynchus coronatus.* Of the Cotingidae, the "campanero" (*Procnias averano* and *P. alba*) and its loud warbling has been mentioned already. It is typical of Guiana. *Rupicola rupicola* only occurs in Guiana devouring fruits as well as insects. *Cephalopterus ornatus*, a bird with very strange plumage "ornaments", occurs in Guiana, as well as the "gold-bird", *Lipaugus vociferans*, and the "trompetero" *Pyroderus scutatus.* These birds prefer fruits to eat. Of the Turdidae *Turdus leucomelas* is a fruit eater. *Coereba flaveola* of the Coerebiade likes pulpy fruits and nectar. The "conoto", *Psarcolius decumanus*, Icteridae, the builder of the pendent nests prefers oranges and bananas. *Gymnomystax mexicanus* eats seeds and grains. *Quiscalus lugubris*, a seed eater, likewise occurs in Guiana. *Psophia crepitans* of the Psophiidae eating fruits lives in Venezuelan Guiana. Of the Thraupidae the "azulejo", the blue bird, *Thraupis virens*, is a very common bird in Venezuela which very much likes fruit. *Tanagra trinitatis* is also a fervant fruit eater. *Cissopis leverina* too lives in Guiana. Furthermore, there are quite a few birds which eat fruits and seeds as well as insects, such as certain fly-catchers and woodpeckers. *Phacellodomus rufifrons*, Furnaridae, devours seeds and insects as well. They are, however, many other fruit and seed eating birds not mentioned here. But of many bird species we do not know whether they occur in the studied area or not.

Leck (1969) found 16 different bird species belonging to 11 families exploiting a Central American fruit tree, *Trichilia cuneata.* Within these bird species there were woodpeckers, flycatchers, honeycreepers, woodcreepers, manakins and others. Haverschmidt (1971) observed 23 bird species in *Guarea guara* trees taking seeds from the splitting fruits in Surinam. Some bird species took the seeds while sitting on a branch, but all the others removed them in flight. Between others, quite a few flycatchers and tanagers are mentioned to eat the seeds. Nineteen visitor species of birds were observed to eat seeds of the understory tree *Guarea glabra* in Panamá, according to Howe and De Steven (1979). Species were listed of the Trogonidae, Ramphastidae, Picidae, Pipridae, Cotingidae, Tyrannidae, Turdidae, Vireonidae, Parulidae, and Thraupidae. The authors observed 12 resident species and 7 North American migrant bird species which fed on the bright orange arilloids. They hypothesize that the fruiting season of *G. glabra* is adaptively synchronized with the northward migration of opportunistically frugivorous North American birds. Snow (1981) presents a long list of food plants of birds. In the neotropical region he considers specialized frugivorous birds certain species of the Cracidae, Steatornithidae (*Steatornis*), Trogonidae (*Trogon*), Capitonidae, Ramphastidae, Cotingidae (*Cotinga, Procnias, Rupicola* and others), Pipridae. Leck (1972) emphasizes that *Cecropia* fruits are a favoured food of many tropical birds and an important fruit resource not only to at least several hundred bird species of the New World, but also for many opportunistic migrants. During his observations in Puerto Rico he observed about 20 different bird species such as parakeets, parrots, woodpeckers, thrushes, warblers, tanagers. *Cecropia* catkins are a preferred food of pigeons, toucans, trogons, fruit crows, flycatchers, woodpeckers, honey creepers, according to Eisenmann (1961).

According to Ridley (1930) Columbidae are the most important dispersers of seeds of any group of birds. They are voracious gleaners of fruits and seeds of *Cecropia peltata* in Panamá (Olsen & Blum 1968). Hall and Swaine (1981) mention frugivorous birds such as hornbills, plantain-eaters and barbets, in a forest of Ghana. Eisemann (1961) observed the above mentioned birds feeding on termites as well as on *Cecropia* catkins which food they favoured without regard whether they are primarily insectivorous or vegetarian. There is apparently a large variety of tropical bird species which eat fruits and seeds, some being obligatory fruit and seed eaters, others omnivores. Many tropical bird species have the ability to feed as omnivores switching among food resources from fruits to insects (e.g. flycatchers). Among the obligatory fruit eaters, on the other hand, we may distinguish between specialists and non-specialized bird species.

It should be emphasized here that the bibliography dealing with bird species in tropical forests is much larger, but that it is very difficult to get the foreign literature in Germany, on the one hand, and that this subject is not of great importance for the studies concerned here, on the other.

A list of all the nocturnal frugivorous tree mammals present at Cabassou, French Guiana, has been elaborated by Charles-Dominique et al. (1981, p. 357).

Aquatic animals

Finally, we should not forget the animals living in the water among which the fishes are of most importance. They are among the most ancient fruit and seed dispersers together with the water currents which have the widest dispersal range over the oceans from continent to continent. Gottsberger (1978) dedicated a special paper to the food of fishes in the Amazon region. Water turtles, iguanas and swimming mammals may also find fallen fruits and seeds in the water to eat them. Fruits of species of *Inga* are one example within many, figs another. Perches and Characinidae have been observed to eat fruit, e.g. figs. A list of all fishes in the stomach of which fruits and seeds have been found in the Amazon area is presented by Gottsberger.

Seasonality of animal life in the forest

Another factor we have to count with is the seasonal life of animals in the forest. Animals are very much dependent on plants concerning their food, on the one hand, and depend on climatic periodocities, on the other hand. Even animals which are not herbivores or fruit eaters may depend to a certain degree on the plants. Sweet rotting fruits or the milky sap leaking out from damaged fruits or barks, for example, may attract insects and these, in their turn, attract insectivorous animals. Herbivores are eaten by carnivorous animals so that the circle is closed making obvious the dependence of animals on plants. Plants, in their turn, are influenced by climatic periodicities, as we have seen, and animal life is, therefore, also dependent on the periodicity of the forest.

Very interesting are the observations on seasonality of tropical bird communities in Panamá by Karr (1976). Rains here begin in late April or early May and continue until December. The dry period then follows from late December through the early part of April. Small animal-dispersed fruits are generally available throughout the year, while large-seeded fruits, eaten by toucans, parrots and cotingids, are more seasonal. Seasonal variations in fruit abundance occur with 2 major peaks: in March-June during a time of increasing rain fall, on the one hand, and in September-October during the later part of the rains. A biomass peak may be observed in the early rainy season, i.e. in May-June. Numbers of insects in and above the canopy are seasonally more variable than insect abundances near the ground. Abundance of spiders, for example, is high in the early part of the rainy season (May-June) and in October. Minor peaks in the number of bird species coincide with the onset of the rainy season. In the high stratum, Karr observed a bird peak in May and a decrease during the dry season when availability of food decreases. In temperate regions, the peak of the breeding season occurs in the late spring and early summer.

For Venezuelan Guiana I could find only some examples of seasonality of animal life which are given in the following. Herons, such as *Ardea candidissima* and *A. garzetta*, have their most beautiful plumage in June, July and August during their mating season. This coincides with the rainy season. Turtles, such as *Emyx tracaxa*, lay their eggs end of February. Before laying eggs, the turtles eat in certain periods fruits with a bitter taste which is transmitted to the eggs. The iguana (*Iguana iguana*) deposits its eggs likewise during the dry season, from February to March. Cracidae, such as *Crax nigra*, the "paují" and other "pavas" (*Penelope purpurascens*) have been observed to mate beginning of April, at the end of the dry season. When the "Southern Cross" is in its culminating point beginning of April, *Crax nigra* starts to warble. *Cacicus*, the bird with the pendant nests, raises its pullets at the beginning of April. *Procnias averano*, the campanero, lets hear its loud warbling in April and May when it is mating time. *Hyla crepitans*, the "singing" frog, on the other hand, starts to crack or warble at the beginning of the rainy season, end of May, sometimes driving people crazy with its concerts.

Although there may be many more examples of periodicity in animal life, it becomes clear that climatic periods play the most important part in the animal life, and in particular the changes of dry and wet seasons. However, when certain animals suffer from the drought, others lay their eggs (reptiles, for example). Some birds have their mating season at the end of the dry season, others during the very wet season. Not all animal species react in the same way, some species prefering certain periods which others avoid. Consequently, a similar pattern arises in animal life as has been demonstrated for plant life. There are plant species which prefer the dry season for dispersal, but others the wet season, and a third group fruits partly in the dry and partly in the wet season. Likewise, animal life is active the year round, but possibly most vigorous at the beginning of the wet season with many exceptions of species which use "seasonal niches" during the year for their activities. The bird peak in May, for example, reported by Karr is certainly related to a fruit peak at the same time, as observed in the studied forest of Venezuelan Guiana. And finally, the seasonality of animals may differ in the different forest strata, as indicated by Karr for insects.

Dispersal methods in dry and humid regions

It is of common knowledge that dispersal strategies of plant species change with the distinct habitats. Wind dispersal prevails in dry regions, while in moist areas zoochory increases. Furthermore, in a warm and moist climate a larger variety of plant and animal species occurs and simultaneously the number of dispersal types augments. Anemochorous fruits and seeds are characteristic, for example, in open grasslands, savannas, steppes, deserts, open areas of high mountains, and mountain peaks. The strong air currents in the high alps and the impoverishment of animal species with increasing height is the reason why anemochory prevails in this region, while zoochory diminishes, and hydrochorous plants are almost absent. Aristeguieta (1967) studied the flora of a dry deciduous forest in the high central savannas (llanos) of Venezuela and came to the conclusion that anemochory is most abundant there. 73% of the species use wind dispersal. Winged fruits and seeds are very frequent, and fruits of Leguminosae predominate, especially of the Papilionaceae family. The region is afflicted by extensive fires during the dry season. Wikander (1980) came to similar results investigating 166 species of the dry deciduous forest of Charallave, a hilly region in the Southern part of the coastal range of Venezuela, where 42% of the species are anemochorous, 30% zoochorous, 19% barochorous, and 9% autochorous. She found that anemochory is more important in the higher more wind-exposed hill tops, whereas zoochory increases towards the lowland and the riverside. Furthermore, the flora of the upper hillside drops its leaves earlier when the dry season arrives than the plants in the lowland and vice versa: the plants of the upper hillside are delayed in the development of the foliage when the rainy season begins. 77% of the species studied have dry fruits, while 32% form fleshy fruits. Within the 42% anemochorous species winged dispersal units preponderate. In the Andine páramos where wind is probably the most important climatic factor, anemochory is the regular method of plant dispersal. The flora of the Andine páramos is rich in Compositae and particularly in *Espeletia* which predominates in the form of different species over large areas. However, when diaspores are in danger to be carried too far by wind, antitelechorous methods may be developed. Certain Compositae of the high South American Andes show a reduction of the pappus which circumstance may be interpreted as an adaptation to the strong storms which would carry the achenes too far (Reiche 1908, Neger 1913, Schimper-Faber 1935). In the so-called "Matas" of the Venezuelan Llanos atelechorous strategies prevail – otherwise these small plant communities could not persist. Wind dispersal acts where wind is effective, and anemochory is an important agent for the colonization of high mountain tops. For the Malayan archipel as early as 1878 Beccari could demonstrate this phenomenon. Pioneer floras which colonize a new spot always show a high percentage of anemochorous species. The first arrivals at an open field, e.g. at a place which was burned down, are represented by anemochorous plants. They may later establish a secondary forest. Secondary forest in the tropics is always distinguished by a variety of anemochorous species. Undoubtedly, new colonization of islands takes place by a majority of anemochorous plants. One of the most famous examples is the island of Krakatau (Treub 1888).

Open areas, in general, are conspicuous by their anemochorous flora. The denser a plant community becomes, the more other dispersal methods increase, while wind dispersal diminishes rapidly. In the humid tropical forest zoochory is most important. In a dense forest, anemochory would be of no or little effect. On the other hand, the variety of animal species is very large and zoochory, therefore, prevails. This fact not only applies to the rain forest, but also to the cloud forest of higher altitudes (approximately from 1200 to 2000 m). Studies in the cloud forest of the "Avila" (rising above Caracas) and of "Rancho Grande" of the Venezuelan coastal range have shown that berries and drupes prevail, particularly in the lower strata. However, as we shall see below, dispersal methods very much depend on the forest layer in which the diaspores develop.

Forest layering and dispersal strategies

Forests are large plant communities in which the microclimate plays an important part. Richards (1952) has treated this subject profoundly for the tropical rain forest. Air movement, for example, is most vigorous in the tree tops, while on the forest floor an extreme stillness of the air reigns. The greatest amount of air movement predominates during the dry season, affecting the tree crowns to a large extent. As a consequence, humidity changes very much in vertical direction of the forest. While the exposed tree crowns may suffer from drought, humidity considerably increases towards the forest floor. The daily and yearly variation of humidity is smaller at the forest floor or in the undergrowth than it is in the upper forest layers. Likewise, temperature in the middle of the day is some 2–3 degrees lower in the undergrowth than among the crowns of the high trees. Illumination changes accordingly. Whereas the exposed trees receive strong irridiation, illumination rapidly decreases towards the forest floor so that the undergrowth only receives 1% or a fraction of the outside daylight. Even the amount of carbon dioxide tends to be slightly richer in the undergrowth than in the crowns of the high trees. It is, therefore, not surprising that we observe a certain layering or stratification within the forest. The tall trees are light demanders, while in the undergrowth the shade type of plants prevails. At the same time, the tall trees are able to endure drought, whereas the plants of the undergrowth are in need of high humidity. We thus find species of very different demands in the distinct stories. These differences in the microclimatic conditions also express themselves in the shape and structure of leaves. As I could show (Roth 1984), the leaves of the upper story trees are of the xeromorphic sun type. They are smaller and have a typical xeromorphic structure, possibly with water-storing tissues, stronger outer epidermis walls, a denser mesophyll with a restricted intercellular system, a larger number of smaller stomata. The sun type of leaves becomes apparent in the higher rate of the proportion palisade parenchyma / spongy parenchyma, in smaller and more elongated palisade cells, and in more numerous but smaller chloroplasts. The undergrowth plants, on the other hand, are of the hygromorphic shade type. They have larger and thinner leaves with delicate cell walls, a large intercellular system, a looser spongy parenchyma, a reduced proportion of the palisade / spongy parenchyma. Occasionally, palisade cells are transformed into funnel-shaped cells with large chloroplasts which have a better capacity to absorb diffuse light. Even the ultrastructure of the chloroplasts differs in the lower strata from those of the upper strata and the chlorophyll proportion a/b increases in the lower strata. During the development of a plant from a seedling to a high tree, the foliage of one and the same individual changes accordingly in outer shape and inner structure. Young plants characteristically develop the shade leaf type, while during growth in height of the tree inner and outer leaf structure progressively becomes modified. Leaf structure thus depends on the height of the developing individual, and young plants have a different leaf structure from grown-up upper story trees of one and the same species. This layering of the forest induced by the vertically changing microclimate also expresses itself in the dispersal strategies of the different species.

It is generally accepted that wind dispersal is more common to the upper story and emergent trees, while in the lower strata zoochory predominates. Some tropical humid forest areas show up to 80% zoochory (Foster 1973). It is not by accident that most of the wind-dispersed Dipterocarpaceae are high or emergent trees. Most of the canopy-level seed production is wind-dispersed in a Malayan Dipterocarp forest. Jones (1956) found more anemochory in the upper stories of a Nigerian rain forest and an increase of zoochory towards the lower layers. 20–30% of the tall trees and lianans have winged or plumed diaspores in a tropical rain forest of Ghana studied by Hall & Swaine (1981). Keay (1957) describes the same relation of wind-dispersal and height of the crowns for a Nigerian forest. Emergent trees are, certainly, most exposed to wind and anemochory is frequent between them. *Ceiba pentandra*, often found as an exposed solitary tree, has dehiscing capsules which release a woolly mass together with the small seeds which are blown to very remote distances. Wikander (1980) found the majority of wind-dispersed species in the upper stories of a dry deciduous forest in Venezuela, but mainly zoochory in the lowest stratum. It has already been mentioned that the same forest develops more anemochorous species on the hillside than in the lowland. Very interesting is the observation of Opler, Baker & Frankie (1975) that shrubby species of *Cordia* are dispersed by birds, tree species of medium height by mammalia and that higher trees use wind-dispersal. Lianas are usually also light demanders and often reach the tree tops. Likewise phanerogamous epiphytes often inhabit the crowns of high trees. It may not surprise, therefore, that lianas (e.g. Bignoniaceae, Sapindaceae) and epiphytes are frequently dispersed by wind. Hall & Swaine (1980) mention *Campylostemon* and *Hippocratea* as climbers of high trees with capsular fruits and wind-dispersed winged seeds. Orchids have dust-seeds which are easily dispersed by wind. Epiphytes of the Bromeliaceae family (e.g. *Tillandsia*) are anemochorous.

Observations of dispersal mechanisms are apparently more numerous in the lower forest stories so that many different kinds of zoochory are described. The microclimate in the forest is of great importance so "that the same area of forest can shelter two quite different faunas, that of the tree tops and that of the undergrowth" (Richards 1952). Interactions between animals and plants are well studied. Whitmore (1975) states for the tropical rain forest of the Far East that animals play a major part in the dispersal of seeds of tropical trees. The most important relation probably exists between plants and birds. Birds are attracted most by bright red and brilliant black colours or by contrasting red and black. Well exposed dangling fruits and seeds facilitate visits of birds as well as of bats. Some plant species develop pendulous fruits with elongated peducles (e.g. mango, Fig. 1). In other cases, the seeds form a long funicle dangling from the dehiscent fruit (e.g. *Swartzis schomburgkii*, Fig. 65). Seeds and their arils often contrast in colour: e.g. shiny black seeds with red arils. Another syndrome of characteristics is represented by plants adapted to bat dispersal. Bat fruits are preferably yellow, brownish or blue. The colours are dull and the odour musty. Mammalia are rather guided by smell than by sight, especially the nocturnal ones. Fruit eating carnivores, particularly civets and mongooses and even bears are seed dispersers in the Far East, according to Whitmore. Monkeys and rodents (squirrels) feed on fruits and seeds. They range within the most important dispersers in some areas. The "rabopelado" (marsupiales) is a very eager fruit eater in

South America. Peccaries feed on fruits and seeds fallen to the ground. While some bird species feed in the uppermost forest layers, others visit the lower trees. Certain mammalia occupy the intermediate strata and others feed on the ground. Adhesive fruits or seeds with hooks and spines or with sticky secretions are more characteristic of herbs or shrubs. Sticky epiornithochorous *Cordia* fruits, for example, with slimy contents develop on shrubs or low trees. Myrmecochory is characteristic of the herbs of the lowest stratum. We thus recognize a certain subdivision of the environment in vertical direction where distinct animal species replace one another so that one could speak of a vertical stratification of biota in which plants and animals are adapted to one another. Stratification of the forest thus reveals itself in the dispersal methods of plants. Regal (1977) came to the interesting result that animal-dispersed and animal-pollinated plants abound today in the subcanopy of moist tropical forests where air movement is reduced and irradiation is patchy, while abiotically dispersed species are among the emergent trees.

Stratification of tropical forests corresponds to an evolutionary process through the "upward struggle of plants for light". Trunks became taller and taller, while the ancient pachycaul tree remained in the undergrowth, and a canopy was raised in this way (Corner 1954). As we have seen already in the present studies, not only anemochory is favoured in high trees, but also autochory. Seeds falling from dehiscing capsules have a better chance to be carried further the higher the parent tree is. Furthermore, light or flat seeds may additionally be carried by the wind. A third type of fruit more frequently found in high trees is the drupe. As compared with the many-seeded berry the drupe represents a derived fruit form in which the number of seeds is reduced often to one. As Janzen (1977) emphasizes "One way to evolutionary engineer a higher fruit/seed ratio is to abort seeds". One-seeded drupes may possibly be more attractive to animals for this reason. However, not only reduction of locules and seeds takes place in the drupe, but also more differentiation into an outer fleshy and edible part as well as into an inner hard and sclerenchymatous part apt to protect the seed mechanically becomes apparent. The drupaceous type may possibly be connected with some seed dormancy, as dormancy occurs in some emergent trees of humid tropical forests, especially those which are hard-coated.

In summary we may say that zoochory prevails in the lower and lowest strata of the forest, while the abiotic method anemochory as well as autochory are favoured in the uppermost strata and in emergent trees. Dispersal methods in the upper strata are more influenced by climatic factors such as wind, drought and insolation. Furthermore, the higher the tree the further the dispersal units may be carried away from the parent plant. Besides the abiotic method and autochory we find drupaceous fruits in many high tree species. Drupes are considered derived by seed reduction. As we have seen before, anemochory is an evolutionally derived dispersal method as well as autochory in the form of dry dehiscent capsules possibly with a reduced number of seeds may be considered derived. In most high tree species we are thus confronted with evolutionally derived dispersal methods, whereas the more ancient zoochorous dispersal method is kept in the lower forest strata. When, during evolutionary processes, tree species gained more and more in height through the upward struggle for light they modified their dispersal methods progressively in the same way as they changed their leaf structure through a transformation of the hygromorphic shade leaf type into the xeromorphic sun leaf type (Roth 1984).

Proportion of fleshy to dry fruits in the different strata

It may be of interest to know the proportion of fleshy and dry fruits in the different forest strata (A-a-aa), as far as information is available. The expression "fleshy" fruit versus "dry" fruit is usually applied in the words widest sense. Berries and drupes eaten by animals are generally designated as fleshy, although the fruit flesh layer may be very thin, while nuts and samaras are considered dry (leathery or woody). In the case of the capsule one usually assumes that the dehiscing fruit is dry. There exist, however, quite a few fleshy capsules, possibly of a more primitive type. The fleshy capsules are deignated as fleshy in this survey, although they are usually not eaten by animals. More complicated still are examples where the "fruit" sensu lato seems to be fleshy and edible, although the ripe gynoecium is represented by a nut, whereas the fleshy parts arise from the perigone, the receptacle, or other accessory parts. This may also be the case of infrutescences where the fruit itself corresponds to a nut, while the infrutescence axis is fleshy. This type is likewise called here "fleshy fruit" for ecological reasons.

Of 322 species taken into consideration here

90 species or 27,95% of the A-story have fleshy fruits and 124 species or 38,50% have dry fruits,
54 species or 16,77% of the a-story have fleshy fruits and 39 species or 12,11% have dry fruits, and
7 species or 2,17% of the aa-story have fleshy fruits and 8 species or 2,48% have dry fruits.

From the above given survey it is obvious that the proportion of species with fleshy fruits to those with dry fruits decreases from the a-story towards the A-story (from 1,38 to 0,72). In other words: the higher the trees become, the drier the fruits. However, we have to take into consideration that the total number of A-story trees very much surpasses that of the a-story trees not to speak of the low number of aa-story trees which is really not representative. On the other hand, the expressions "fleshy" or "dry" very much used in ecological studies do not supply the same conclusive evidence as the consideration of the different fruit types (see p. 167).

Monochory and polychory

Roughly speaking we distinguish certain principal types of dispersal, as there are: Achory (atelechory), barochory, autochory, zoochory (with endo-, syn-, and epizoochory), hydrochory, and anemochory. Within zoochory we encounter most specialization through the great variety of dispersal agents (ichthyochory, saurochory, ornithochory, mammaliochory, myrmecochory). Endozoochory, synzzochory, and epizoochory may become mixed up in certain more complicated cases. Epizoochory may work well when fruits or seeds are furnished with spines, burrs, hooks, glands, slimy secretions. Endozoochory is characterized by a mutual combination of "supply and demand". Hydrochory may be realized through rain drops only or through rain wash. Submerged or floating diaspores may be transported by rivers, lakes or the ocean. Guppy made a profound contribution to this subject. Very small diaspores (dust diaspores) may be carried by the wind for their light weight. Other anemochorous diaspores are plumed or winged, appear as inflated balloons, or in the form of tumbleweeds and wind-ballistics.

However, these observations have a different look when practice begins. Many dispersal units are dispersed by 2 or more agents, being diplochorous or polychorous. Moreover, some of the above mentioned categories are ill definded. When are we able to speak of real "autochory"? Should some activity be implied in this dispersal type? *Ecballium*, for example, has a very active strategy of dispersal. *Dorstenia* shoots with projectiles using a turgor mechanism, while *Hura crepitans* throwing its seeds as far as 45 meters has a xerochastic method. These and many other examples are mentioned in my book "Fruits of Angiosperms". But what shall we do with the majority of dehiscing fruits (capsules)? Is their dehiscence always as spectacular and effective as in the above mentioned examples? Do all autochorous plants throw their dispersal units far away? A large number of them certainly opens their capsules slowly just letting fall out the seeds. The seeds may then fall to the ground close to the mother plant. Many capsules have a delayed dehiscence not opening before fallen to the ground. In which category should we put them? May we still speak of autochory? Fruits dehiscent on the ground may later be eaten by animals. Many examples of achory, barochory and zoochory are apparently catalogued within the autochorous types. Moreover, regarding the great variety of fruits in the tropics, we are happy in most cases to know whether a fruit is dehiscing or not, but we are not able to say whether it opens effectively enough to really disperse its seeds in this way. Small seeds may be carried away by the wind when the capsule opens, but we still speak of an autochorous dispersal type.

Barochory, on the other hand, means that heavy dispersal units fall to the ground by their own weight possibly to roll away from the mother plant for a certain distance that depends on the inclination of the ground on which the mother plant grows and of the weight of the dispersal unit. Barochory is considered a primitive dispersal method not seldom observed in tropical humid forests. The globular fruits of the caulicarpous *Couroupita guianensis*, Lecythidaceae, are very hard and considerably heavy, but I have never seen them roll away very far from the mother plant. They may rot on the forest floor and finally attract animals when half decomposed. The very large and very attractive infrutescences of *Pandanus* which are colourfully painted with red, orange or yellow fall to the ground when ripe and there certainly atract animals which eat them. Wikander and Gonzalez (1978) mention 2 cases of barochory (*Acacia macrantha* and *Cassia emarginata*) with indehiscent legumes which fall to the ground by their weight (weighing about 1,7 and 6,4 g respectively), while the seeds already begin to germinate within the decomposing fruit. Maybe this is an adaptation to achory. Examples of this kind lead us to the conclusion that barochory and achory are at least related to one another and that barochory is not a very specialized type of dispersal.

It is rather probable that in the tropical humid forest several dispersers and possibly more than one dispersal method are used by many plants. Many dehiscing autochorous fruits have seeds with an aril or a caruncle to be dispersed by animals (e.g. Euphorbiaceae). Autochory is here combined with zoochory. Fruits with winged seeds usually dehisce first to release the seeds which are dispersed by wind. Autochory and anemochory are apparently combined with one another in this case. Many winged fruits and seeds or dispersal units with hairs are dispersed by wind as well as by water. Rain is apparently a very important climatic factor in the rain forest and rain wash may be a method of transportation for any fruit or seed on the ground not too heavy to be washed away. Epizoochoric fruits and seeds may use any adequate "vehicle" that is passing by. Animals of very different kinds can enjoy the sweet pulp of a fig, mango, paw paw, mammey apple, soursap, sapote, cashew "apple", beach grape or the like. On the other hand, many dispersal mechanisms using a single method or a single disperser are specialized. Hydrochorous dispersal – considered relatively ancient – may be assisted by zoochory (e.g. ichthyochory). Wind dispersal is dependent on air currents and the special shape and weight of the dispersal unit. Anemochory is a derived and specialized form of dispersal in which the diaspore gradually adapted to the dispersal method by reduction of the specific weight and the enlargement of the surface through formation of wings or long hairs, by transformation of the perianth, perigone, calyx or bracts into flying organs. The point of gravity of the dispersal unit has much influence on the way of gliding as well as has the turbulence of the air (strong wind, cyclone etc.) on the distance to which the diaspore is dispersed. The higher the tree and the faster the air movement the further the diaspore may be blown, as a rule. The same is true of the seeds dispersed in an autochorous way. However, some winged dispersal units are very large (e.g. *Centrolobium, Gyranthera caribensis*) and merely fall to the ground close to the mother plant or are dispersed only over short distances. Arrived at the ground they may be transported further by a rain wash or by animals.

In summary, in many cases we have to count with several dispersal possibilities of one and the same dispersal unit: e.g. different animals are able to disperse the fruit or seed (birds, bats, mammals) or autochorous dispersal is combined with zoochory, or barochory is connected with zoochory (the diaspore is dispersed when fallen to the ground). Furthermore, a delayed zoochorous dispersal is realized when fruits fall to the ground (achory, barochory) and after rotting are dispersed by animals, when the pulp inside becomes free (*Enterolobium cyclocarpum*). In certain cases, even fruits and seeds are dispersed separately (by scatterhoarding of the fruits and following seed dispersal). Many distinct dispersal possibilities consequently exist and a clear classification of a certain dispersal type in a certain dispersal category is sometimes difficult.

Species diversity and series of species

Richards (1945) was the first to point to the great number of closely allied species occurring side by side in the tropical rain forest. Continuous series not only concern allied species, but also taxa of higher ranks, such as genera and families. Federov (1966) mentions natural series or groups of affinity in families such as the Burseraceae (*Protium*), Elaeocarpaceae (*Sloanea*), Lecythidaceae (*Eschweilera*), Mimosaceae (*Parkia, Pithecellobium*), Myrtaceae (*Eugenia*), Sapotaceae, which are also of interest in the present study. Federov reasons that the multitude of allied species may be distributed among different strata and synusiae and that previous ecological isolation provides the possibility of the co-existence of series of species within the same area. But he concludes that an overwhelming majority of allied species are found within the same stratum, not seldom in the upper stratum among emergent trees.

A great variety of species can coexist in the tropical rain forest each species forming a small population from the very origin but stable with respect to the number of constituent individuals. According to calculations of Black et al. (1950) population density in the Amazonian region is about one individual per hectare (100 × 100 m). There exists apparently an inverse relationship between the number of species and the number of individuals. Very few species attain a density of 40–50 up to 200 or more individuals per hectare. The species with a larger population density are usually of lower strata, according to Fedorov, while the lowest population density is characteristic of the stratum of emergent trees and the uppermost storey. As a possible explanation, I would like to add here that this is the stratum of most competition. As most trees flower seldom and more or less irregularly, cross pollination is scanty and self-pollination presumably prevails (biotic isolation). However, several families with series of allied species have dioecious trees, such as the Burseraceae, Euphorbiaceae, Guttiferae, and others (see Ashton 1969). Self-pollination is a condition favourable for the rapid origin of species by means of genetic drift. Mutant genes accumulate in populations and contribute to the relatively rapid orgin of series of closely allied species. The allied species show characters that are neither beneficial nor deletorious. A series of species may thus be regarded as a single disintegrating species in statu nascendi.

Ashton (1969) suggests that clumping (of species) is most pronounced in those families where means of dispersal do not exist or are unreliable, e.g. in the Apocynaceae, Leguminosae and in animal-dispersed families such as the Meliaceae, Burseraceae, Sapotaceae. As complexity of the forest increases, the numbers of biotic niches into which evolution can take place increases. A variety of different climates exists within the forest canopy, according to Ashton, into which plants can be adapted. As the forest hightens, a multitude of microclimates and microhabitats appear (Corner 1954). Ashton furthermore emphasizes that a great number of allied species occurring side by side in many genera and families is a particular feature of South-East Asian rain forest, not, for instance, occuring in Central America. However, this statement cannot be supported here, as we have already seen. Another possibility of explaining the presence of series of allied species arises through the refuge theory. Species diversity may be caused by a large number of factores, according to Prance, such as isolation into allopatric populations, adaptation to particular ecological niches, predator pressure on seeds and seedlings (Janzen), phenological isolation, and others.

Refuge theory

According to Haffer (1969), Vuilleumier (1971), and Prance (1978), repeated climatic changes during Pleistocene and Post-Pleistocene resulted in restrictions and reexpansions of the rain forest in the lowland areas of tropical South America. During dry phases, the forest covered only small disjunct patches, while the rest of the area was mostly occupied by savanna. These isolated regions acting as refuges for tropical forest trees became the places of speciation. Each time a continuously dispersed species was broken up into allopatric populations occupying areas with a slightly different ecology, the opportunity of speciation increased. During dry phases, local selection pressures allowed differentiation among populations of a species inhabiting different forest refugia. In humid phases, the forest expanded and a complex pattern of species diversity resulted.

These new species should be more or less similar to one another differing in few not very important characteristics so that series of species arise comprising a larger number of similar species. Such a group may be recognized in *Miconia* where the species mainly differ by their fruiting times spread out over the entire year (Snow 1965). In the area studied of Guiana, a great number of such groups or series of species may be recognized within many families, such as the Rosaceae (the genus *Licania*), Apocynaceae (*Aspidosperma*), Bignoniaceae (*Tabebuia*), Burseraceae (*Protium*), Euphorbiaceae (*Sapium*), Guttiferae (*Rheedia*), Lauraceae, Lecythidaceae (*Eschweilera*), Mimosaceae (*Inga*), Sapotaceae (*Pouteria*), and others. As could be demonstrated in my studies, the allied species belonging to a series of species are not distinguished by their dispersal mechanisms. But although the allied species develop the same or very similar dispersal units, their rates of distribution may differ very much. While most species show very low rates with only a few individuals, in the forest studied, some have very high rates with 1.000 to several 1.000 individuals. These differences can certainly not be explained by the dispersal mechanism which is the same in all representatives belonging to the same series or group. The reason for the higher rates of distribution could be found in a better germination power, higher resistance to adverse conditions and against predators, a stronger radical system, faster growth, better reproduction at shorter intervals, and other factors – but this could not be examined in detail here. On the other side, it is possible that historical events are the reason for the abundance in individuals of some species and the reduced number in individuals of other similar species of the same group or series. This is the more probable, as the shield of Guiana represents the oldest part of the South American continent. The species with the lowest numbers of individuals could have arisen from more or less marginal "islands" of the disjunct forest area, while the species with a large number of individuals could come from larger refuge areas of more or less central position. The result of the present studies would thus be that the dispersal methods are of minor importance for the number of individuals present in the modern primary tropical forest, while historic-geological events would be of more significance. "Dispersal is but *one* of the factors determining the presence of plants". Although the refuge theory is only one of many possibilities to explain this phenomenon, in the varying numbers of individuals among allied species it certainly becomes evident that dispersal mechanisms are not responsible for these differences.

Number of individuals and height categories in homogeneous groups

In certain families, several species of the same genus or even of different genera show almost the same type of fruits and seeds and the same dispersal mechanisms, however, the number of individuals of each species is very different. In some of these homogeneous groups only one species stands out for its high number of individuals, while the other species have only few individuals, but more often – two or several species show high or higher numbers of individuals, while the remaining species are represented by few individuals. Examples are described in the following:

Within the Apocynaceae all species studied of *Aspidosperma* as well as *Himatanthus articulatus* develop dehiscent capsules with winged seeds which are dispersed by wind. However, although the dispersal mechanism is the same in all 5 species of *Aspidosperma* as well as in *Himatanthus*, the number of individuals in the forest is quite different. The highest number is reached by *A. excelsum* with 1.760 individuals. The lowest numbers are 0, 3 and 59. Other numbers lie in between (550 and 650). This fact cannot be explained by the dispersal mechanism.

The 6 species of Bignoniaceae have capsules with winged seeds (2 species of *Jacaranda* and 4 species of *Tabebuia*). The number of individuals is very low in 4 species, while one species has 111 individuals and another one reaches 704 individuals.

Four species of *Cordia*, Boraginaceae, have zoochorous drupes. But their numbers of individuals are very different, oscillating between 0, 164, 263 and 624.

All 12 species of Burseraceae (7 species of Protium, 3 species of *Tetragastris*, one species of *Trattinickia* and "Maramo") develop dehiscing drupes or drupaceous capsules which are mainly dispersed by animals. However, the numbers of individuals are quite different. While 6 species are represented by a number of individuals below 100, others have 155, 293 and 408 individuals respectively. Two species, however, are represented by very high numbers of individuals, namely 1.157 and 2.669.

The 6 species of *Sloanea*, Elaeocarpaceae, have capsular fruits. Of 3 species we know that the seeds are furnished with an orange aril. It is quite possible that the other 3 species likewise have seeds with arils. 2 species have a very similar number of individuals (167 and 168), while the number of individuals of the other species oscillates between 1, 8, 47, and 66.

21 species of the Euphorbiaceae have capsular fruits and are autochorous. However, their numbers of individuals are very different. While most species have n number of individuals below 100, some have 180, 190, 349, and 423 individuals. But 2 species are represented by very high numbers: 1.706 and 3.679 individuals, respectively.

Within the Guttiferae, 3 species have capsular fruits and are autochorous. However, 2 species have only 76 and 264 species, while one species is represented by as many as 1.893 individuals.

The 14 species of Lauraceae develop all drupes which are zoochorous. Their number of individuals oscillates between a few individuals, more than 100, 207, and 300 individuals.

Of the 11 species of Lecythidaceae all develop capsular pyxidia. However, 2 species are anemochorous having winged seeds. The other 9 species likewise have deshiscing pyxidia, but the seeds are dispersed by animals. While the number of individuals is low to moderate in 5 of the zoochorous species (between 1 and 247), 4 species have a very high number of individuals: 803, 1.025, 2374, and 3.446.

The 16 species of *Inga*, Mimosaceae, have indehiscent zoochorous pods. Their numbers of individuals are very low, except some species with more than 200 individuals and 2 species with 404 and 409 individuals. Most of the auto-zoochorous species of the Mimosaceae with dehiscent pods have a low number of individuals except one species with 2.534 individuals.

All 8 autochorous species of the Caesalpiniaceae with dehiscent pods have a minimal number of individuals except one species with 434 individuals.

Of the 3 autochorous species of the Papilionaceae one species has 0 individuals, another 188, while the third is represented by as many as 1.262 individuals.

Most of the 7 auto-zoochorous species of Meliaceae have very low numbers of individuals, except 2 species with 885 and 1.376 individuals respectively.

All 14 species of Myrtaceae have drupes and are autochorous. Most of them have very low numbers of individuals. One has 140, another 243, and a third 328 individuals. The highest number is reached by 506 individuals in one species.

The 15 species of Rosaceae are all zoochorous and develop drupes which are very similar in their outer aspect and inner structure (species of *Couepia, Hirtella, Licania, Parinari*, and *Prunus*). However, their numbers of individuals are very different. While 10 species remain below 100 individuals, 3 species have between 101 and 343 individuals, another species is represented by 679 individuals, and only one has as many as 3.258 individuals.

The 9 species of Rutaceae have all capsules and are autochorous. They belong to different height categories. However, only one species reaches 203 individuals, while most of the species remain below 37 individuals.

The 21 species of Sapotaceae are very much

alike in their dispersal mechanisms. Most of them have drupes or few-seeded berries dispersed by animals (birds and bats). However, their numbers of individuals are very different: 7 species remain below 100 individuals, 10 species have between 100 and 500 individuals, 3 species have between 614 and 795 individuals, and only one species is represented by as many as 1.658 individuals.

These examples show first of all that the dispersal mechanism is not so important for the number of individuals present in the forest, as van der Pijl has pointed out already "Dispersal is but one of the factors determining the presence of plants". And one should add here: that dispersal seems to be one of the least important factors for the presence of plants in the primary forest studied, and possibly in most primary forests of this type. In about 17 families we could observe homogeneous groups. All types of dispersal mechanisms are represented in these groups: 2 families are anemochorous, 5 families are autochorous, 3 families are auto-zoochorous, and 7 families are zoochorous. Formation of homogeneous groups, therefore, does not depend on the dispersal type. It is true that the height of the tree may influence the number of individuals present in the forest, as certain dispersal mechanisms are more advantageous in a definite forest layer, as we have seen. Anemochory and autochory are more advantageous in the upper forest layers, zoochory in the lower forest layers. However, within the Apocynaceae all 6 anemochorous species belong to height category A, and only one species has an outstanding number of individuals (1.760). Height differences are, consequently, not the reason of differences in the number of individuals. Within the 6 anemochorous Bignoniaceae the species of lesser height (a) has most individuals (704). Of the 4 zoochorous species with drupes of the Boraginaceae all belong to height category A, and only one species reaches 624 individuals. Of the 12 zoochorous species of the Burseraceae one half is of height category A and the other half of height category a. Of the 2 species with an outstanding number of individuals one belongs to height category A (2.669 individuals), while the other is of height category a (1.157 individuals). Of the 6 species of Elaeocarpaceae, 4 belong to height category A and 2 to a. One species of height category A reaches 167 individuals and another one of height category a 168 individuals. Of the 21 autochorous species of Euphorbiaceae, 10 belong to height category A, 9 to a, and one to aa. One species with most individuals belongs to height category A (3.679 individuals) and another one to a with 1.706 individuals. The 3 autochorous species of Guttiferae are all of height category a. Nevertheless, one species reaches as many as 1.893 individuals. Of the 14 zoochorous species of Lauraceae, 6 are of height category A and 5 of a. None of the species reaches a very high number of individuals. Of the 8 auto-zoochorous species of Lecythidaceae, all are of height category A. The highest numbers of individuals are reached by species of height category A (3.446, 2.374, 1.025). Of the 16 zoochorous species of Mimosaceae, 6 belong to height category A, 8 to a, and 2 to aa. The highest numbers of individuals are reached by 2 species belonging to height category A (404 and 409). Of the 8 autochorous species of Caesalpiniaceae, 6 are of height category A, and 2 of a. The species with most individuals belongs to height category a. All 3 autochorous species of Papilionaceae belong to height category A. However, the difference in the number of individuals is enormous: 0, 188 and 1.262, respectively. Of the 7 auto-zoochorous species of Meliaceae 2 belong to height category A, and 5 to a. The species with most individuals (1.376) belongs to a, followed by a species with 885 individuals of height category A. Of the 14 zoochorous species of Myrtaceae, 4 belong to height category A and 10 to a. The species with most individuals (506) is of height category A, followed by 2 species of height category a with 328 and 243 individuals, respectively. Of the 15 zoochorous species of Rosaceae 12 are of height category A and 2 of a. The species with most individuals is a high tree of heigh category A (with 3.258 individuals). Of the 9 autochorous species of Rutaceae, 5 are of height category A, one of a, and 2 of aa. However, none of the species reaches a high number of individuals. Of the 21 zoochorous species of Sapotaceae, 15 belong to height category A, and 5 to a. The species with most individuals is of height category A (1.658 individuals), followed by another one of height category A (795 individuals), and one of height category a (722 individuals).

In summary we could say that the height categories do not play a very important part within homogeneous groups concerning the number of individuals. However, there are many other factors which could influence the number of individuals present in the forest, as there are: Damage by predators, resistance of the tree, rapidity of growth, extension of the root system, germination power of the seeds, competition with other species, adaptability of species, availability of pollinators, number of fruits and seeds produced per year, intervals between fruiting times, original habitat of the species and others. We do not know whether the members of a homogeneous group differ in these characteristics. Assuming that they have developed under the same environmental conditions, the differences should not be too great. We would thus expect of a homogeneous group that the number of individuals of all pertaining species is relatively low and that there are only slight fluctuations in the number of individuals from species to species. Such a group is represented by the Lauraceae and the Rutaceae, for example. While in the Rosaceae the numbers of individuals are spread out over a larger scale (up to 679 individuals), one species stands out for its very high number of individuals (3.258). One could thus speak here of a certain "dominance". This species covers 67,91% of the total number of individuals. Within the Apocynaceae one species is likewise dominant concerning the number of individuals reaching 58,23% of the anemochorous individuals. One dominant species concerning the number of individuals is also found in the Bignoniaceae covering 78,30%. Of the 4 zoochorous species with drupes of the genus *Cordia*, 59,37% of the individuals are covered by one species. Within the Burseraceae, a very homogeneous group, 2 species have outstanding numbers of individuals, one species reaching 48,98%, the other 21,23%, while the next following species only covers 12,80%. Within the genus *Sloanea*, Elaeocarpaceae, one species reaches 36,56% and the second in place 36,54%. Of the 24 autochorous species of Euphorbiaceae, one species covers 57,43%, and the second in place 26,63% of the total number of individuals. The next following species only reaches 2,96%. Here, the dominance of 2 species becomes very obvious. Of the auto-zoochorous Lecythidaceae, one species covers 42,25% of the total individuals, and the next following species 29,11%. Of the 16 zoochorous species of *Inga*, Mimosaceae, one species covers 22,03%, and the next following 21,76%. On the third place follows a species reaching 15,46%. Of the 7 auto-zoochorous species of Meliaceae, one species covers 55,84%. Of the 14 zoochorous species of Myrtaceae, one species reaches 32,56%, a second species 21,10%, and a third 15,63% of the total individuals. Of the 21 zoochorous species of Sapotaceae, one species covers as much as 25,31%, the next following species 12,13%, and a third species 9,37%.

From this survey the following conclusions may be drawn: Where only one species is dominant, this may cover as much as 78% of the total individuals. Where 2 or 3 species have an outstanding number of individuals, the percentages diminish accordingly. We have already asked the question: How may this dominance in individuals be explained and the answer was that many factors influence the number of individuals by which a species is represented in the forest. It is certainly impossible to study all these factors in detail for all species present in the forest of Venezuelan Guiana. Modern authors suggest that both competition and predation influence and control most the existence and abundance of plant species. Leigh (1983) emphasized that insect pests do not allow any one of the species to become too common (see also the many papers of Janzen). We have already seen that species characteristic of another habitat are usually represented by few individuals in the rain forest. Furthermore it could be shown that very resistant species (e.g. against drought or those supporting a poor soil) frequently do not show an elevated number of individuals. Likewise, species with a very fast growth or those developing a large amount of flowers or flowering every year, as a rule, do not stand out for their high numbers of individuals. However, exceptions always exist and a few were mentioned above.

Dispersal mechanisms, on the other hand, seem to play a subordinate part in the primary humid forest concerning the number of individuals present. Usually sufficient seedlings develop. Considering the age of most primary forest trees – which oscillates between a few hundred to several thousand years – the survival of the species is assured even with a not very effective dispersal mechanism.

The tropical rain forest seems to have established long ago concerning the number of invividuals, more or less maintaining the "status quo" at present time with relatively little changes. We thus could approach the problem from another side, suggesting that the responsibility for the present status lies in certain historic-geological events. In this connection a hypothesis much discussed at the present may be proposed, the refuge theory. If the homogeneous groups have developed in the way the refuge theory proposes, i.e. through contraction and expansion of the forest and formation of new species in disjunct areas, dominance in individuals of one or a few species within a homogeneous group would be easier to explain. The question is, however: are the dominant species of today the original species of the humid tropical forest or did they evolve during the contraction of the forest arising in a slightly changed environment and later had – during the expanding phase of the forest – a better chance to develop more individuals, e.g. in certain niches? Both possibilities could have been realized and for both examples exist. Of the eight autochorous species of Caesalpiniaceae studied one occupies a dominant position concerning the number of individuals, covering about 95%, and this species is characteristic of lowlands and the deciduous forest besides the rain forest. But as a rule one should assume that the original species of the rain forest prevail in the number of individuals, unless they are at the declining line already (e.g. *Caryocar nuciferum*). However, all these considerations remain speculations and the actual data available are too few to draw any definite conclusions. Besides, so many factors are interwoven in the whole spectrum of the rain forest community so that several constituents may become determining in the dominance of certain species.

Evolutionary considerations

Our phylogenetic considerations are founded on the suggestion that the early Angiosperm flora was composed of relatively low stature trees or shrubs protected by higher Gymnospermous trees.

During the Carboniferous era, the first forests were composed of trees which reached a height of 30–40 m (e.g. *Lepidodendron, Sigillaria*) accompanied by a flora of smaller tree ferns (Marattiales) and Pteridosperms of shrubby or liana habit. The undergrowth was formed by herbaceous Sphenophyllae, Hepaticeae and Musci. Simultaneously with the Carboniferous forests the Pteridospermous shrubby flora with *Glossopteris* as a guiding form developed in South Africa, India, Australia, and the South of South America. It is accepted that a Gymnosperm forest originally dominated in the tropics composed of Conifers (*Araucaria*), Cycadatae, Bennettitatae, and Gingkoatae, accompanied by tree ferns and Equisetatae, as well as stragglers of the Pteridospermae. Particularly in the lower Cretaceous, the Bennettitatae with 30–40.000 species and partly with a palm-like habit were abundantly represented. *Ginkgo* reached a height of up to 40 meters. The Cordaitidae were trees which reached up to 30 m in height. *Sequoiadendron giganteum* and other mammoth trees of the Pinidae reached even 100 meters or more. The Bennettitatae with the genus *Cycadeoidea* had their culminating point in the lower Cretaceous. *Cycadeoidea maxima* produced giant flowers with 600 seeds, while *Monanthesia* (Mexico) developed flowers in each leaf axil. Apparently, we have to count with a luxuriant tree flora in tropical forests of these periods, luxuriant more in stature than in the number of species. Only in the lower Cretaceous the Angiosperms appear finding a prosperous area for development. Although the direct origin of the Angiosperms is till obscure, being a fervent point of controversy for many authors, we follow here the ideas of Bancroft (1930) who suggested that a low-growing Pteridosperm of the type of *Glossopteris* from which the Caytoniales are derived is the origin of the Angiosperms. Even this is a theory, it is very likely that the first Angiosperms had a more shrubby habit or were low-growing trees.

When, as suggested here, the early Angiosperms were shrubs or small trees which – in the struggle for light – progressively heightened, it becomes clear, why high trees use more derived dispersal mechanisms, such as anemochory, autochory through dry dehiscing capsules or develop drupes with a reduced seed number, as compared with smaller trees of the undergrowth which have maintained the more original zoochorous dispersal types. Furthermore, the progressive transformation of the hygromorphic shade leaf of small trees into the xeromorphic sun leaf of high trees can easily be explained in this way (Roth 1984).

During evolution fruits with arillate seeds transformed into more xeromorphic dispersal units, according to Corner (1949). Large arillate seeds – originally undetached – have no dormancy and cannot withstand desiccation. Dry dispersal units such as the anemochorous samaras or winged seeds, on the other hand, or dry autochorous capsules prevail in the upper forest layers and in drier more open habitats. Anemochory and autochory, as observed by me in the upper forest strata, are "late", according to van der Pijl (1966). Both dispersal types cover a high percentage in dry and arid climates. Dry dehiscent fruits with dry seeds are derived, as drying out and dehiscence are relative progressions. Furthermore, the detachment of seeds (e.g. in autochorous fruits) is a secondary characteristic, according to Corner (1949). Originally seeds remained undetached on the fruit, possibly for the display of the aril. The formation of a separating layer, e.g. in the form of a meristem or of two or more layers distinguished by their cell size and cell wall thickness has to be considered a derived character, not only in fruits and seeds, but also on leaves and in the bark (Roth 1977 b, 1981). Additionally, the fruits became smaller and the seeds less numerous during evolution (Corner 1954). A further derived fruit form is, therefore, represented by the drupe in which not only a reduction of seeds took place, but also a further differentiation into a fleshy mesocarp and possibly a hard endocarp. Through a thicker endocarp the seeds may assume a better viability so that "mechanical" dormancy results.

Forest stratification and the evolution of the heightening forest

As we have seen in the last chapter, the theory maintained here is that the first Angiosperms were represented by small trees or shrubs which developed well protected below the high crowns of Gymnosperms and tree ferns. Higher and emergent Angiosperm trees evolved secondarily by and by when the Angiosperms conquered the scenery. This theory very well reconciles with the forest layering, as pointed out by me for leaf structure in the different strata (Roth 1984). In my book "Stratification of tropical forests, as seen in leaf structure" I was able to show that a transformation of the hygromorphic shade leaf into the xeromorphic sun leaf takes place during the development of the heightening trees whereby the "sensible height" ranges about between 5 - 10 (15) meters. At this height, the hygromorphic shade leaf frequently begins to transform into a xeromorphic sun leaf, depending on the sensibility of the species as well as on certain environmental conditions. This process has, however, to be considered continuous and progressive. With the heightening of the plant the microclimate progressively changes in so far, as the humidity decreases, while air movement and illumination increase in intensity. However, as could be shown for *Gyranthera caribensis* (Roth 1984), the transformation of the hygromorphic shade leaf into the xeromorphic sun leaf may also take place under invariable environmental conditions without a notable growth in lenght of the young plant so that an endogenous rhythm seems to be imprinted in the genetic code expressing itself during the development of the plant. A progressive transformation of the hygromorphic shade leaf into the xeromorphic sun leaf induced by environmental factors already became a genetic property and is, at least, obvious where an "anticipated planning" became reality. In some species, the young plants begin to develop certain structural xeromorphic peculiarities or characteristics of the sun leaf, at a height where the majority of the species still retains the characteristics of the hygromorphic shade leaf. The genetically fixed inner rhythm is originally induced by changes in the microclimate, i.e. by the progressive heightening of the plant. Although the rhythm of the structural transformation of the leaves is possibly genetically laid down, it is generally activated by the changing microclimate during the heightening of the plant. An endogenous rhythm of comparable nature probably represents the annual rhythm in wood and bark of certain temperate tree species which, although translocated to a homogeneous tropical environment, continue with the formation of annual rings. Zalenskij's law which states that more xeromorphic peculiarities and characteristics of the sun leaf become evident on leaves with the increasing height of leaf insertion on the plant seems to have its origin in evolution. It is quite remarkable that this law is more pronounced in woody species such as *Fagus, Carpinus, Acer, Tilia* (Schramm 1912), in *Eucalyptus globulus* (Magnus 1876 and many other authors cited by Napp-Zinn 1974) and in almost all 232 tree species of Venezuelan Guiana studied by me (Roth 1984), while many exceptions occur in the Graminae and in herbaceous dicotyledons. Zalenskij's law can thus be explained by the progressive heightening of the Angiosperm forest during evolution.

Surprisingly, stratification of the forest not only becomes evident in the leaf structure of trees, but also reveals itself in the evolution of the dispersal units. Lower shade-tolerant tree species not infrequently show heavy gravity-dispersed seeds or fruits, a dispersal mechanism considered more primitive. As the forest heightened during evolution and the higher and emergent trees appeared, autochorous or anemochorous dispersal mechanisms began to develop. Anemochory frequently found in drier and more open regions is considered more derived, while zoochory dominating in the lower forest strata is regarded as more primitive and ancient. It is, furthermore, suggested here that dry dehiscent capsules of the autochorous type are likewise derived, at least in certain families. In the Leguminosae, van der Pijl suggests that the follicle dehescing at the ventral suture and retaining the arillate seeds for the display to birds is more primitive than the legume dehiscing with 2 valves and releasing the seeds in an autochorous way. Autochory is a dispersal mechanism more adequate for a drier environment which is also found in temperate regions, being particularly characteristic of herbs. The herbaceous habit, however, seems to be derived from the arboreous habit. As animal life is most vigorous in the lower and middle strata of the forest, it is not surprising that species of the uppermost strata as well as emergent trees shifted from the biotic dispersal mechanisms towards more abiotic devices. In high and emergent trees which are more exposed to wind, drought and insolation anemochory and autochory seem to be more advantageous and both dispersal mechanisms also occur in more open and drier habitats. Possibly, it has to be distinguished between phylogenetically more primitive autochorous types, on the one hand, and more derived ones, on the other hand (see above).

The stratification of the tropical forest could thus be explained by the evolution of the Angiosperms which developed from more primitive lower trees with hygromorphic shade leaves and zoochorous seeds or fruits by heightening of their stems into upper canopy trees with xeromorphic sun leaves and, often, more abiotic dispersal mechanisms, such as involved in anemochory or autochory. In order to develop more elongated stems, these had to be furnished with more tensile and compressive strength, the activity of the cambium had to be extended, the problems of water transport had to be solved, structural conditions had to be changed in favour of the improvement of transpiration and photosynthesis. From the phylogentic point of view, the highest trees have thus reached the last phase of ecological niche formation in the heightening forest. The "critical height" of 5 - 10 (15) meters of many tree species in the tropical forest at which transformation of the hygromorphic shade leaf into the xeromorphic sun leaf takes place, may possibly be considered the original height of the ancient Angiosperms.

Comparison of bark, leaf and fruit structure in the families studied

Comparing the characteristics of fruit and seed structure with those elaborated for bark (Roth

1981) and leaves (Roth 1984) interesting observations can be made concerning the different families studied. Some families are definitely homogeneous in their characters relating to fruits and seeds as well as to bark and leaf structure. One of these families is the Lecythidaceae which forms a very homogeneous group with a very regular bark in which stratification of soft and hard bast prevails. The hard bast is composed of fibers. Likewise the structure of the leaves is very homogeneous. Most leaves are of the sun type and show some xeromorphic characters (high trees!). The Lecythidaceae are distinguished to a large degree by a lack of peculiarities. As emphasized already, the fruits of the Lecythidaceae are pyxidia with a dehiscing lid and the seeds have often a fleshy funicle eaten by animals. Particularly the species of *Eschweilera* coincide in their fruit and seed morphology as well as in their leaf and bark structure. A series of similar species seems to be realized in the genus *Escweilera.*

Another homogeneous group is represented by the Bignoniaceae and in particular by the genus *Tabebuia.* The bark has a very regular appearance being stratified, whereby the hard bast consists in fibers. Likewise the leaves have several structural characteristics in common. Disc-shaped glandular hairs are most outstanding in the genus *Tabebuia.* The representatives of the Bignoniaceae studied are all high trees with capsular fruits and winged seeds dispersed by wind.

A further homogeneous family is the Lauraceae. The bark is distinguished by alternating hard bast plates. The hard bast is composed of fibers. The leaves are of the sun type with some xeromorphic features. Oil idioblasts are an outstanding peculiarity. The fruits correspond to zoochorous drupes.

The Boraginaceae show a stratified pattern in the bark which, however, is not as regular as in other families. Fibers in the hard bast are common to all species studied. In the leaf structure, the features vary little from species to species. There is usually one palisade layer. Large stomata and stellate hairs are special characters. The trees are all high and most of them develop drupes.

More or less homogeneous are also the Polygonaceae studied concerning bark and leaf structure. In the bark, sclereids as well as fibers are developed, and crystals are an outstanding peculiarity. The leaves are distinguished by their large size, a well-developed palisade parenchyma and large druses. The fruits of the species of *Coccoloba* studied represent nuts.

The bark of the Sapotaceae studied is characterized by alternating plates of hard bast, composed of fibers, and by secretory cells arranged in tangential rows. The leaves are large and coriaceous, and of the xeromorphic type. Tannins are abundant in the cells. The leaf structure is somewhat plump and primitive. The fruits are drupes or berries dispersed by birds and bats.

However, not all families are homogeneous in their features concerning fruit and seed as well as bark and leaf structure. The Apocynaceae form a heterogeneous group concerning their bark structure. There may be fibers or sclereids in the hard bast or the hard bast may be completely absent. A variation in all features of bark structure could be observed. Although the species studied are all high trees, the original leaf type seems to be the hygromorphic shade leaf which progressively adapted to drought and insolation in the upper canopy by formation of longer palisade cells, hiding of the stomata, development of protuberances in the epidermis and possibly by formation of hairs and sclereids. The leaf structure is thus subject to a more or less common scheme. All studied species of *Aspidosperma* develop capsules with winged seeds.

Another family in which not all structural features are homogeneous is the Moraceae. Although the bark structure is of an irregular type, the Moraceae form a homogeneous group concerning bark characters. Leaf structure, on the other hand, is diversified, as variations in the epidermis structure, in the formation of the stomata and the mesophyll occur, while cystoliths and a multilayered epidermis are family characters. However, the fruit morphology is more or less homogeneous, as most species form fleshy infrutescences eaten by animals.

The Melastomaceae form a homogeneous group concerning their bark structure which is irregular and possibly primitive. The bark is, furthermore, delicate. Concerning leaf structure, *Mouriria* has very xeromorphic leaves, while the other species studied are different. Fruit formation in the Melastomaceae is homogeneous, as zoochorous berries are developed.

Heterogeneous in their bark structure as well as in their leaf structure are the Euphorbiaceae. Different types of bark formation are distinguished. The leaves are of the hygromorphic shade type with distinct pecularities. In their fruit structure, on the other hand, most Euphorbiaceae are homogeneous developing dry autochorous capsules.

The delicate bark of the Myrtaceae studied is distinguished by a regular stratification and by septate crystal strands. In their leaf structure, however, the species differ. The leaves may be small or large. Most species are represented by smaller trees. The fruits are zoochorous drupes.

In certain families bark structure and leaf type coincide very well. In the Nyctaginaceae the bark is thin and a hard bast absent. The leaves are of the hygromorphic shade type. The species studied are characteristic of a humid and shady environment. However, the fruits are represented by zoochorous nuts.

A thick rhytidome and xeromorphic sun leaves, on the other hand, are peculiar of the Myristicaceae studied. The fruits are capsular and auto-zoochorous.

Lacistema aggregatum has a thick bark and leathery leaves of the sun type with xeromorphic features. Another coincidence between bark and leaf structure is to observe within the Rubiaceae. *Alseis* develops an aerenchyma in the rhytidome and a loose spongy parenchyma in the leaves.

As could be expected, some families form homogeneous groups concerning their bark and leaf structure, as well as their fruit and seed morphology, others are homogeneous only in fruit morphology, while they may differ either in bark or leaf structure or in both.

Usefulness of tropical fruits and seeds

The fruits mainly serve for seed dispersal and – when zoochorous – contain attractants for animals. Juicy fruits contain a large amount of water. As nutritious contents the carbohydrates, usually in the form of sugars, occupy the first place. Fatty oils are relatively seldom in fruits and the fruit with the highest fat content is the avocado. Oily droplets of different size fill most of the parenchymatous cells of this large berry. The protein content of fruits is usually low. Under certain circumstances, the germinating seed may make a profit by the nutritious contents of the fruit. Vitamins, enzymes, and hormones are further important components of fruits (see also Roth 1977 b).

Seeds, on the other hand, are mainly storage organs storing food either in the endo or perisperm or in the embryo itself. They store carbohydrates in the form of starch, sugars or even cellulose (date), they may be rich in proteins and fats, but are poor in water. Besides these nutritious components fruits and seeds contain the so-called secondary compounds which we have mentioned already in the chapter "Chemical protection".

There is a vast number of tropical tree species and herbs which bear valuable fruits and seeds, but very little is known of their contents, particularly considering the secondary compounds. However, the natives know to use these indigenous fruits and seeds very well applying them not only as a food, but also for many kinds of diseases and other purposes. They inherit an ancient tradition overcome from their ancestors – in South America particularly from the Indians. It would be very easy to write a voluminous book on tropical useful plants enumerating all the useful components of tropical wild plants. This book would also be very beneficial to show the world what will be lost when tropical forests are finally destroyed by the advancing "progress" of technics. There are thousands and thousands of species from which precious wood, fibers, cork, filling material, gums, resins, perfumes, mucilages, saponins, tannins, natural colours, and ink could be obtained, not to speak of the chemical substances present in tropical plants which – extracted – could be used pharmacologically. No one who has not visited yet the tropical rain forests can imagine the enormous variety of secondary compounds inherent in tropical forest species. In the following I may just mention a few examples.

There is first of all a large variety of edible fruits and seeds which are unknown in other countries, but are of very agreable taste. However, they can only be processed in the native countries. Besides *Bertholletia excelsa* there are other Lecythidaceae with edible nuts of the genus *Lecythis*. Fruits of Sterculiaceae are rich in mucilage and tannins. Although we know and use only *Theobroma cacao*, other genera such as *Guazuma* produce similar mucilaginous and astringent fruits. Some fruits are used as "receptacles" and substitutes of bottles and even as music instruments ("maracas", "marimba"), e.g. from *Crescentia cujete* ("calabash"). Woolly hairs used as a filling material similar to the "capok" of *Ceiba pentandra* are also found in the fruits of other Bombacaceae, such as *Ochroma*. The Sapotaceae are famous for their latex content and there are quite a few "nisperos" besides *Achras sapota* in the same family which supply chewing gums. Other species of Sapotaceae yield gutta-percha, balatá, shea butter, and oils. Resins are abundant in many tropical families such as Guttiferae, Leguminosae, Burseraceae, Flacourtiaceae. Saponins are not seldom in fruits. Saponins with hemolytic properties are known of *Melia azedarach*. Enormous is the number of seeds and fruits which contain mucilage (see also Grubert 1974). Secondary compounds such as alkaloids, glycosides (cyanogenic glycosides), acids, tannins and others have already been mentioned (see chapter "Chemical protection").

However, fruits and seeds as well as other parts, such as leaves, bark, roots and wood, of many tropical plant species contain secondary compounds the chemical constitution of which is not studied yet, but only the curative properties or the poisonous effects on man and animals are known. Of many tropical plants it is said that they have purgative and laxative properties, that they are emetocathartic, vermifugal, sudorific or febrifugal. Others are antirheumatic, antidysenteric, diuretic, tonic or astringent. Some are effective against diarrhea, pectoral diseases and poisonous snake bites, and few help even against leper, syphilis, and coagulating poisons. Others are curative for the skin or cure intestinal disorders. Some produce vomiting, convulsions, and signs of drunkenness, while others cure these sicknesses. Some are used for abortion, others as insecticides, or fish poisons (*Piscidia* received its name from it). And, finally, the heart "poisons" or cardiac stimulants are manyfold. Many of these chemical compounds are dosis-dependent, having a curative effect in small quantities, but are poisonous in large quantities. Furthermore, enzymes with digestive effect could be used in medicine. A large variety of examples is mentioned in Pittier (1926/1979) as well as in Blohm (1962) and in the literature cited there (see also Roth 1977).

CHAPTER 5

FINAL CONCLUSIONS

The studied area comprising about 155,5 ha or 1.555.000 m^2 accomodates 67.777 individuals, trees (and shrubs) of a stem diameter not less than 10 cm at breast height. Individuals of species calculated for 1 ha occur at densities between 0,0064 up to 23,65 (e.g. *Chaetocarpus schomburgkianus*). Black, Dobzhansky, & Pavan (1950) who studied population densities in Amazonian forests give values of individuals per species between 0 and 16 per hectare. However, the population density of a half of the species is less than one individual per hectare. Comparing the above given data for 155,5 ha in Venezuelan Guiana, the category "less than one individual" goes down to 0,0064 individuals per ha which is less than 1/100 of an individual per ha. It thus results from this calculation that one hectare as sampling area is much too small for the evaluation of actual population densities and that a large amount of species and individuals is simply not covered in this way.

In total 58 families with 375 species and 66.977 individuals were studied concerning their dispersal mechanisms. Of the 375 species studied 233 belong to height category A (30 m and above), 107 to height category a (10–29 m), and only 19 to aa (below 10 m). Of the 66.977 individuals studied 45.543 belong to height category A, 18.814 to height category a, and 2.262 individuals to height category aa (Table I). Within the families with a large number of species and individuals we find the Lecythidaceae, Euphorbiaceae, Sapotaceae, Leguminosae (Mimosaceae, Papilionaceae, Caesalpiniaceae), Burseraceae, Rosaceae, Apocynaceae, Guttiferae, Meliaceae, Myrtaceae, Sapindaceae, Boraginaceae, Lauraceae. Some families are distinguished by very high trees such as the Apocynaceae, Boraginaceae, Lecythidaceae, Papilionaceae, Rosaceae, Bignoniaceae, and others.

Of all the trees and shrubs studied, 50 species with 8.088 individuals are anemochorous, 66 species with 13.845 individuals are autochorous (with dehiscing capsules), 57 species with 20.829 individuals are auto-zoochorous (with dehiscing capsules but additional zoochorous devices), and 196 species with 23.962 individuals are exclusively zoochorous (Table II).

In percentages

13,33% of the species or 12,07% of the individuals are anemochorous

17,50% of the species or 20,67% of the individuals are autochorous

15,20% of the species or 31,09% of the individuals are auto-zoochorous

52,53% of the species or 36,00% of the individuals are zoochorous

The percentages of anemochorous and autochorous species are relatively high as compared with other tropical humid forests studied (e.g. by Foster 1973 or Hall & Swaine 1981).

Anemochorous are mostly high to very high trees so that anemochory is almost restricted to the uppermost forest layers (Bignoniaceae, Apocynaceae, Vochysiaceae). As the uppermost story is most exposed to wind and insolation, wind dispersal is most adequate in this layer. Autochory with dry dehiscing fruits is also much more common among high trees (Euphorbiaceae, *Sloanea*, Caesalpiniaceae). Autochory seems to be more advantageous when the seeds fall from a great height becoming removed further from the parent plant.

Besides, high trees often develop derived fruit forms such as single or few-seeded drupes (Rosaceae, Sapotaceae, Lauraceae, Boraginaceae, Burseraceae, and others), nuts or samaras. A reduction of the seed number becomes thus obvious towards the higher forest strata. Auto-zoochorous or zoochorous species are mostly trees of the lower forest layers, except the above mentioned examples with derived fruit forms. Thus a certain layering of the forest becomes evident concerning fruit types and their dispersal mechanisms. This layering depends, at least partly, on the microclimate of the forest.

Within the Leguminosae we observe the following changes in the forest studied (supposing that the Mimosaseae are more primitive and the Papilionaceae more derived):

Mimosaceae	Caesalpiniaceae	Papilionaceae
no of species and no of individuals	declining →	
no of species with indehiscent pods	declining →	
no of anemochorous species	increasing →	
more polyspermy	→	more monospermy
no of species of height category A	increasing →	

Concerning the fruiting times (Table III) we observe that anemochorous species mainly fruit during the dry season (February) or at the beginning of the wet season (May) when the high winds arise. Due to the starting rains May should be very advantageous for the germination of seeds. Fruiting times of autochorous species are more spread out over the year, although a certain preference of May/June, August and December is conspicuous. Autochorous species thus fruit in rainy periods as well as in drier or dry months, covering a broader spectrum with different seasons. A fruiting peak in the dry season (in December extending into January and February) is apparent. The fruiting times of auto-zoochorous species are more or less spread out over the entire year, although a certain concentration of fruiting species in February and May/June is perceptible. Within the zoochorous species a striking concentration of fruiting species becomes evident in February. Although fruiting is spread out over the entire year, a second peak becomes obvious in May/June extending over July and August. A third concentration of fruiting species is visible at the end of the year (mainly December). Those species fruiting during the wet season may be adapted to the more vigorous animal life in the forest, whereas fruiting during the dry season may satisfy animals when other food resources as well as water become scarce. Many birds and other animals may start already during the dry months with their mating season or already with sexual reproduction, nesting or laying eggs.

Fruiting times of the species studied are influenced by the climate: the regular alternation of the dry and the wet season seems to be the driving factor. Anemochorous species seem to have adapted best to this climatic change, displaying 2 fruiting peaks, one in the dry season (February) and the other at the beginning of the wet season (May). Concerning fruiting times and climatic periods, we may distinguish between 3 different fruiting types: one type prefering the dry season (many anemochorous species), another type prefering the wet season, and a third one the dry as well as the wet season. It is supposed here that the "dry fruiting" types are mainly represented by dry dehiscing or anemochorous dispersal units, the "wet fruiting" types by fleshy drupes or berries, and the "dry-wet fruiting" types by intermediate forms.

A second important factor influencing the fruiting times is the animal life. This is most vigorous during the wet season, extending over a large part of the year. Certain zoochorous plant species, however, are adapted to the special necessities of determined animal species, e.g. those fruiting during the dry season. The staggering of fruiting times over the year by closely related species may be influenced by animal life. But finally, also animal life is mostly dependent on the seasons so that a complete circle is closed again.

Snow (1965) distinguishes a north-tropical seasonality and a south-tropical seasonality. North of the equator the dry season from January to May determines the fruiting peak of the north-tropical species, while south of the equator the main dry season in September-October exercises its influence over the fruiting peak of the south-tropical species. In the intermediate area where 2 dry seasons may be observed (January-May and September-October) species may show 2 fruiting peaks. Species indigenous of the northern hemisphere should thus follow the north-tropical pattern, those coming from the southern hemisphere the south-tropical pattern, while species indigenous of intermediate regions (e.g. British Guiana) should follow the pattern with 2 wet and 2 dry seasons, and possibly show 2 fruiting peaks. In the material studied by me, many species have a single fruiting peak, others have 2, and some even 3. One could speculate that species indigenous of the north-tropical region keep their north-tropical rhythm in the form of an endogeneous (?) rhythm, and the species of the south-tropical region likewise keep the south-tropical rhythm with a single fruiting peak, while those of the intermediate region show 2 peaks. However, the fruiting times of species occurring in Mexico generally do not coincide with those observed in Venezuelan Guiana. To a large extent, the seasonal periodicity of fruiting seems to depend on the local climatic factors, some species prefering the dry season for fruiting, others the wet season, and a third group fruiting in the dry as well as in the wet season.

Using the data of Rollet concerning fruiting times, only little differences are perceptible in the 3 different strata of the forest studied. Fruiting times are spread out over the year in all strata (A-a-aa), although an accumulation of fruiting trees in February is evident in the a-story. A second peak of fruit maturation occurs in May with the highest number of fruiting individuals, followed by February. While in February 18 fruiting species with 4.482 individuals were observed in the a-story (Table IV), in May occured 17 species with 5.233 individuals. It has, however, to be taken into account that the number of individuals of a certain species is counted twice or even 3 or more times when fruiting extends over several months. In the highest (A) story, a marked concentration of fruiting trees is found in February (49 species with 21.152 individuals), while in May 31 species with 13.589 individuals are fruiting. The least number of individuals is found in July (1.371). About one half of the species and individuals in the highest story is adapted to fruiting during the dry period (November, December, January, February, March), in total 102 species with 42.388 individuals. A second peak in May and June comprises species which prefer the wet season for fruiting. During the wet season (May/October) 112 species with 42.810 individuals are fruiting in the A-story. These values are comparable with the data obtained in the a-story: somewhat less than one half of the species (42) with 12.432 individuals fruit during the dry season, while somewhat more species (48) with somewhat less individuals (11.894) fruit during the wet season. The number of species and individuals is thus well balanced during the dry and the wet season.

When the families studied with their special dispersal types are related to the fruiting times, the tight relations of the anemochorous species with the climate become obvious again. The preferred months of the anemochorous species are February of the dry season and May, when the wet season starts with rains and heavy winds. Anemochorous species growing in drier habitats likewise prefer the dry season for fruiting and species growing in dry habitats are frequently anemochorous.

Within one and the same family we observe that fruiting times are more or less spread out over the year. This is partly due to the fact that different dispersal methods prevail within a family. But also in families where one and the same dispersal method prevails, e.g. zoochory or autochory, the different species not infrequently fruit at different times. This becomes most evident in species belonging to the same genus (e.g. *Sloanea* of the Eleaocarpaceae). Two or more species of the same genus usually fruit at very different times so as to avoid one another in fruiting. In many examples we observe a certain variation of an original "fruiting theme" through extension, adding or shifting of fruiting months. By delay or anticipation of the fruiting time, by adding of further fruiting months or by shifting to other fruiting months as well as by extension of the fruiting, the basic pattern may be changed and new dispersal possibilities may arise for the shifting species. One for many example represent the Rosaceae or the Lecythidaceae. As an example it is assumed that a certain species fruits in February. By delay of fruiting a shifting to March may result or by anticipation of fruiting a third species may already start to fruit in January. A fourth species may simply extend fruiting from February to March. A fifth species may add a month, e.g. of the wet season, thus fruiting in February and August. In this way, particularly zoochorous species may find other dispersers in other or additional months. This type of fruiting pattern imitates a pattern of mutual avoidance. However, we have to emphasize that not only zoochorous species avoid one another (Janzen), but also species with other dispersal mechanisms, such as anemochorous or autochorous methods where selection of fruiting times through the influence of animals does not play any part.

A wide range of fruiting extended over several months of the year, and possibly covering some months of the wet as well as of the dry season may be advantageous under certain circumstances. But within these examples we find some species with a high number of individuals as well as those with a low number. Different "niches" may be responsible for the distinct fruiting times of species of the same genus or family, for adding of fruiting months and the shifting to other months, for delay, anticipation or extension of the fruiting. A further possibility to explain these phenomena is the extension and contraction of the forest in earlier periods connected with the formation of refuges.

Examining the geographical distribution of the species studied, as well as their habitats and fruiting times, some comparisons with Mexico and Surinam could be made. However, Mexico and the area studied in Venezuelan Guiana have only few common species, whereas Surinam and Venezuelan Guiana have a great many species in common. But species very common in Surinam are frequently rare in the area studied of Venezuelan Guiana and vice versa. Species which grow well in other habitats, such as the semideciduous or deciduous forest, the secondary forest or the savanna, usually have a low number of individuals in the humid forest of Venezuelan Guiana. Although the species may have a good tolerance of different habitats and soils, a strong resistance to drought and insolation with an ample altitudinal range, growing well in disturbed areas, they may be poorly represented in the humid forest studied. Resistance, rapid growth or easy adaptation to different climatic conditions are frequently not important characteristics of a species to increase the number of individuals in the area in question.

Spondias mombin, for example, a resistant tree with a good distribution rate in the secondary forest, also growing in the deciduous forest, and widely planted in villages, shows only 36 individuals in the forest studied, while *Tapura guianensis*, a typical wide-spread forest species of Amazonia, according to Prance (1973), with considerable variation in leaf size and shape, possibly reflecting isolation of various populations during drier periods, is represented by as many as 1.527 individuals. *Pentaclethra macroloba* with 2.534 individuals had a much larger extension in prehistoric times, but now occurs in disjunct areas. The toxic seeds which are well protected against predation show an excellent germination capacity. The tree grows well on low fertility soils and is dominant in the tropical wet forest of La Selva, Costa Rica.

In many cases we have to count with several dispersal possibilities of one and the same species, e.g. different animals disperse the fruits or seeds (birds, bats, mammals) or autochorous dispersal is combined with zoochory, or barochory is connected with zoochory. A delayed zoochorous dispersal is realized when fruits fall to the ground by barochory, for example, and after rotting are dispersed by animals when the pulp inside becomes free (Leguminosae). Even fruits and seeds may be dispersed separately, by scatterhoarding and following seed dispersal. Hydrochory and zoochory may be combined with one another when fruits fall into the water and later are eaten by fish. Rain wash may often be an additional dispersal mechanism when small fruits or seeds are concerned. Many plant species are thus polychorous and a clear distinction of dispersal types may sometimes be difficult.

Dispersal strategies differ according to the forest layer where they occur. Zoochory prevails in the lower and lowest strata, while more abiotic methods such as anemochory and autochory are favoured in the uppermost strata and in emergent trees. Seeds of autochorous species may be dispersed in an abiotic way in so far as the wind may carry or the rain may wash them away. Dispersal methods in the upper strata are more influenced by climatic factors, such as wind, drought and insolation. The higher the tree, the further the dispersal units may be carried away from the parent tree. Besides, we find drupaceous fruits in high tree species which are considered derived fruit forms through reduction of seeds. Anemochory as well as autochory (in the form of dry dehiscent capsules possibly with a reduced number of seeds) are likewise regarded as derived. In most high tree species we are thus confronted with evolutionally derived dispersal methods, whereas the more ancient zoochorous dispersal method is kept in the lower forest strata. When during evolutionary processes tree species gained more and more in height through the "upward struggle for light", they modified their dispersal methods progressively in the same way as they changed their leaf structure through a transformation of the hygromorphic shade leaf type into the xeromorphic sun leaf type (see Roth 1984).

It is a commonplace that a great variety of species can coexist in the tropical rain forest, each species forming a small population from the very origin but stable with respect to the number of constituent individuals. According to Black et al. (1950) population density in the Amazon region is about 1 individual per hectare (100 × 100 m). Very few species attain a density of 40 - 50 up to 200 individuals per hectare. The species with a larger population density are usually among the lower strata, according to Fedorov (1966), while the lowest population density is characteristic of the stratum of emergent trees and the uppermost story. Fedorov considers self-pollination a condition favourable for the rapid origin of species by means of genetic drift. Ashton (1969) suggests that "clumping" of species is most pronounced in those families where means of dispersal do not exist or are unreliable. He cites the Apocynaceae, Leguminosae and animal-dispersed families such as the Meliaceae, Burseraceae, Sapotaceae. Ashton furthermore emphasizes that a great number of allied species occurring side by side in many genera and families is a particular feature of southeast Asian rain forest not, for instance, occurring in Central America. However, the results obtained here neither support the theory of Fedorov nor that of Ashton for the area of Venezuelan Guiana studied. As could be shown before, the species with the highest numbers of individuals frequently come from the highest story (A), while the lowest population density should be found in this stratum, according to Fedorov. The statement of Ashton that the dispersal methods in the above cited families are unreliable, is open to discussion. That clumping of species also occurs in northern South America is a fact. One explanation of the series of species could be the complexity of the tropical forest and the large number of biotic niches into which evolution can take place. It is quite evident that the different strata in the forest serve as different niches where new species can move. Different degrees of humidity and illumination, of air movement, of pollination and dispersal possibilities may influence speciation in the different strata. In this case one should suppose that series of species of a certain genus should belong to different forest strata. But in the forest studied the series of species frequently belong to the same stratum, mainly the A-stratum (e.g. Lecythidaceae, Rosaceae, Apocynaceae). Furthermore, the species with the highest numbers of individuals often belong to the same stratum (A), not to different strata. There are examples where 2 dominant species of a series belong to different strata (A and a), but in the majority of examples 2 or 3 dominant species come from the same stratum, usually from the A-stratum. No special dispersal methods seem to be favoured in the formation of series of species, as Ashton suggests (see also Mac-Arthur 1969). Among the series of species there are bird and bat-dispersed Sapotaceae and zoochorous Rosaceae, autochorous Euphorbiaceae, anemochorous Apocynaceae, auto-zoochorous Lecythidaceae. However the series of species may have developed, attention is focussed here on the most striking fact, the "dominance" concerning the number of individuals of a single or a few species within a homogeneous group.

In about 17 families studied, several species of the same genus or of different genera of the same family show the same dispersal mechanisms with the same fruit and seed types (forming homogeneous groups or series of species) but the numbers of individuals in the forest studied very much differ in the different species. In some of these homogeneous groups only one species stands out for its high number of individuals, while the other species have low numbers of individuals, but more often 2 or several species show higher numbers of individuals, while the remaining species are represented by few individuals. Examples presented in this book show that the dispersal type is one of the least important factors responsible for the existence of homogeneous groups. As can clearly be seen, all types of dispersal mechanisms are represented in the homogeneous groups: in 2 families they are anemochorous, in 5 families they are autochorous, in 3 families they are auto-zoochorous, and in 7 families they are zoochorous. Neither do the height categories play a very important part in the formation of homogeneous groups. Within the Apocynaceae, for example, all 6 anemochorous species belong to height category A, but only one species has an outstanding number of individuals (1.760). Of the 12 zoochorous species of Burseraceae 2 have outstanding high numbers of individuals (2.669 and 1.157 respectively). One of these species is of height category A, the other of a. Of the 21 autochorous species of Euphorbiaceae, one species reaches 1.706 individuals, the other 3.679 individuals. One species belongs to height category A, the other to a. Of the 8 auto-zoochorous species of Lecythidaceae, 3 species of height category A reach very high numbers of individuals (1.025, 2.374, 3.446). Eight species altogether belong to height category A. Of the 15 zoochorous species of Rosaceae, one species has 3.258 individuals, however, most species are very high trees (A). Of the 21 zoochorous species of Sapotaceae, 3 species reach high numbers of individuals (722, 795, 1.658). Most species of the Sapotaceae belong to height category A. Concerning the number of individuals, the height categories do not play an outstanding part within the homogeneous groups, although we cannot deny that a larger number of tree species of the highest category (A) have elevated numbers of individuals.

Assuming that the species of a homogeneous group have developed under the same environmental conditions, the differences in the habit of the distinct species should not be too great and it would be expected that the number of individuals of all allied species is relatively low with only slight fluctuations from species to species. Such a group is represented by the Lauraceae or the Rutaceae. However, in many families one, two or three species are "dominant". Where only one species is dominant, this may cover as much as 78%, of the total number of individuals. The tropical rain forest seems to have established long ago concerning the number of individuals, more or less maintaining the "status quo" with relatively little changes so that dispersal methods play a subordinate part in this respect. The responsibility for the present status may possibly lie in historic-geological events. If the homogeneous groups have developed in disjunct areas, as the refuge theory proposes, dominance of species could be explained either by the prevalence of the original species of the humid tropical forest or by the development of new species better adapted to certain niches. When 2 or more species are dominant, the latter possibility should be suggested.

The refuge theory (Haffer, Vuilleumier, Prance) offers a certain possibility to explain the origin of homogeneous groups or series of species in a genus or family (e.g. Rosaceae with *Licania*, Apocynaceae with *Aspidosperma*, Bignoniaceae with *Tabebuia*, Burseraceae with *Protium*, Euphorniaceae with *Sapium*, Guttiferae with *Rheedia*, Lecythidaceae with *Escweilera*, Mimosaceae with *Inga*, Sapotaceae with *Pouteria* and others). Although the species belonging to such a group have the same dispersal mechanisms, only one or a few species show an elevated number of individuals, while the other species are represented by very low numbers of individuals. The reason for the higher rates of distribution of certain species could be based on better germination power, higher resistance to adverse conditions and against predators, a stronger radical system, faster growth, better reproduction at shorter intervals, but these characteristics could not be examined in detail here. Several examples of special resistance, very good germination power, excellent growth have been presented in this book, but the number of individuals of these species is not elevated, as a rule. It is, however, possible that historic-geological events are the reason for the abundance in individuals of some species and the reduced number of individuals in other similar species of the same series. The species with the lowest numbers of individuals could have been arisen from more or less marginal "islands" of the disjunct forest areas, while the species with the largest numbers of individuals could come from larger refuge areas of more or less central position. In the modern primary tropical forest the dispersal methods would thus be of minor importance for the number of individuals of a certain species, while historic-geological events would have been more significant. However, this suggestion is only a speculation or working hypothesis for future studies.

The transformation of the hygromorphic shade leaf into the xeromorphic sun leaf taking place during the heightening of the trees shows a "sensible height" of about 5–10 (15) meters. At this height the hygromorphic shade leaf begins about to transform into a xeromorphic sun leaf, depending on the sensibility of the species. With the heightening of the plant the microclimate progressively changes, as the humidity decreases, while air movement and illumination increase in intensity. Although the rhythm of the structural transformation of the leaves is possibly genetically laid down, it is generally activated by the changing microclimate during the heightening of the plant. In *Gyranthera caribensis*, for example, this transformation even takes place under unvariable environmental conditions without a notable growth in length of the plant so that an endogeneous rhythm seems to be imprinted in the genetic code of the plant making itself obvious through an "anticipated planning" in dwarf plants of *Gyranthera* (Roth 1984). In some species, the young plants seem to develop certain structural xeromorphic pecularities of the sun leaf at a height where the majority of the species still retains the characteristics of the hygromorphic shade leaf. The already genetically fixed inner rhythm is originally induced by changes in the microclimate. Zalenskij's law which states that more xeromorphic peculari-ties and characteristics of the sun leaf become evident with the increasing height of leaf insertion on the plant, seems to be based on evolution. It is quite remarkable that this law is more pronounced in woody species, while many exceptions occur in the Gramineae and in herbaceous dicotyledons. Zalenskij's law can thus be explained by the progressive heightening of the Angiosperm forest during evolution.

Stratification of the forest not only becomes evident in the leaf structure of trees, but also reveals itself in the evolution of the dispersal units. Lower shade-tolerant tree species not infrequently show heavy gravity-dispersed seeds or fruits or have zoochorous dispersal mechanisms considered more primitive. As the forest heightened during evolution and the higher and emergent trees appeared, anemochorous and autochorous dispersal mechanisms began to develop. Anemochory frequently found in drier and more open habitats is regarded as more derived. Autochorous species of Leguminosae occupy similar habitats to those of anemochorous species (e.g. very dry areas, steppes), according to Buchwald (1895). Furthermore, it is suggested here that the dry dehiscent capsules of the autochorous type are likewise derived, as compared with the zoochorous dispersal mechanisms. In the Leguminosae, van der Pijl suggests that the follicle dehiscent at the ventral suture and retaining the arillate seeds for the display to birds is more primitive than the legume dehiscing with 2 valves and releasing the seeds in an autochorous way. Autochory is a dispersal mechanism more adequate for a drier environment which is also found in temperate regions, being particularly characteristic of herbs. The herbaceous habit, however, seems to be derived from the arboreous habit. As animal life is most vigorous in the lower and middle strata of the forest, it is not surprising that species of the lower forest stories are zoochorous, while high and emergent trees, more exposed to wind, drought and insolation, adapted to wind-dispersal or to autochory with dry dehiscent capsules, as autochory seems to be more advantageous in higher wind-exposed trees or in drier and more open habitats. Possibly, one has to distinguish between phylogenetically more primitive capsular dispersal types (e.g. with fleshy fruits retaining the seeds), on the one hand, and more derived types with dry dehiscing capsules releasing the seeds, on the other. The stratification of the tropical humid forest could thus be explained through the evolution of the Angiosperms which developed from more primitive lower trees with hygromorphic shade leaves and zoochorous fruits or seeds by heightening of their stems into upper canopy trees with xeromorphic sun leaves and anemochorous or autochorous dispersal mechanisms. From the phylogenetic point of view, the highest trees have thus reached the last phase of ecological niche formation in the heightening forest. The "critical height" of 5–10 (15) meters of many tree species in the tropical humid forest at which the transformation of the hygromorphic shade leaf into the xeromorphic sun leaf takes place, may possibly be considered the original height of ancient Angiosperm trees or shrubs. My phylogenetic considerations are founded in the suggestion that the early Angiosperm flora was composed of relatively low stature trees or shrubs protected by higher Gymnospermous trees or tree ferns. During the struggle for light the Angiospermous trees heightened progressively and derived dispersal mechanisms, such as anemochory and autochory with dry dehiscent capsules, or zoochorous drupes with a reduced number of seeds developed in the higher forest strata. At the same time the transformation of the hygromorphic shade leaf into the xeromorphic sun leaf took place (Roth 1984).

A comparison of bark, leaf and fruit structure in the families studied shows that some families are definitely homogeneous in all the characters studied. One of these families is the Lecythidaceae with a very regular stratified bark structure and with leaves of the xeromorphic sun type (high trees). The Lecythidaceae are distinguished to a large degree by a lack of peculiarities. The fruits are pyxidia with a dehiscing lid and the seeds have often a fleshy funicle eaten by animals. Particularly in the genus *Eschweilera* a series of species seems to be present. Another homogeneous group is the Bignoniaceae with the genus *Tabebuia*. The bark is very regular and stratified and disc-shaped glandular hairs are an outstanding character in the leaves of *Tabebuia*. The representatives of the Bignoniaceae studied are all high trees with capsular fruits and winged seeds dispersed by wind. A further homogeneous family is the Lauraceae with alternating hard bast plates, leaves of the sun type and oil idioblasts. The fruits correspond to zoochorous drupes. The bark of the Sapotaceae is very characteristic with alternating hard bast plates, secretory cells arranged in tangential rows and large coriaceous leaves of a somewhat plump and primitive structure. The fruits are drupes or berries dispersed by birds and bats.

There are, however, also heterogeneous families concerning bark, leaf and fruit structure, such as the Apocynaceae, Moraceae, Euphorbiaceae, Myrtaceae and others. In certain families bark and leaf structure coincide very well. In the Nyctaginaceae the bark is thin and a hard bast absent, while the leaves are of the hygromorphic shade type. The fruits are zoochorous nuts surrounded by an anthocarp. The species studied are characteristic of a humid and shady environment. A thick rhytidome and xeromorphic sun leaves are peculiar of the Myristicaceae. The fruits are capsular and auto-zoochorous. *Lacistema aggregatum* has a thick bark and leathery leaves of the sun type with xeromorphic features. *Alseis* of the Rubiaceae develops an aerenchyma in the rhytidome and a loose spongy parenchyma in the leaves.

As could be expected, some families form homogeneous groups concerning their bark and leaf structure, as well as their fruit and seed morphology, others are homogeneous only in fruit and seed morphology, while they may differ either in bark or in leaf structure or in both.

Concerning the morphology and inner structure of fruits and seeds, certain families are distinguished by their special dispersal units, such as the Bignoniaceae or Apocynaceae by dry capsules (or mericarps) with winged seeds which are dispersed by wind, the Euphorbiaceae by dry dehiscent capsules and dry seeds, the Sapotaceae by fleshy berries or drupes and large seeds with a conspicuous hilar region, the Lecythidaceae by large woody pyxidia (*Eschweilera*) and large seeds with a fleshy funicle. Some families are known for their aromatic or resinous secretions included in secretory cavities or cells such as the Anacardiaceae, Annonaceae, Flacourtiaceae, Burseraceae, Rutaceae, and Guttiferae, by laticifers (Sapotaceae) or by mucilage (Sterculiaceae). Some families are homogeneous concerning their fruit and seed formation, while others are heterogeneous developing drupes, berries, samaras or capsules at the same time. As homogeneous families we may mention the Bignoniaceae, Apocynaceae, Burseraceae (*Protium*), Rosaceae (drupe), Elaeocarpaceae (*Sloanea*), Euphorbiaceae, Lauraceae, while the Boraginaceae, Flacourtiaceae, Guttiferae and Sapindaceae, for example, represent heterogeneous families. However this categorization only refers to the species studied here. Some fruits and seeds show very interesting surface structures in the form of knobs, protuberances or hooks (Fig. 3, 22, 45, 58, 81, 94, 95, 96, 99, 100). Large protuberances occur on *Matelea maritima*, although they have no dispersal function on the fruit, as the seeds are dispersed by wind (Fig. 3). A palisade layer occurs in the epicarp of species of *Rheedia* (Figs. 47 and 49) as well as of *Ouratea guianensis* (Fig. 80). Parenchymatous cells with weavy walls are a peculiarity of *Drypetes variabilis* (Fig. 37) and of *Torrubia cuspidata* (Fig. 78) and possibly serve to strenghten the tissue. Dead tracheidal cells with pits occur in *Laetia procera* (Fig. 41) and in *Terminalia amazonia* (Fig. 32 center). They develop from parenchyma cells and in the case of *Terminalia* are air-filled to reduce the specific weight of the samara. Collenchymatous layers may be composed of cells with angular thickenings (Fig. 20 above) or form a cartilaginous tissue with irregular wall thickenings (Fig. 20 below, Fig. 46). Sclereids may adopt peculiar shapes and sclerenchymatous layers may characterize certain species by their special formation or position. Sclereids may either have an annular aspect, as seen in transection (Fig. 93 above, Fig. 92), or may occur scattered or in small groups within the parenchyma (Figs. 38 center, 37), while ramified sclereids are characteristic of *Aspidosperma excelsum* and *A. marcgravianum* (Figs. 9, 10). Crossing fiber layers are characteristic of many endocarp formations (Figs. 13, 38, 42, 43, 50, 86, 87, 101). Peculiar are secretory cavities developing within the vascular bundles and decomposing during fruit maturation (Fig. 106). The ribs on the pericarp of *Touroulia guianensis* strengthened with a sclerenchymatous layer alternate with parenchymatous "pockets" filled with oil droplets (Figs. 82, 83, 84). The spiny fruit of *Apeiba echinata* resembles a sea urchin (Fig. 100), while the fruit of *A. tibourbou* develops long hairs resembling a wick. Long stiff hairs surround the fruit of *Sloanea grandiflora* giving the impression of a human head (Fig. 34). Very seldom a periderm develops in the pericarp. However, *Catostemma communis*, Bombacaceae, makes an exeption with a very large stratified periderm comprising up to 50 cell layers or more. The cork cells have U-shaped wall thickenings. As several periderms develop in the pericarp, a scaly rhytidome arises comparable to that observed in barks of the stem (Figs. 15, 16, 17). Stratified lenticels may be observed in *Carapa* sp. (Fig. 71). A conspicuous ventral meristem is present in the periscarp of *Gustavia augusta* (Fig. 60). As could be shown by Roth (1977 b), a ventral meristem not seldom occurs at the ventral side of carpels.

The outer shape of fruits is most variable. Characteristically pear-shaped are the drupes of the Rosaceae (*Licania*) (Fig. 85). Capsules show different ways of dehiscence. Besides septicidal and/or loculicidal dehiscence (which may be combined) the mesoscarp may separate from the endocarp in a tangential dehiscence (*Clusia*, Fig. 44). Most interesting is the dehiscence of *Cochlospermum* (Figs. 24, 25, 26). The exo- and mesoscarp first separate from the endocarp by tangential splitting, then the united exo- and mesoscarp dehisces loculicidally, and finally follows a partial septicidal opening of the endocarp. In this way the seeds are released comparatively slowly. *Cochlosper-*

mum sp. is a species characteristic of the semideciduous and tropophilous forest. The fine structure of capsules may be elaborate, e.g. in *Aristolochia* grandiflora (Fig. 4, below) where the vascular bundles bring about a ladder structure. Very variable is the shape of the leguminous fruit. It may be helicoidal as in *Enterolobium cyclocarpum* (Fig. 62, above) or helicoidally twisted (Fig. 62, below). Some samaras are single-winged (Fig. 67, above) others are completely surrounded by a ring-like wing (Fig. 68, above). A very regular arrangement of scales in the form of spirostichies is obvious in the fruit of *Mauritia* (Fig. 81, below). Accessory fruit parts may bring about cup-shaped structures such as the cupule of Lauraceae (Fig. 51, above), while the fruit of *Triplaris* adopts the shape of a flying shuttlecock through 3 enlarged sepals.

Not seldom the seeds imprint their position and shape on the inside of the pericarp (Fig. 70, above for *Trichilia*). The arrangement of the seeds is peculiar in many fruits (Fig. 4, above) e.g. in *Swietenia* where the elongated wind-dispersed seeds hang down from the central column. Arillate dangling seeds occur in *Swartzia schomburgkii* (Fig. 67, below). The seeds of *Cochlospermum* (Figs. 24, 26) have a helicoidal structure being surrounded by a wreath of hairs. For further information of fruit structures, dehiscence types and other details see Roth (1977 b).

Finally, I want to draw attention to the usefulness of tropical fruits and seeds and the manifold possibilities of their application. Besides their nutritious values (water, carbohydrates, fatty oils, and proteins) vitamins, enzymes and hormones are important components of fruits. Seeds, on the other hand, being mainly storage organs of nutritious substances are poor in water. Apart from the nutritious contents the so-called secondary compounds are of high importance. Very little is known of the chemical composition of valuable fruits and seeds of the vast number of tropical tree species, shrubs and herbs. However, the natives very well know to make use of the indigenous fruits and seeds not only as a food, but also for many kinds of diseases, not to speak of other purposes. Restricting the examples only to fruits and seeds, there are thousands and thousands of species from which fibers, filling materials, latex, gums, resins, mucilage, saponins, tannins, natural colours, perfumes and the like could be obtained. Well known is the fabrication of receptacles and music instruments from woody fruits. Of many secondary compounds the chemical constitution is not studied yet, but only the curative properties or the poisonous effects are known. Fruits and seeds of many tropical plant species have purgative or laxative properties, or their contents are emetocathartic, vermifugal, sudorific or febrifugal. Other substances are antirheumatic, antidysenteric, diuretic, tonic or astringent, not to forget the many cardiac stimulants. Some fruits and seeds are effective against diarrhea, pectoral diseases, poisonous snake bites or help against leper, syphilis, others are curative for the skin or cure intestinal disorders. Some are used for abortion, others as insecticides or fish poisons. The list could be extended ad libitum (see Pittier 1926/1979, Blohm 1962).

BIBLIOGRAPHY

André, E. (1964): Un naturalista en la Guayana. Collección Cuatricentenario. Caracas.

Aristeguieta, L. (1967): El bosque seco caducifolio de los altos llanos centrales. Bol. Soc. Ven. Cienc. Nat. Vol. 27: 128-307 and 395-438.

Aronow, L. and F. Kerdel-Vegas (1965): Seleno-cystathionine, a pharmacologically active factor in the seeds of *Lecythis ollaria*. Nature 205: 1185-1186.

Ashton, P.S. (1969): Speciation among tropical forest trees: Some deduction in the light of recent evidence. Biol. J. Linn. Soc. 1: 155-196.

Ashton, P.S. (1976): An approach to the study of breeding systems, population structure and taxonomy of tropical trees. Linn. Soc. Symp. 2: 35-42.

Ashton, P.S. (1977): A contribution of rain forest research to evolutionary theory. Ann. Missouri Bot. Garden 64: 694-705.

Augspurger, C.K. (1983a): Seed dispersal of the tropical tree, *Platypodium elegans*, and the escape of its seedlings from fungal pathogens. J. Ecol. 71: 759-771.

Augspurger, C.K. (1983b): Offspring recruitment around tropical trees: Changes in cohort distance with time. Oikos 40: 189-196.

Axelrod, D.I. (1952): Variables affecting the probabilities of dispersal in geologic time. Bull. Am. Mus. Nat. Hist. 99: 177-188.

Baker, H.G. (1970): Evolution in the tropics. Biotropica 2: 101-111.

Baker, H.G. (1972): Seed weight in relation to environmental conditions in California. Ecology 53: 997-1010.

Baker, H.G., K.S. Bawa, G.W. Frankie, and P.A. Opler (1983): Reproductive biology of plants in tropical forests. In: F.B. Golley (ed.) Tropical rain forest ecosystems. Elsevier Sci. Publ. Co., New York.

Baker, H.G. and G.L. Stebbins (eds.) (1965): The genetics of colonizing species. Acad. Press, New York - London.

Bankroft, H. (1930): The arborescent habit in Angiosperms (a review). New Phytologist 29: 153-169 and 227-275.

Beck v. Mannagetta, G. (1913): Frucht und Same. Handwörterbuch der Naturwissenschaften 4. Fischer; Jena.

Beer, T. and M.D. Swaine (1977): On the theory of explosively dispersed seeds. New Phytologist 78: 681-694.

Bell, E.A. and D.H. Janzen (1971): Medical and ecological considerations of L-Dopa and 4-HTP in seeds. Nature 229: 136-137.

Berg, R.Y. (1966): Seed dispersal of *Dendromecon*: its ecologic, evolutionary and taxonomic significance. Amer. J. Bot. 53: 61-73.

Berg, R.Y. (1975): Fruit, seed, and myrmecochorous dispersal in *Micrantheum* (Euphorbiaceae). Norw. J. Bot. 22: 173-194.

Black, G.A., T. Dobzhansky, and C. Pavan (1950): Some attempts to estimate species diversity and population density of trees in Amazonian forests. Bot. Gaz. 111: 413-425.

Blohm, H. (1962): Poisonous plants of Venezuela. Wissenschaftliche Verlagsgesellschaft M.B.H., Stuttgart.

Boelcke, O. (1964): Estudio morfológico de las semillas de Leguminosas Mimosoideas y Caesalpinioideas de interés agronómico en Argentina. Darwiniana 7: 240-322.

Borzi, A. (1903): Biologia dei semi di alcune specie d'Inga. Atti Real. Acad. Lincei 12: 131-140.

Bravato, M. (1974): Estudio morofológico de frutos y semillas de las Mimosoideae (Leguminosae) de Venezuela. Acta Bot. Venez. 9: 317-361.

Brodie, H.J. (1951): The splash cup dispersal mechanisms in plants. Canad. J. Bot. 29: 224-234.

Brodie, H.J. (1955): Springboard dispersal mechanisms operated by rain. Canad. J. Bot. 33: 156-167.

Brokaw, N.V.L. (1983): Tree falls: Frequency, timing, and consequences. In: E.G. Leigh, A.S. Rand and D.M. Windsor (eds.), The ecology of a tropical forest (seasonal rhythms and long-term changes).

Buchwald, J. (1895): Die Verbreitungsmittel der Leguminosen des tropische Afrika. Engler's Bot. Jb. 19: 494-561.

Burkart, A. (1943): Las leguminosas argentinas (silvestres y cultivadas). ACME, Buenos Aires. (2nd. ed. 1952).

Burrows, F.M. (1973): Calculation of primary trajectories of plumed seeds in steady winds with variable convection. New Phytologist 72: 647-664.

Burrows, F.M. (1975a): Calculation of the primary trajectories of dust seeds, spores and pollen in unsteady winds. New Phytologist 75: 389-403.

Burrows, F.M. (1975b): Wind-borne seed and fruit movement. New Phytologist 75: 405-418.

Cant, J.G.H. (1979): Dispersal of *Stemmadenia donnel-smithii* by birds and monkeys. Biotropica 11: 122.

Carlquist, S. (1966): The biota of long-distance dispersal. II. Loss of dispersability in Pacific Compositae. Evolution, N.Y. 20: 30.

Carlquist, S. (1969): Toward acceptable evolutionary interpretation of floral anatomy. Phytomorphology 19: 332-362.

Charles-Dominique, P., M. Atramentowicz, M. Charles-Dominique, H. Gerard, A. Hladik, C.M. Hladik, and M.F. Prévost (1981): Les mammifères frugivores arbicoles nocturnes d'une forêt guyanaise: inter-relations plantes – animaux. Rev. Ecol. (Terre et Vie) 35: 341-435.

Cheke, A.S., W. Nanakorn and Ch. Yankoses (1979): Dormancy and dispersal of seeds of secondary forest species under the canopy of a primary tropical rain forest in northern Thailand. Biotropica 11: 88-95.

Glegg, D.O., E.E. Conn, and D.H. Janzen (1979): Developmental fate of the cyanogenic glucoside linamarin in Costa Rican wild lima bean seeds. Nature 278: 343-344.

Colquhoun, M.K. and A. Morley (1943): Vertical zonation in woodland bird communities. J. Anim. Ecol. 12: 75-81.

Conn, E.E. (1979): Cyanide and cyanogenic glycosides. In: Rosenthal and Janzen (eds.), Herbivores: their interaction with secondary plant metabolites. Academic Press, New York.

Connell, J.H. (1978): Diversity in tropical rain forests and coral reefs. Science 199: 1302-1310.

Constance, L. (ed.). (1963): Amphitropical relationship in the herbaceous flora of the Pacific coast of North and South America. A symposium. Quart. Rev. Biol. 38: 109-177.

Corner, E.J.H. (1949): The durian theory or the origin of the modern tree. Ann. of Bot., N.S. 13: 367-414.

Corner, E.J.H. (1951): The leguminous seed. Phytomorphology 1: 117-150.

Corner, E.J.H. (1952): Durians and dogma. Indones, Jour. Nat. Sci. 108: 141-145.

Corner, E.J.H. (1954): The evolution of tropical forest. In: J. Huxley, A.C. Hardy, and E.B. Ford (eds.), Evolution as a process. Allen and Unwin, London, p. 34-46.

Corner, E.J.H. (1967): On thinking big. Phytomorphology 17: 24-28.

Corner, E.J.H. (1976): The seeds of Dicotyledons. 2 vols. Cambridge University Press, Cambridge – London – NY – Melbourne.

Cremer, K.W. (1966): Dissemination of seed from *Eucalyptus regnans*. Australian Forestry 30: 33-37.

Croat, T.B. (1974): A case for selection for delayed fruit maturation in *Spondias* (Anacardiaceae). Biotropica 6: 135-137.

Crocker, W. and L.V. Barton (1953): Physiology of seeds. Chronica Botanica Company, Waltham, Mass.

Davis, T.A. (1966): Morphology of horns in exceptional fruits of six palm species. Phytomorphology 16: 314-326.

Eisenberg, J.F., and R.W. Thorington (1973): A preliminary analysis of a neotropical mammal fauna. Biotropica 5: 150-161.

Eisenmann, E. (1961): Favorite foods of neotropical birds: flying termites and *Cecropia* catkins. Auk 78: 636-638.

Endress, P.K. (1973): Arils and aril-like structures in woody Ranales. New Phytol. 72: 1159-1171.

Engler, A. (1964): Syllabus der Pflanzenfamilien II. (Angiospermen). Borntraeger, Berlin.

Estrada, A. and R. Coates – Estrada (1984): Fruit eating and seed dispersal by howling monkeys (*Alouatta palliata*) in the tropical rain forest of Los Tuxtlas, Mexico. Amer. J. Primatology 6: 79-91.

Estrada, A. and T.H. Fleming (eds.) (1986): Frugivores and seed dispersal. – Tasks for vegetation science. Dr. W. Junk Publ., Dordrecht.

Ewel, J.J. and A. Madriz (1968): Zonas de vida de Venezuela. MAC, Caracas.

Faegri, K. and L. van der Pijl (1966): The principles of pollination ecology. Pergamon Press, Oxford. (2nd. edit. 1971).

Fahn, A. and E. Werker (1972): Anatomical mechanisms of seed dispersal. Seed Biology I (Ed. T.T. Kozlowski). Academic Press, New York – London.

Fahn, A. and M. Zohary (1955): On the pericarpial structure of the legumen, its evolution and relation to dehiscence. Phytomorphology 5: 99-111.

Fedorov, A.N.A. (1966): The structure of the tropical rain forest and speciation in the humid tropics. J. Ecol. 54: 1-11.

Fleming, T.H. (1979): Do tropical frugivores compete for food? Amer. Zool. 19: 1157-1172.

Fleming, T.H. and E.R. Heithaus (1981): Frugivorous bats, seed shadows, and the structure of tropical forests. Biotropica 13 (supplement): 45-53.

Foster, R.B. (1973): Seasonality of fruit production and seed fall in a tropical forest ecosystem in Panama. Ph. D. Thesis, Duke University.

Foster, R.B. (1983): The seasonal rhythm of fruit fall on Barro Colorado Island. In: E.G. Leigh Jr., A.S. Rand, and D.M. Windsor (eds.), The ecology of a neotropical forest: seasonal rhythms and long-term changes. Smithonian Institution Press, Washington, D.C. p. 151-172.

Foster, S.A. and C.H. Janson (1985): The relationship between seed size, gap dependence, and successional status of tropical rain forest woody species. Ecology 66: 773-780.

Frank, H.K. and L. Betancourt (1980): Die Paranuss. Teil 1: Herkunft, Gewinnung, chemische und physikalische Eigenschaften. Deutsche Lebensmittel-Rundschau 76: 7-11.

Frank, H.K., L. Betancourt and Mirtha Uboldi (1980): Die Paranuss. Teil 2: Verderb und Bildungsbedingungen für Aflatoxine. Deutsche Lebensmittel-Rundschau 76: 47-50.

Frankie, G.W., H.G. Baker, and P.A. Opler (1974): Comparative phenological studies of trees in tropical wet and dry forest in the lowlands of Costa Rica. J. Ecol. 62: 881-919.
Freeland, W.J. and D.H. Janzen (1974): Strategies in herbivory by mammals: the role of plant secondary compounds. Amer. Naturalist 108: 269-289.
Frohne, D. and U. Jensen (1979): Systematik des Pflanzenreichs unter besonderer Be?tcksichtigung chemischer Merkmale und pflanzlicher Drogen. 2. Aufl. G. Fischer, Stuttgart-New York.
Futuyama, D.J. and M. Slatkin (1983): Coevolution. Sinauer, Sunderland, Mass.
Gadgil, M. (1971): Dispersal: Population consequences and evolution. Ecology 52: 253-261.
Garwood, Nancy C. (1983): Seasonal rhythm of seed germination in a semideciduous tropical forest. In: E.G. Leigh, A.S. Rand, and D.M. Windsor, The ecology of a tropical forest: seasonal rhythms and long-term changes. Oxford University Press, Oxford.
Gautier-Hion, A., J.M. Duplantier, R. Quris, F. Feer, C. Sourd, J.-P. Decoux, G. Dubost, L. Emmons, C. Erard, P. Hecketsweiler, A. Moungazi, C. Roussilhon, and J.M. Thiollay (1985): Fruit characters as a basis of fruit choice and seed dispersal in a tropical forest vertebrate community. Oecologia 65: 324-337.
Gentry, A.H. (1974): Flowering phenology and diversity in tropical Bignoniaceae. Biotropica 6: 64-68.
Gentry, A.H. (1976): Bignoniaceae of southern Central America: Distribution and ecological specificity. Biotropica 8: 117-131.
Gentry, A.H. (1982): Patterns of neotropical plant species diversity. In: M.K. Hecht, B. Wallace and G.T. Prance (eds.), Evolutionary Biology, Vol. 15. Plenum Press, New York and London, pp. 1-84.
Gilbert, L.E. and P.H. Raven (eds.) (1975): Coevolution of animals and plants. Symposium V. First Internat. Congress of systematics and evolutionary biology in Boulder, Colorado, August 1973. Univ. Texas Press, Austin and London.
Gindel, I. (1960): Biological function of fruit. Nature 187: 42-44.
Glander, K.E. (1979): Feeding associations between howling monkeys and basilik lizards. Biotropica 11: 235-236.
Gómez-Pompa, A., C.V. Vasquez-Yanes, and S. Guevara (1972): The tropical rain forest: a non-renewable resource. Science 177: 762-765.
Gottsberger, G. (1978): Seed dispersal by fish in the inundated regions of Humaitá, Amazonia. Biotropica 10: 170-183.
Green, D.S. (1980): The terminal velocity and dispersal of spinning samaras. Amer. J. Bot. 67: 1218-1224.
Gregory, P.H. (1973): The microbiology of the atmosphere. 2nd. ed. Leonard Hill, Aylesbury.
Grubert, M. (1974): Studies on the distribution of myxospermy among seeds and fruits of Angiospermae and its ecological importance. Acta Biol. Venez. 8: 315-551.
Guevara, S. and A. Gómez-Pompa (1972): Seeds from surface soils in a tropical region of Veracruz, Mexico. J. Arnold Arbor. 53: 312-335.
Gunn, Ch. R. and J.V. Dennis (1976): World guide to tropical drift seeds and fruits. Demeter Press. Quadrangle Book Co. New York Times, New York.
Guppy, H.B. (1917): Plants, seeds and currants in the Westindies and Azores. Macmillan, London.
Haffer, J. (1969): Speciation in Amazonian forest birds. Science 165: 131-137.
Hall, J.B. and M.D. Swaine (1980): Seed stock in Ghanaian forest soils. Biotropica 12: 256-263.
Hall, J.B. and M.D. Swaine (1981): Distribution and ecology of vascular plants in a tropical rain forest (Forest vegetation in Ghana). Geobotany 1. Dr. W. Junk Publishers. The Hague-Boston-London.
Hallê, F., R.A.A, Oldeman, and P.B. Tomlinson (1978): Tropical trees and forests. An architectural analysis. Springer-Verlag, Berlin-Heidelberg-New York.
Hamilton, W.D. and R.M. May (1977): Dispersal in stable habitats. Nature 269: 578-581.

Harborne, J.B. (1977): Introduction to ecological biochemistry. Academic Press, London-New York-San Francisco.

Harborne, J.B. (1979): Flavonoid pigments. In: Rosenthal and Janzen (eds.). Herbivores: Their interaction with secondary plant metabolites. Academic Press, New York.

Harper, J.L. (1977): Population biology of plants. Academic Press, London-New York-San Francisco.

Harper, J.L., P.H. Lovell, and K.G. Moore (1970): The shapes and sizes of seeds. Ann. Rev. Ecol. and Syst. 1: 327-356.

Harrison, J.L. (1962): The distribution of feeding habits among animals in a tropical rain forest. J. Animal. Ecol. 31: 53-64.

Hartshorn, G.S. (1972): The ecological life history and population dynamics of *Pentaclethra macroloba*, a tropical wet forest dominant, and *Stryphnodendron excelsum*, an occasional associate. Diss. Coll. of forest Resources, University Washington.

Hatheway, W.H. and H.G. Baker (1970): Reproductive strategies in *Pithecellobium* and *Enterolobium* — further information. Evolution 24: 253-254.

Haverschmidt, F. (1971): Large number of birds exploiting a fruit tree in Surinam. Wilson Bull. 83: 104-105.

Hedberg, O. (1969): Evolution and speciation in a tropical high mountain flora. Biol. J. Linn. Soc. 1: 135-148.

Heel, W.A. van (1973): Flowers and fruits in Flacourtiaceae. I. *Scaphocalyx spathacea* Ridl. Blumea 21: 259-279.

Heel, W.A. van (1979): Flowers and fruits in Flacourtiaceae. IV. *Hydnocarpus* spp., *Kiggelaria africana* L., *Casearia* spp., *Berberidopsis corallina* Hook. F. Blumea 25: 513-529.

Heithaus, E.R. (1982): Coevolution between bats and plants. In: T.H. Kunz (ed.), Ecology of bats, Plenum Press, New York and London, pp. 327-368.

Heithaus, E.R., T.H. Fleming, and P.A. Opler (1975): Foraging patterns and resource utilization in seven species of bats in a seasonal tropical forest. Ecology 56: 841-854.

Herrera, C.M. (1981): Are tropical fruits more rewarding to dispersers than temperate ones? Amer. Naturalist 118: 896-907.

Herrera, C.M. (1982a): Defense of ripe fruit from pests: its significance in relation to plant-disperser interactions. Amer. Naturalist 120: 218-241.

Herrera, C.M. (1982b): Seasonal variation in the quality of fruits and diffuse coevolution between plants and avian dispersers. Ecology 63: 773-785.

Herrera, C.M. (1984): Avian interference of insect frugivory: an exploration into the plant-bird-fruit pest evolutionary triad. Oikos 42: 203-210.

Hertel, H. (1966): Structure, form, movement. Reinhold Publ. Corp. New York.

Hopkins, H.C., and M.J.G. Hopkins (1983): Fruit and seed biology of the neotropical species of *Parkia*. In: S.L. Sutton, T.C. Whitmore & A.C. Chadwik (eds.). Tropical rain forest: Ecology and management. pp. 197-209.

Howe, H.F. (1977): Bird activity and seed dispersal of a tropical wet forest tree. Ecology 58: 539-550.

Howe, H.F. (1979): Fear and frugivory. Amer. Naturalist 114: 925-931.

Howe, H.F. (1980): Monkey dispersal and waste of a neotropical fruit. Ecology 61: 944-959.

Howe, H.F. (1983a): Fruit production and animal activity in two tropical trees. In E.G. Leigh, A.S. Rand, and D.M. Windsor (eds.), The ecology of a tropical forest: seasonal rhythms and long-term changes. Smithonian Institution Press, Washington, D.C.

Howe, H.F. (1983b): Annual variation in a neotropical seed-dispersal system. In: S.L. Sutton, T.C. Whitmore, and A.C. Chadwick), (eds.) Tropical rain forest: Ecology and management pp. 211-227. Blackwell, Oxford.

Howe, H.F. and G.F. Estabrook (1977): On intraspecific competition for avian dispersers in tropical trees. Amer. Naturalist 111: 817-832.

Howe, H.F. and G.A. van de Kerckhove (1979): Fecundity and seed dispersal of a tropical tree. Ecology 60: 180-189.

Howe, H.F. and G.A. van de Kerckhove (1980): Nutmeg dispersal by tropical birds. Science 210: 925-927.

Howe, H.F. and G.A. van de Kerckhove (1981): Removal of wild nutmeg (*Virola surinamensis*) crops by birds. Ecology 62: 1093-1106.

Howe, H.F. and R.B. Primack (1975): Differential seed dispersal by birds of the tree *Casearia nitida* (Flacourtiaceae). Biotropica 7: 278-283.

Howe, H.F. and W.M. Richter (1982): Effects of seed size on seedling size in *Virola surinamensis*, a within and between tree analysis. Oecologia 53: 347-351.

Howe, H.F., E.W. Schupp, and G.C. Westley (1985): Early consequences of seed dispersal for a neotropical tree (*Virola surinamensis*). Ecology 66: 781-791.

Howe, H.F. and J. Smallwood (1982): Ecology of seed dispersal. Ann. Rev. Ecol. Syst. 13: 201-228.

Howe, H.F. and D. de Steven (1979): Fruit production, migrant bird visitation, and seed dispersal of *Guarea glabra* in Panamá. Oecologia 39: 185-196.

Hoyos, J. (1976): Los árboles de Caracas. Sociedad de Ciencias Naturales, Caracas.

Hoyos, J. (1978): Flora tropical ornamental. Sociedad de Ciencias Naturales, La Salle, Caracas.

Hubbell, S.P. (1980): Seed predation and the coexistence of tree species in tropical forests. Oikos 35: 214-229.

Janson, C.H. (1983): Adaptation of fruit morphology to dispersal agents in a neotropical forest. Science 219: 187-189.

Janzen, D.H. (1967): Synchronization of sexual reproduction of trees within the dry season in Central America. Evolution 21: 620-637.

Janzen, D.H. (1969): Seed-eaters versus seed-size, number, toxicity and dispersal. Evolution 23: 1-27.

Janzen, D.H. (1970): Herbivores and the number of tree species in tropical forests. Amer. Naturalist 104: 501-528.

Janzen, D.H. (1971a): Escape of *Cassia grandis* L. beans from predators in time and space. Ecology 52: 964-979.

Janzen, D.H. (1971b): Seed predation by animals. Ann. Rev. Ecology and Systematics 2: 465-492.

Janzen, D.H. (1972a): Escape in space by *Sterculia apetala* seeds from the bug *Dysdercus fasciatus* in a Costa Rican deciduous forest. Ecology 53: 350-361.

Janzen, D.H. (1972b): Association of a rainforest palm and seedeating beetles in Puerto Rico. Ecology 53: 258-261.

Janzen, D.H. (1972c): Protection of *Barteria* (Passifloraceae) by *Pachysima* ants (Pseudomyrmecinae) in a Nigerian rain forest. Ecology 53: 885-892.

Janzen, D.H. (1973): Host plants as islands. II. Competition in evolutionary and contemporary time. Amer. Naturalist 107: 786-790.

Janzen, D.H. (1974): Tropical blackwater rivers, animals, and mast-fruiting by the Dipterocarpaceae. Biotropica 6: 69-103.

Janzen, D.H. (1975a): Intra- and interhabitat variations in *Guazuma ulmifolia* (Sterculiaceae) seed predation by *Amblycerus cistelinus* (Bruchidae) in Costa Rica. Ecology 56: 1009-1013.

Janzen, D.H. (1975b): Ecology of plants in the tropics. Inst. Biol. Stud. Biol. 58. 66 pp. E. Arnold, London.

Janzen, D.H. (1975c): Behavior of *Hymenaea courbaril* when its predispersal seed predator is absent. Science 189: 145-147.

Janzen, D.H. (1976a): The depression of reptile biomass by large herbivores. Amer. Naturalist 110: 371-400.

Janzen, D.H. (1976b): Two patterns of pre-dispersal seed predation by insects on Central American deciduous forest trees. Linn. Soc. Symp. Series 2 «Tropical trees» (ed. J. Burley and B.T. Styles): 179-188.

Janzen, D.H. (1976c): Why do bamboos wait so long to flower. Linn. Soc. Symp. 2: 135-139.

Janzen, D.H. (1977a): Intensity of predation on *Pitchecellobium saman* (Leguminosae) seeds by *Merobruchus columbinus* and *Stator limbatus* (Bruchidae) in a Costa Rican deciduous forest. Tropical Ecology 18: 162-176.

Janzen, D.H. (1977b): Promising directions of study in tropical animal-plant interactions. Ann. Miss. Bot. Gard. 64: 706-736.

Janzen, D.H. (1977c): Why fruits rot, seeds mold, and meat spoils. Amer. Naturalist 111: 691-713.

Janzen, D.H. (1977d): Variation in seed weight in Costa Rican *Cassia grandis* (Leguminosae). Tropical Ecology 18: 117-186.

Janzen, D.H. (1977e): How southern cowpea weevil larvae (Bruchidae: *Callosobruchus maculatus*) die on nonhost seeds. Ecology 58: 921-927.

Janzen, D.H. (1978a): A bat-generated fig seed shadow in rain forest. Biotropica 10: 121.

Janzen, D.H. (1978b): Seeding patterns of tropical trees. In: Tomlinson, P.B. and M.H. Zimmerman (eds.), Tropical trees as living systems, pp. 83-128. Cambridge University Press, Cambridge.

Janzen, D.H. (1978c): Complications in interpreting the chemical defenses of trees against tropical arboreal plant-eating vertebrates. In: G.G. Montgommery (ed.), The ecology of arboreal folivores, pp. 73-84. Smithonian Institution Press, Washington, D.D.

Janzen, D.H. (1979a): How many parents do the wasps from a fig have? Biotropica 11: 127-129.

Janzen, D.H. (1979b): New horizons in the biology of plant defenses. In: G.A. Rosenthal and D.H. Janzen (eds.), Herbivores, their interaction with secondary plant metabolites, pp. 331-350. Academic Press, New York.

Janzen, D.H. (1980a): When is it coevolution? Evolution 34: 611-612.

Janzen, D.H. (1980b): Specificity of seed-attacking beetles in a Costa Rican deciduous forest. J. of Ecol. 68: 929-952.

Janzen, D.H. (1982): Variation in average seed size and fruit seediness in a fruit crop of a guanacaste tree (Leguminosae: *Enterolobium cyclocarpum*). Amer. J. Bot. 69: 1169-1178.

Janzen, D.H. (1983): Dispersal of seeds by vertebrate guts. In: D.J. Futuyama and M. Slatkin (eds.), Coevolution, pp. 232-262. Sinauer, Sunderland, Mass.

Janzen, D.H. (1984): Dispersal of small seeds by big herbivores: foliage is the fruit. Amer. Nat. 123: 338-353.

Janzen, D.H. and M.L. Higgins (1979): How hard are *Enterolobium cyclocarpum* (Leguminosae) seeds? Brenesia 16: 61-67.

Janzen, D.H., H.B. Juster, and I.E. Liener (1976): Insecticidal action of the phytohemagglutinin in black beans on a bruchid beetle. Science 192: 795-796.

Janzen, D.H., H.B. Juster, and E.A. Bell (1977): Toxicity of secondary compounds to the seed-eating larvae of the bruchid beetle *Callosobruchus maculatus.* Phytochemistry 16: 223-227.

Janzen, D.H., G.A. Miller, J. Hackforth-Jones, C.M. Pond, K. Hooper, and D.P. Janos (1976): Two Costa Rican bat-generated seed shadows of *Andira inermis* (Leguminosae). Ecology 57: 1068-1075.

Janzen, D.H. and D.E. Wilson (1974): The cost of being dormant in the tropics. Biotropica 6: 260-262.

Jordano, P. (1983): Fig-seed predation and dispersal by birds. Biotropica 15: 38-41.

Kaden, N.N. and M.E. Kirpicnikov (1965): A possible contemporary system of fruit terminology. Taxon 14: 218-223.

Karr, J.R. (1976): Seasonality, resource availability, and community diversity in tropical bird communities. Amer. Naturalist 110: 973-994.

Keay, R.W.J. (1957): Wind-dispersed species in a Nigeria forest. J. Ecol. 45: 471-478.

Knöppel, H.A. (1970): Food of Central Amazonian fishes. Contribution to the nutrient-ecology of Amazonian rain-forest-streams. Amazoniana 2: 257-352.

Koelmeyer, K.O. (1959): (1959/60): The periodicity of leaf change and flowering in the principal communities of Ceylon. Ceylon Forester 4: 157-189, 308-364.

Kozlowski, T.T. (1972): Seed Biology. Vol. I. Academic Press, New York and London.

Laan, P.A. van der (1951): A note about the differences between *Derris* and *Lonchocarpus* as an insecticide. Indones. J. Nat. Science 107: 99-101.

Land, H.C. (1963): Birds at a tropical feeding tree. Wilson Bull. 75: 199-200.

Larcher, W. (1980): Physiological Plant Ecology. Springer, Berlin-Heidelberg-New York. 2nd. ed.

Leck, C.F. (1969): Observations of birds exploiting a Central American fruit tree. Wilson Bull. 81: 264-269.

Leck, C.F. (1971): Overlap in the diet of some neotropical birds. Living bird 10: 89-106.

Leck, C.F. (1972): Observations of birds at *Cecropia* trees in Puerto Rico. Wilson Bull. 84: 498-500.

Lee, D.W. (1979): Biological activity of seed proteins in Malesian legumes. Biotropica 11: 214-218.

Lee, T.D. (1984): Effects of seed number per fruit on seed dispersal in *Cassia fasciculata* (Caesalpiniaceae). Bot. Gaz. 145: 136-139.

Leigh, E.G. (1983): Why are there so many kinds of tropical trees? In: E.G. Leigh, A.S. Rand, and D.M. Windsor (eds.), The ecology of a tropical forest (seasonal rhythms and long-term changes). Oxford University Press, Oxford.

Leigh, E.G., A.S. Rand, and D.M. Windsor (1982): The ecology of a tropical forest (seasonal rhythms and long-term changes). Smithonian Institution Press. Washington, D.C.

Leighton, M. and D.R. Leighton (1983): Vertebrate responses to fruit seasonality within a Bornean rain forest. In: S.L. Sutton, T.C. Whitmore & A.C. Chadwick (eds.). Tropical rain forest: Ecology and management. pp. 181-196.

Leite, A.M.C. and J.M. Rankin (1981): Ecología de sementes de *Pithecolobium racemosum* Ducke. Acta Amazonica 11: 309-318.

Levey, D.J. and T.C. Moermond (1984): Fruit choice in neotropical birds: the effect of distance between fruits on preference patterns. Ecology 65: 844-850.

Levin, D.A. (1974): The oil content of seeds: an ecological perspective. Amer. Naturalist 108: 193-206.

Liebermann, D., J.B. Hall, M.D. Swaine, and M. Liebermann (1979): Seed dispersal by baboons in the Shai Hills, Ghana. Ecology 60: 65-75.

Liener, I.E. (1979): Phytohemagglutinins. In: Rosenthal, G.A. and D.H. Janzen (eds.), Herbivores: their interaction with secondary plant metabolites. Academic Press, New York, p. 567-598.

Longman, K.A. and J. Jenik (1974): Tropical forest and its environment. Longman, London and New York.

Mac-Arthur, R.H. (1969): Patterns of communities in the tropics. Biol. J. Linnean Soc. London 1: 19-30.

Maguire, B. (1970): On the flora of the Guyana Highland. Biotropica 2: 85-100.

Maguire, B. et col. (1981): The Botany of the Guyana Highland. Part XI. Memoirs of the New York Bot. Garden, Vol. 32.

Manasse, R.S. and H.F. Howe (1984): Competition for dispersal agents among tropical trees: influences of neighbors. Oecologia 59: 185-190.

Margalef, R. (1968): Perspectives in ecological theory. University of Chicago Press, Chicago-London.

McClure, H.E. (1966): Flowering, fruiting and animals in the canopy of a tropical rain forest. Malay. Forester 29: 182-203.

McCutchen, C.W. (1977): The spinning rotation of ash and tulip tree samaras. Science 197: 691-692.

McDiarmid, R.W., R.E. Ricklefs, and M.S. Foster (1977): Dispersal of *Stemmadenia donnell-smithii* (Apocynaceae) by birds. Biotropica 9: 9-25.

McKey, D. (1975): The ecology of coevolved seed dispersal systems. In: Gilbert, L.E. and P.H. Raven (eds.), Coevolution of animals and plants, pp. 159-191. University of Texas Press, Austin.

McKey, D. (1978): Soils, vegetation, and seed-eating by Black *Colobus* monkeys. In: G.G. Montgommery (ed.), Ecology of arboreal folivores: 423-437. Smithonian Institution Press, Washington D.C.

McKey, D. (1979): The distribution of secondary compounds within plants. In: G.A. Rosenthal and D.H. Janzen (eds.), Herbivores: their interaction with secondary plant metabolites, pp. 55-133. Academic Press, New York.

Medina, E., H.A. Mooney, and C. Vázquez-Yánez (1984): Physiological ecology of plants of the wet tropics. Tasks for vegetation science 12 (eds. H. Lieth and H.A. Mooney).

Medway, F.L.S. (1972): Phenology of a tropical rain forest in Malaya. Biol. J. Linn. Soc. 4: 117-146.

Mittermeier, R.A. and M.G.M. van Roosmalen (1981): Preliminary observations on habitat utilization and diet in eight Surinam monkeys. Folia Primatologica 36: 1-39.

Moermond, T.C. and J.S. Denslow (1983): Fruit choice in neotropical birds: effects of fruit type and accessibility on selectivity. J. of Animal Ecol. 52: 407-420.

Mohana Rao, P.R. (1974): Nutmeg seed: its morphology and developmental anatomy. Phytomorphology 24: 262-273.

Mohana Roa, P.R. (1976): Seed and fruit anatomy of *Pterospermum acerifolium* (Stercucliaceae). Phytomorphology 26: 363-369.

Montilla, R. (1977): El Llano. Soc. Venez. Ciencias Naturales, Caracas.

Moreno, Patricia (1977). Latencia y viabilidad de semillas de árboles tropicales. Interciencia 2: 298-302.

Mori, S. (1970): The ecology and uses of the species of *Lecythis* in Central America. Turrialba 20: 344-350.

Mori, S.A. and Jacqueline A. Kallunki (1976): Phenology and floral biology of *Gustavia superba* (Lecythidaceae) in Central Panamá. Biotropica 8: 184-192.

Morton, E.S. (1973): On the evolutionary advantages and disadvantages of fruit eating in tropical birds. Amer. Naturalist 107: 8-22.

Nadelmann, H. (1889): Über die Schleimendosperme der Leguminosensamen. Ber. Dtsch. Bot. Ges. 7: 248-255.

Neger, F.W. (1913): Biologie der Pflanzen auf experimenteller Grundlage. Enke-Verlag, Stuttgart.

Norberg, R.A. (1973): Autorotation, self-stability, and structure of single-winged fruits and seeds (samaras) with comparative remarks on animal flight. Biol. Rev. 48: 561-596.

Olson, S.L. and K.E. Blum (1968): Avian dispersal of plants in Panamá. Ecology 49: 565-566.

Opler, P.A., H.G. Baker, and G.W. Frankie (1975): Reproductive biology of some Costa Rican *Cordia* species (Boraginaceae). Biotropica 7: 234-247.

Oppenheimer, J.R. and G.E. Lang (1969): *Cebus* monkeys: effect on branching of *Gustavia* trees. Science 165: 187-188.

Osmaston, H.A. (1965): Pollen and seed dispersal in *Chlorophora excelsa* and other Moraceae, and in *Parkia filicoidea* (Mimosaceae), with special reference to the role of the fruit bat, *Eidolon helvum*. Commonwealth Forestry Review 44: 96-104.

Parkin, J. (1953): The Durian theory – a criticism. Phytomorphology 3: 80-88.

Pellegrini, O. (1952): Affinità systematiche fra Pruneae e Leguminosae dimostrate dai tessuti omologhi dei pericarpi. Delpinoa 5: 5-48.

Pennington, T.D. and J. Sarukhan (1968): Arboles tropicales de México. Inst. Nac. Invest. Forestales, Mexico.

Perry, D.R. (1986): Die Kronenregion des tropischen Regenwaldes. Spektrum der Wissenschaft, Sonderdruck.

Pfeiffer, H. (1891): Die Arillargebilde der Pflanzen-Samen. Engler's Bot. Jb. 13: 492-540.

Pijl, L. van der (1952): Ecological variations on the theme pod. Indones. J. Nat. Sci. 108: 6-12.

Pijl, L. van der (1955): Sarcotesta, aril, pulpa and the evolution of the angiosperm fruit. I, II. Proc. Ned. Acad. Wet. (C) 58: 154-161, 307-312.

Pijl, L. van der (1956): Classification of the leguminous fruits according to their ecological and morphological properties. Proc. K. Ned. Acad. Wet. (C) 59: 301-313.

Pijl, L. van der (1957): On the arilloids of *Nephelium, Euphorbia, Litchi* and *Aesculus*, and the seeds of Sapindaceae in general. Acta Bot. Neerl. 6: 618-641.

Pijl, L. van der (1966): Ecological aspects of fruit evolution. Proc. K. Ned. Acad. Wet. (C) 69: 597-640.

Pijl, Van der der (1969): Evolutionary action of tropical animals on the reproduction of plants. Biol. J. Linn. Soc. 1: 85-96.

Pijl, L. van der (1972): Principles of dispersal in higher plants. Springer, Berlin-Heidelberg-New York.

Pino J. Isabel M. (1979): Morfología y anatomía relacionada a la taxonomía de plantas superiores. Seminario de grado, UCV, Caracas.

Pittier, H. (1926): Manual de las plantas usuales de Venezuela (y su suplemento 1939), 2nd. ed. (1970) Fundación Eugenio Mendoza, Caracas. 620 pp.

Plummer, G.L. and Catherine Keever (1963): Autumnal daylight weather and camphor-weed dispersal in the Georgia Piedmont region. Bot. Gaz. 124: 283-289.

Prance, G.T. (1973): Phytogeographic support for the theory of Pleistocene forest refuges in the Amazon Basin, based on evidence from distribution patterns in Caryocaraceae, Chrysobalanaceae, Dichapetalaceae and Lecythidaceae. Acta Amazonica 3: 5-28.

Prance, G.T. (1976): The pollination and androphore structure of some Amazonian Lecythidaceae. Biotropica 8: 235-241.

Prance, G.T. (1978): The origin and evolution of the amazon flora. Interciencia 3: 207-222.

Prance, G.T. and S.A. Mori (1978): Observations on the fruits and seeds of neotropical Lecythidaceae. Brittonia 30: 21-33.

Prévost, M.F. (1983): Les fruits et les graines des espèces végétales pionnières de Guyane. Rev. Ecol. (Terre et Vie) 38: 121-145.

Proctor, V.W. (1968): Long-distance dispersal of seeds by retention in digestive tract of birds. Science 160: 321-322.

Radford, A.E., W.C. Dickison, J.R. Massey, and C.R. Bell (1974): Vascular Plant Systematics. Harper and Row Publ., New York-London.

Ranga Swamy, N.S. (1967): Morphogenesis of seed germination in Angiosperms. Phytomorphology 17: 477-487.

Regal, P.J. (1977): Ecology and evolution of flowering plant dominance. Science 196: 622-629.

Rehr, S.S., D.H. Janzen, and P.P. Feeny (1973): L-dopa in legume seeds: a chemical barrier to insect attack. Science 181: 81-82.

Reiche, K. (1907): Grundzüge der Pflanzenverbreitung in Chile. Die Vegetation der Erde VIII.

Rhoades, D.F. (1979): Evolution of plant chemical defense against herbivores. In: G.A. Rosenthal and D.H. Janzen (eds.), Herbivores: their interaction with secondary plant metabolites. Academic Press, New York. p. 4-54.

Richards, P.W. (1945): The floristic composition of primary tropical rain forest. Biol. Rev. 20: 1-13.

Richards, P.W. (1952): The tropical rain forest. University Press, Cambridge.

Richards, P.W. (1969): Speciation in the tropical rain forest and the concept of the niche. Biol. J. Linn. Soc. 1: 149-153.

Ridley, H.N. (1905): On the dispersal of seeds by wind. Ann. of Bot. 19: 351-363.

Ridley, H.N. (1930): The dispersal of plants throughout the world. Reeve, Ashford.

Röhl, E. (1959): Fauna descriptiva de Venezuela. Nuevos Gráficos, 4th ed., Caracas.

Rollet, B. (1967a): Estudio del bloque experimental Rio Grande (5000 héctares de selva pluvial en la Guayana Venozolana). 58 pp. MAC, Direccion de Recursos Naturales Renovables, Caracas.

Rollet, B. (1967b): Inventario forestal en la Guayana Venezolana. MAC, Informe 3, 2a. Caracas.

Rollet, B. (1969a): Etudes quantitatives d'une forêt dense humide sempervirente de plain de la Guayane Vénézuélienne. Thèse Doct., Toulouse.

Rollet, B. (1969b): Etudes quantitatives d'une forêt dense humide sempervirente de plaine en Guayane Vénézuélienne. Trav. Lab. Forest. Toulouse, T. 1, Vol. 8, Art. 1: 1-36.

Rollet, B. (1969c): La Régéneration naturelle en forêt dense humide sempervirente de plaine de la Guayane Vénézuéliénne. Bois et Forêts des Tropiques 124: 19-38.

Roosmalen, M.G.M. van der (1977): Surinaams Vruchtenboek. Inst. Syst. Bot. Utrecht.

Rosenthal, G.A. and E. Bell (1979): Naturally occurring, toxic non-protein amino acids. In: G.A. Rosenthal and D.H. Janzen (eds.), Herbivores: their interaction with secondary plant metabolites. Academic Press, New York.

Rosenthal, G.A. and D.A. Janzen (1979): Herbivores: their interaction with secondary plant metabolites. Academic Press, New York.

Roth, Ingrid (1974): Desarrollo y estructura anatómica del merey: *Anacardium occidentale* L. Acta Bot. Venez. 9: 197-223.

Roth, Ingrid (1976): Estructura interna de los domacios foliares en *Tococa* (Melastomaceae). Acta Biol. Venez. 9: 221-258.

Roth, Ingrid (1977a): Über die anisokotyles Keimlinge von *Gyranthera carbibensis*, Bombacaceae. (Struktur der Samenschale, des Makro- und Mikrokotyledos und der Primärblätter). Bot. Jahrb. Syst. 97: 515-547.

Roth, Ingrid (1977b): Fruits of Angiosperms. In: H.J. Braun, S. Carlquist, P. Ozenda, I. Roth (eds.). Encyclopedia of Plant Anatomy. Borntraeger, Berlin-Stuttgart.

Roth, Ingrid (1980): Blattstruktur von Pflanzen aus feuchten Tropenwäldern. Bot. Jahrb. Syst. 101: 489-525.

Roth, Ingrid (1981): Structural patterns of tropical barks. In: H.J. Braun, S. Carlquist, P. Ozenda, and I. Roth (eds.). Encyclopedia of Plant Anatomy. Borntraeger, Berlin-Stuttgart.

Roth, Ingrid (1984): Stratification of tropical forests as seen in leaf structure. In: H. Lieth (ed.), Tasks for vegetation science 6. Dr. W. Junk Publ., The Hague-Boston-Lancaster.

Roth, Ingrid and Norma Carrasquel (1969): The peeling off of the epidermis in fruits of *Jacquinia loeflingii* Carrasquel. Acta Bot. Venez. 6: 139-145.

Roth, Ingrid and Ingrid Clausnitzer (1972): Desarrollo y anatomía del fruto y de la semilla de *Carica Papaya* L. (Lechosa). Acta Bot. Venez. 7: 187-206.

Roth, Ingrid and Helga Lindorf (1971a): Desarrollo y anatomía del fruto y de la semilla de *Theobroma cacao* L. Acta Bot. Venez. 6: 261-295.

Roth, Ingrid and Helga Lindorf (1971b): Anatomía y desarrollo del fruto y de la semilla del café. Acta Bot. Venez. 6: 197-238.

Roth, Ingrid and Helga Lindorf (1972a): Desarrollo y anatomía del fruto y de la semilla de la granada (*Punica granatum* L.). Acta Bot. Venez. 7: 143-162.

Roth, Ingrid and Helga Lindorf (1972b): Anatómia y desarrollo del fruto y de la semilla de *Achras zapota* (Níspero). Acta Bot. Venez. 7: 121-141.

Roth, Ingrid and Helga Lindorf (1972c): Desarrollo y anatomía del fruto y de la semilla de *Citrus*. Acta Bot. Venez. 9: 149-176.

Roth, Ingrid and Helga Lindorf (1974a): La interpretación morfológica de la semilla de las Rubiaceae y en especial del café. Acta Bot. Venez. 9: 141-147.

Roth, Ingrid and Helga Lindorf (1974b): Desarrollo y anatomía del fruto y de la semilla de *Myristica fragrans* van Houtt. Acta Bot. Venez. 9: 149-176.

Rudd, V.E. (1965): The American species of *Ormosia* (Leguminosae). Contr. U.S. National Herb. 32: 279-384.

Ryan, C.A. (1979): Proteinase inhibitors. In: Rosenthal, G.A. and D.H. Janzen (eds.), Herbivores: their interaction with secondary plant metabolites, pp. 599-618. Academic Press, New York.

Salisbury, E.J. (1942): The reproductive capacity of plants. Bell, London.

Schimper, A.F.W. (1935): Pflanzengeographie auf physiologischer Grundlage. 3rd. ed. Revised by C.F. von Faber. Jena.

Schnee, L. (1960): Plantas comunes de Venezuela. Revista Fac. de Agronomía UCV, No. 3. Caracas.

Schneider, S. (1935): Untersuchungen über die Samenschleudermechanismen verschiedener Rhoeadales. Jahrb. wiss. Bot. 81: 663-704.

Schoener, T.W. (1968): Sizes of feeding territories among birds. Ecology 49: 123-141.

Scott, A.C. and T.N. Taylor (1983): Plant/animal interactions during the Upper Carboniferous. Bot. Rev. 49: 259-307.

Seigler, D.S. (1979): Toxic seed lipids. In: Rosenthal, G.A. and D.H. Janzen (eds.), Herbivores: their interaction with secondary plant metabolites. Academic Press, New York.

Sheldon, J.C. and F.M. Burrows (1973): The dispersal effectiveness of the achene-pappus units of selected Compositae in steady wind with convection. New Phytol. 72: 665-675.

Sheldon, J.C. and J.T. Lawrence (1973): Apparatus to measure the rate of fall of wind-dispersed seeds. New Phytol. 72: 677-680.

Silander, J.A. jr. (1978): Density-dependent control of reproductive success in *Cassia biflora*. Biotropica 10: 292-296.

Silvertown, J.W. (1980): The evolutionary ecology of mast seeding in trees. Biol. J. Linnean Soc. 14: 235-250.

Singh, R.P. (1954): Structure and development of seeds in Euphorbiaceae: *Ricinus communis* L. Phytomorphology 4: 118-123.

Sinnot, E.W. and I.W. Bailey (1915): Changes in the fruit type of Angiosperms coincident with the development of the herbaceous habit. Science, N.S. 41: 179.

Skutch, A.F. (1980): Arils as food of tropical american birds. Condor 82: 31-42.

Sláma, K. (1979): Insect hormones and antihormones in plants. In: Rosenthal, G.A. and D.H. Janzen (eds.), Herbivores: their interaction with secondary plant metabolites. Academic Press, New York.

Smith, C.C. (1975): The coevolution of plants and seed predators. In: L.E. Gilbert and P.H. Raven (eds.), Coevolution of animals and plants: 53-77. University of Texas Press, Austin.

Smythe, N. (1970): Relationships between fruiting seasons and seed dispersal methods in a neotropical forest. Amer. Naturalist 104: 25-35.

Snodderly, D.M. (1979): Visual discriminations encountered in food foraging by a neotropical primate: implications for the evolution of color vision. In: E.H. Burtt (ed.), The behavioural significance of color, pp. 238-281. Garland Press, New York & London.

Snow, D.W. (1962): The natural history of the oilbird, *Steatornis caripensis*, in Trinidad, W.I. II. Population, breeding ecology and food. Zoologica 47: 199-221.

Snow, D.W. (1966): A possible selective factor in the evolution of fruiting seasons in tropical forest. Oikos 15: 274-281.

Snow, D.W. (1971): Evolutionary aspects of fruit eating by birds. Ibis 113: 194-203.

Snow, D.W. (1981a): Coevolution of birds and plants. In: P.L. Forey (ed.), The evolving biosphere, pp. 169-178. Cambridge University Press, Cambridge.

Snow, D.W. (1981): Tropical frugivorous birds and their food plants: A world survey. Biotropica 13: 1-14.

Sousa, S.M. (1969): Influencia de las aves en la vegetación de la laguna del Majahual en los Tuxtlas, Ver. Bol. Soc. Bot. Mexico 30: 97-112.

Southwood, T.R.E. (1985): Interactions of plants and animals: patterns and processes. Oikos 44: 5-11.

Sporne, K.R. (1954): Statistics and the evolution of dicotyledons. Evolution 8: 55-64.

Stebbins, G.L. (1971): Adaptive radiation of reproductive characteristics in the Angiosperms. II. Seeds and seedlings. Ann. Rev. Ecol. and Syst. II: 237-260.

Stopp, K. (1958): Die verbreitungshemmenden Einrichtungen in der Süd-afrikanischen Flora. Botan. Studien 8. Jena.

Stanley, S.M. (1982): Macroevolution and the fossil record. Evolution 36: 460-473.

Stebbins, G.L. and F.J. Ayala (1981): Is a new evolutionary synthesis necessary? Science 213: 967-971.

Steven, D. de, and F.E. Putz (1984): Impact of mammals on early recruitment of a tropical canopy tree, *Dipteryx panamensis*, in Panama. Oikos 43: 207-216.

Stopp, K. (1962): Antitelechore Einrichtungen bei den Gattungen *Sesamum, Rogera* und *Psilocaulon*. Beitr. Biol. Planzen 37: 63-76.

Struhsaker, T.T. (1977): Palm-nut smashing by *Cebus* a. *apella* in Colombia. Biotropica 9: 124-126.

Stubblebine, W., J.H. Langenheim, and D. Lincoln (1978): Vegetative response to photoperiod in the tropical leguminous tree *Hymenaea courbaril* L. Biotropica 10: 18-29.

Suessenguth, K. (1938): Neue Ziele der Botanik. München.

Swain, T. (1976): A. d'A. Bellairs and C.B. Cox (eds.) Morphology and biology of reptiles, pp. 107-122. Academic Press, New York.

Swain, T. (1979): Tannins and lignins. In: Rosenthal, G.A. and D.H. Janzen (eds.), Herbivores: their interaction with secondary plant metabolites. Academic Press, New York, p. 657-682.

Swaine, M.D. and T. Beer (1977): Explosive seed dispersal in *Hura crepitans* L. (Euphorbiaceae). New Phytologist 78: 695-708.

Takhtajan, A. (1959): Evolution der Angiospermen. G. Fischer, Jena.

Tammes, P.M.L. (1951): Relationship between relative humidity, temperature and growth of micro-organisms in the tropics. Indones. J. Nat. Sci. 107: 45-46.

Thompson, J.N. and M.F. Willson (1978): Disturbance and the dispersal of fleshy fruits. Science 200: 1161-1163.

Thomson, R.B. (1927): Evolution of the seed habit in plants. Trans. R. Soc. Canada 21: 229-272.

Torres, A. (1968): Algunas plantas leñosas de Cumaná. Escuela de Ciencias, Cumaná.

Treub, M. (1888): Notice sur la nouvelle flore de Krakatau. Ann. Jard. Bot. Buitenzorg 7: 213-223.

Uhl, C. and K. Clark (1983): Seed ecology of selected Amazon basin successional species. Bot. Gaz. 144: 419-425.

Ulbrich, E. (1928): Biologie der Früchte und Samen (Karpobiologie). Springer, Berlin.

Uphof, J.C.T. (1942): Ecological relations of plants with ants and termites. Bot. Review 8: 563-598.

Vandermeer, J.H. (1974): Relative isolation and seed predation in *Calliandra grandiflora*, a Mimosaceous legume from the highlands of Guatemala. Biotropica 6: 267-268.

Vandermeer, J.H. (1975): A graphical model of insect seed predation. Amer. Naturalist 109: 147-160.

Vásquez-Yanez, C., A. Orozco, G. François and L. Trejo (1975): Observations on seed dispersal by bats in a tropical humid region in Veracruz, Mexico. Biotropica 7: 73-76.

Vuilleumier, B.S. (1971): Pleistocene changes in the fauna and flora of South America. Science 173: 771-780.

Walter, H. (1961): Die Vegetation der Erde in ökologischer Betrachtung. Fischer, Stuttgart.

Walter, H. and H. Flohn (1973): Die Vegetation der Erde in Öko-physiologischer Betrachtung. 3rd. ed. Vol. I. Die tropischen und subtropischen Zonen. G. Fischer, Stuttgart.

Wheelwright, N.T. (1985a): Competition for dispersers, and the timing of flowering and fruiting in a guild of tropical trees. Oikos 44: 465-477.

Wheelwright, N.T. (1985b): Fruit size, gape width, and the diets of fruit-eating birds. Ecology 66: 808-818.

Wheelwright, N.T., W.A. Haber, K.G. Murray, and C. Guindon (1984): Tropical fruit-eating birds and their food plants: a survey of a Costa Rican lower montane forest. Biotropica 16: 173-192.

Wheelwhright, N.T. and G.H. Orians (1982): Seed dispersal and animals: contrasts with pollen dispersal, problems of terminology, and constraints on coevolution. Amer. Naturalist 119: 402-413.

Whitmore, T.C. (1975): Tropical rain forests of the Far East. Clarendon Press, Oxford.

Whitmore, T.C. (1983): Secondary succession from seed in tropical rain forests. Forest. Abst. 44: 767-779.

Wikander, Tatiana (1980): Mecanismos de dispersión de frutos y semillas del bosque tropical decíduo por la sequía de Chrallave. Trabajo de ascenso. Caracas.

Williams, M.A.J. (1975): Late Pleistocene tropical aridity synchronous in both hemispheres? Nature 253: 617-618.

Williams, R.O. (1951): The useful and ornamental plants in Trinidad and Tobago. Guardian Comm. Print., Trinidad and Tobago.

Willson, M.F. and J.N. Thompson (1982): Phenology and ecology of color in bird-dispersed fruits, or why some fruits are red when they are "green". Canad. J. Bot. 60: 701-713.

Wunderlich, Rosalie (1967): Some remarks on the taxonomic significance of the seed coat. Phytomorphology 17: 301-311.

Zim, H.S. and H.M. Smith (1953): Reptiles and amphibians, a guide to familiar American species. Simon and Schuster, New York.

Zohary, M. (1950): Evolutionary trends in the fruiting head of Compositae. Evolution 4: 103-109.

Fig. 1. Above left: *Carica papaya*, Right: *Theobroma cacao*, both showing caulicarpy Below, left: *Mangifera indica* with flagellocarpy. Right: Infrutescence of *Pandanus*.

Fig. 2. Above: *Ochroma,* still closed and dehisced fruits with "kapoklike" locular hairs. Below: *Luffa cylindrica, intact fruits and reticular vascular system supplying the "luffa sponge".*

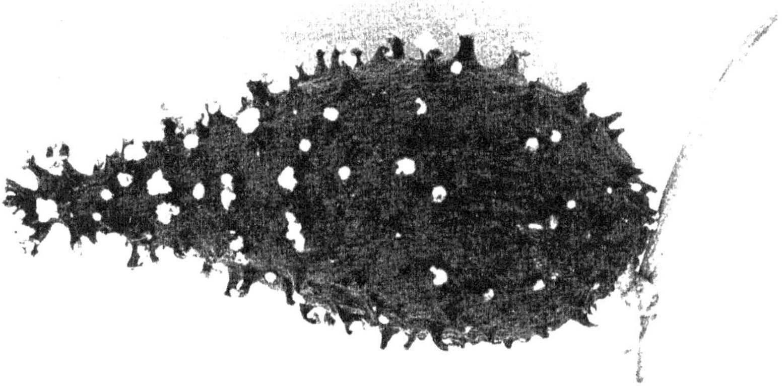

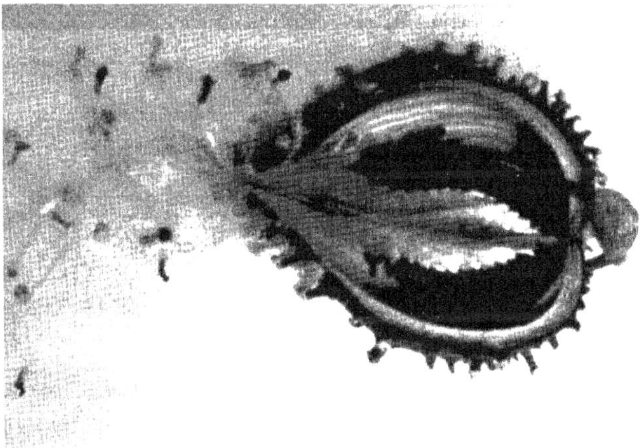

Fig. 3. Fruit of the herbaceous climber *Matelea maritima*, intact fruit with surface protuberances and halved fruit exposing placenta and black seeds with hairs.

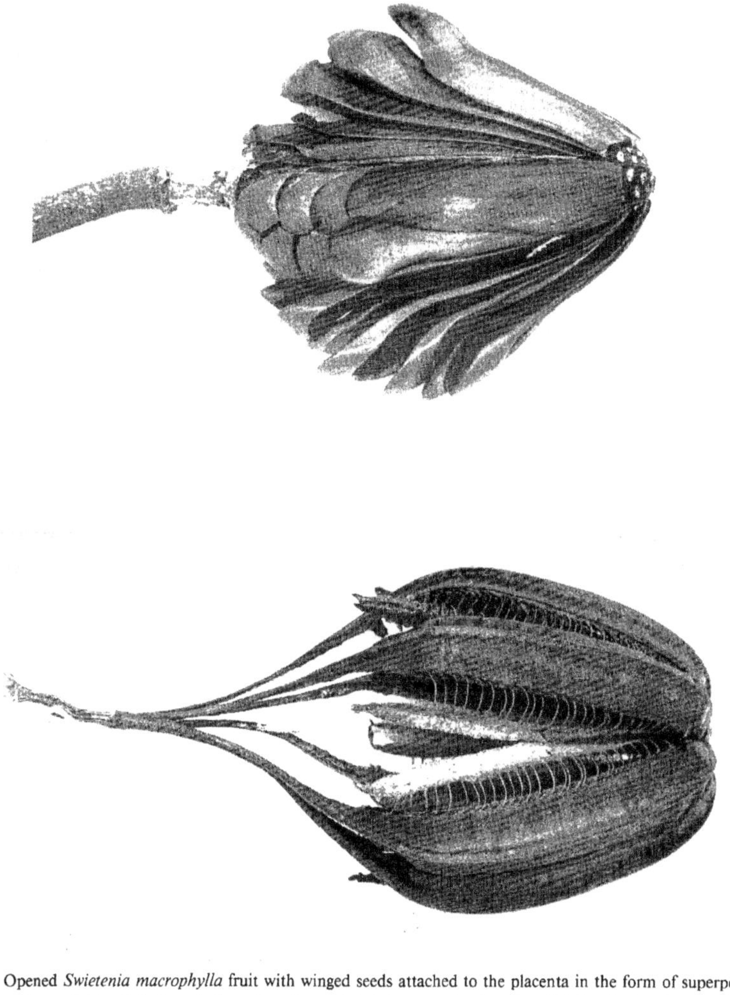

Fig. 4. Above: Opened *Swietenia macrophylla* fruit with winged seeds attached to the placenta in the form of superposed rows. Below: *Aristolochia grandiflora,* dehisced fruit with "windows" penetrated by vascular bundles. Within the locules the flattened seeds are arranged like coins in a roll of money.

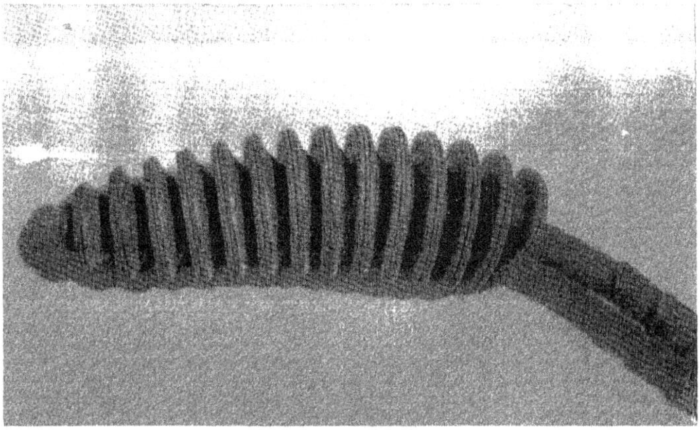

Fig. 5. Above: Infrutescence of *Cyclanthus*. Carpels forming a "staircase pattern". Below: *Allamanda* fruit with 2 follicular mericarps.

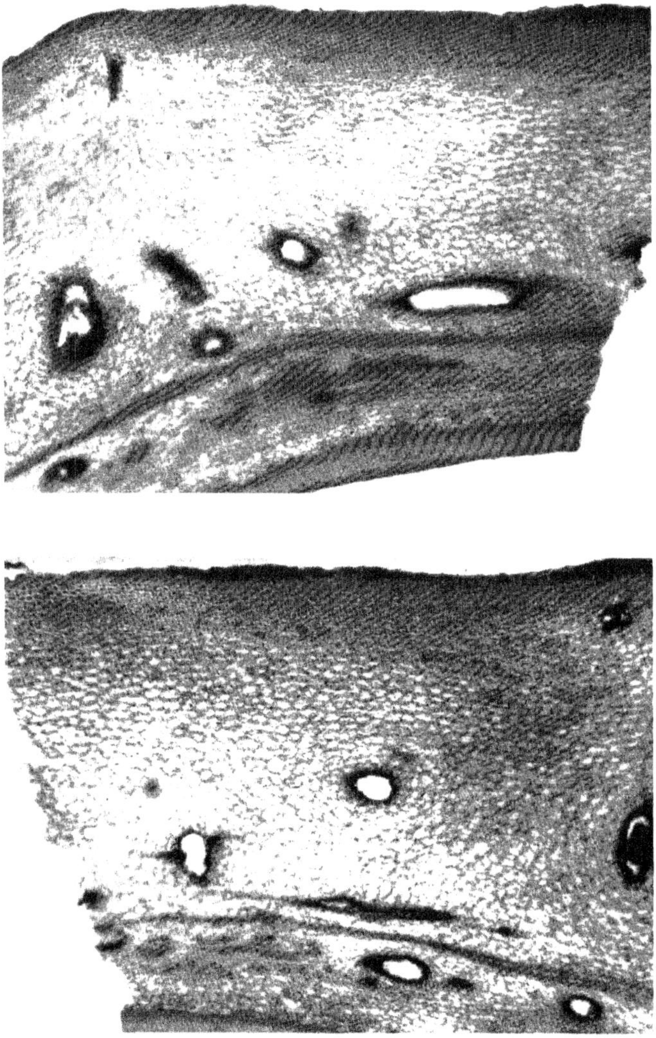

Fig. 6. Tapirira guianensis, Anacardiaceae, t.s. The pericarp is mainly composed of parenchymatous cells in which large secretory cavities are embedded. The outer part of the fruit wall is collenchymatous, while the endocarp is fibrous.

Fig. 7. Above and left: Aggregate fruit of *Xylopia nitida* var. *nerviosa,* Annonaceae. The follicles contain 2-5 seeds. Below: *Anaxagorea,* head-like aggregate fruit composed of a multitude of follicles.

Fig. 8. Aspidosperma megalocarpon, Apocynaceae. Above: The 2 halves of the mericarp. Below: Mericarp Halve with winged seed.

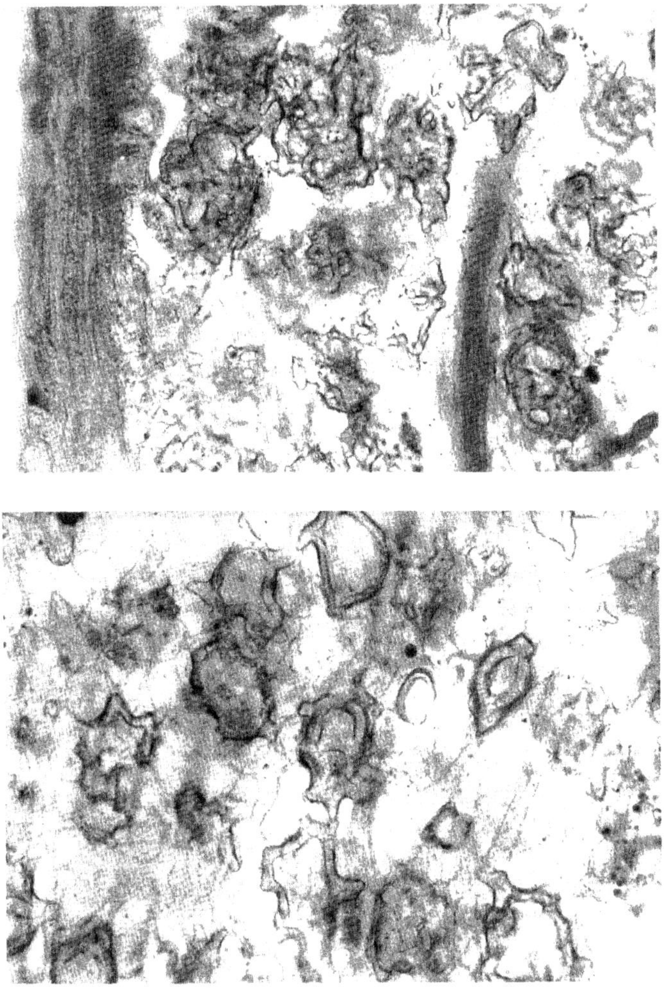

Fig. 9. Aspidosperma megalocarpon, parenchymatous fruit wall, as seen in t.s., in which sclereids of irregular shape are frequently scattered. Above: Vascular bundles in longitudinal section.

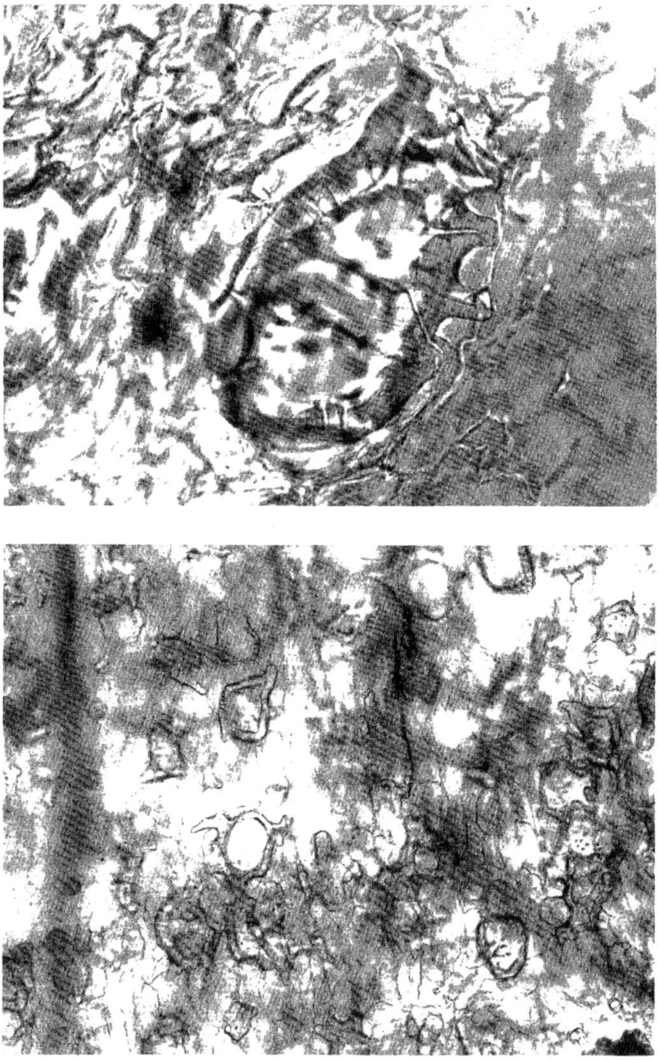

Fig. 10. Above: *Aspidosperma excelsum,* Apocynaceae, large sclereid embedded in the parenchyma of the fruit wall. Below: *Aspidosperma marcgravianum,* sclereids of different shape.

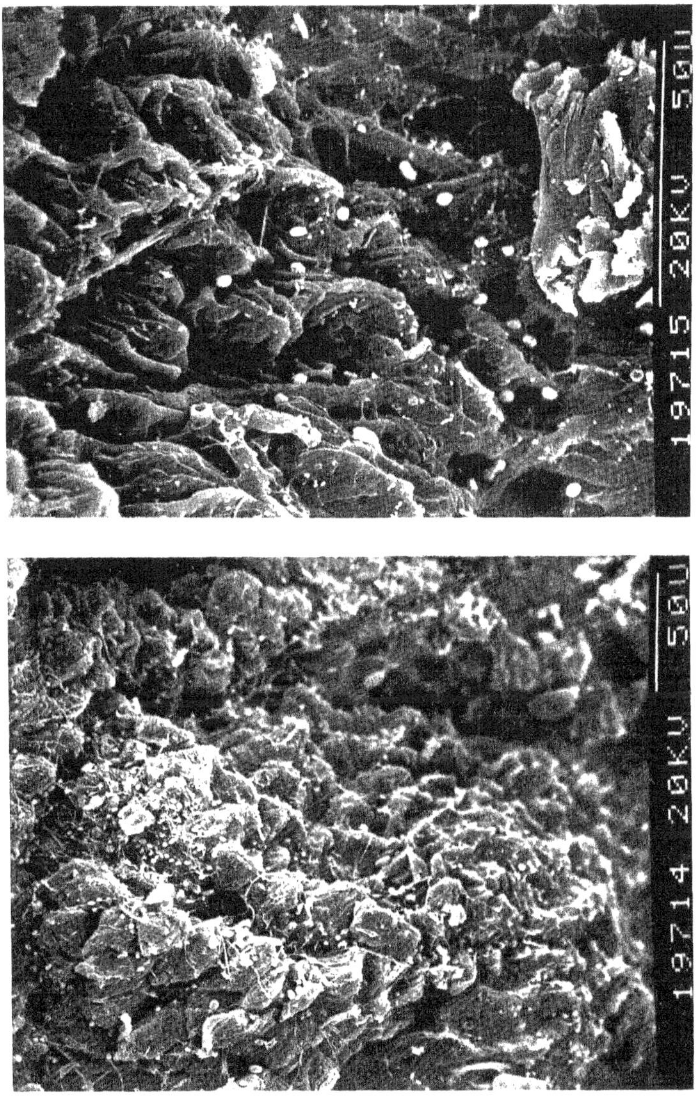

Fig. 11. Aspidosperma excelsum, spiny fruit surface, as seen in the scanning microscope.

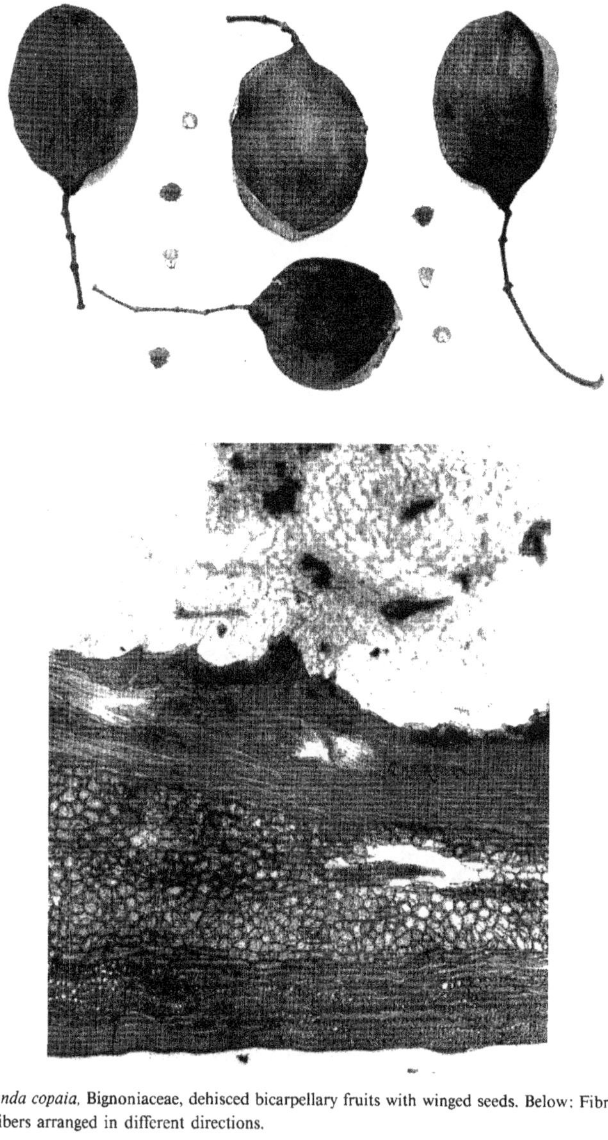

Fig. 12. Above: *Jacaranda copaia*, Bignoniaceae, dehisced bicarpellary fruits with winged seeds. Below: Fibrous endocarp of *Jacaranda copaia* with groups of fibers arranged in different directions.

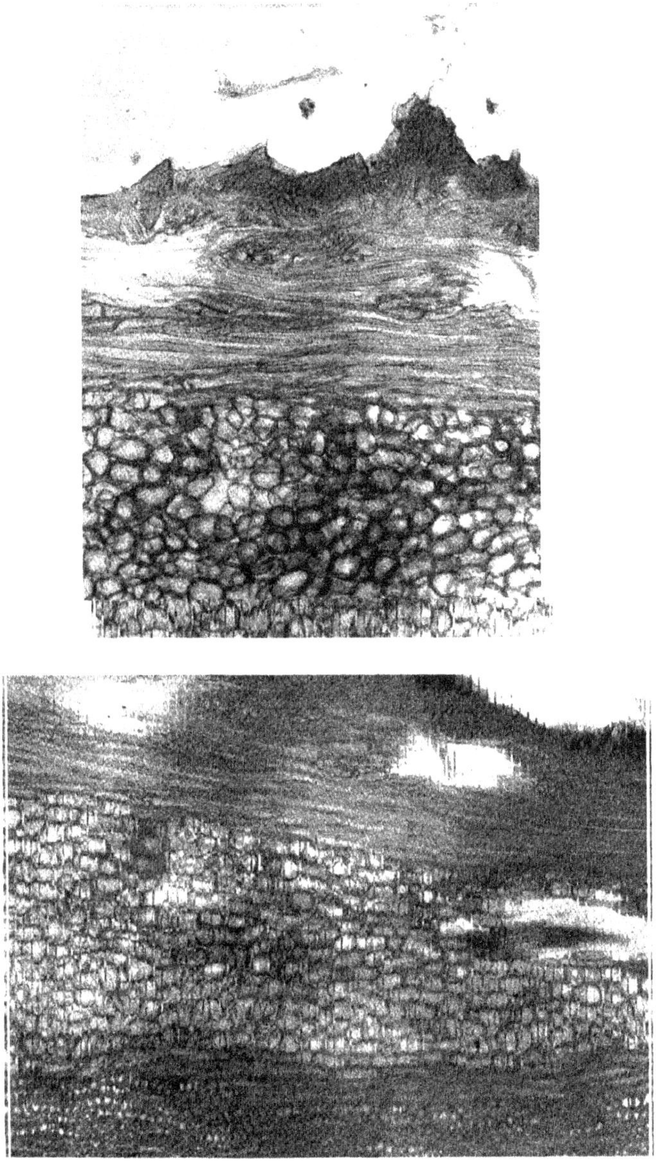

Fig. 133. Jacaranda copaia. Above: Outer part of endocarp protruding into the parenchymatous mesocarp. Below: Endocarp fibers partly sectioned longitudinally and partly transversely. Vascular bundles are included in parenchymatous isles of the endocarp.

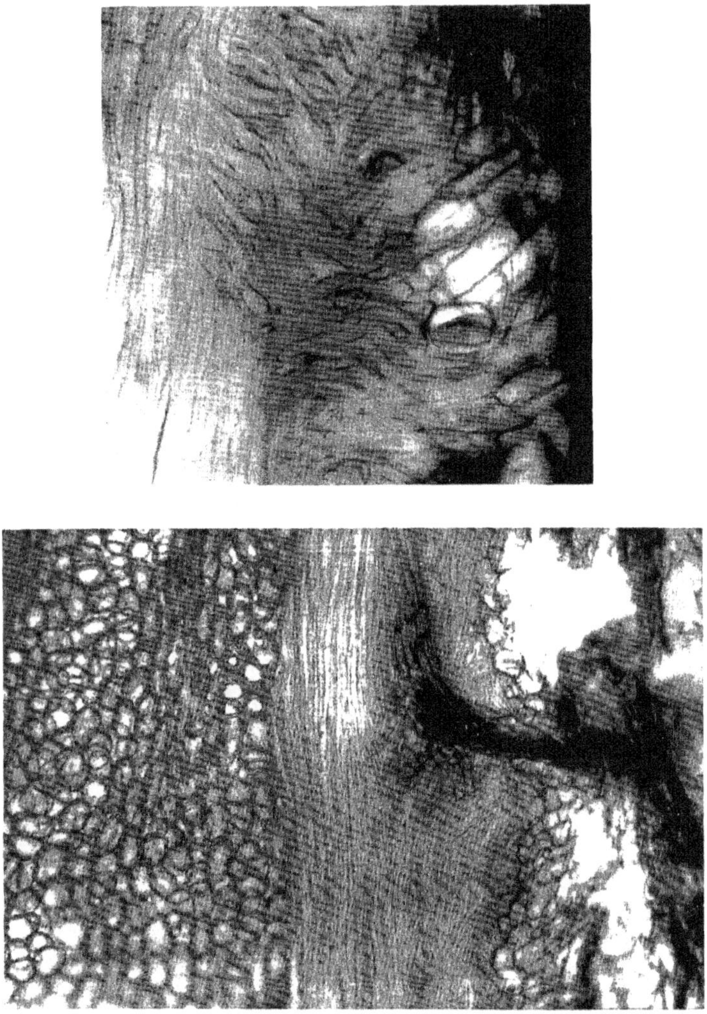

Fig. 14. Jacaranda copaia. Above: Part of vascular bundle cut transversely. Below: Ramification of vascular bundle entering the mesocarp.

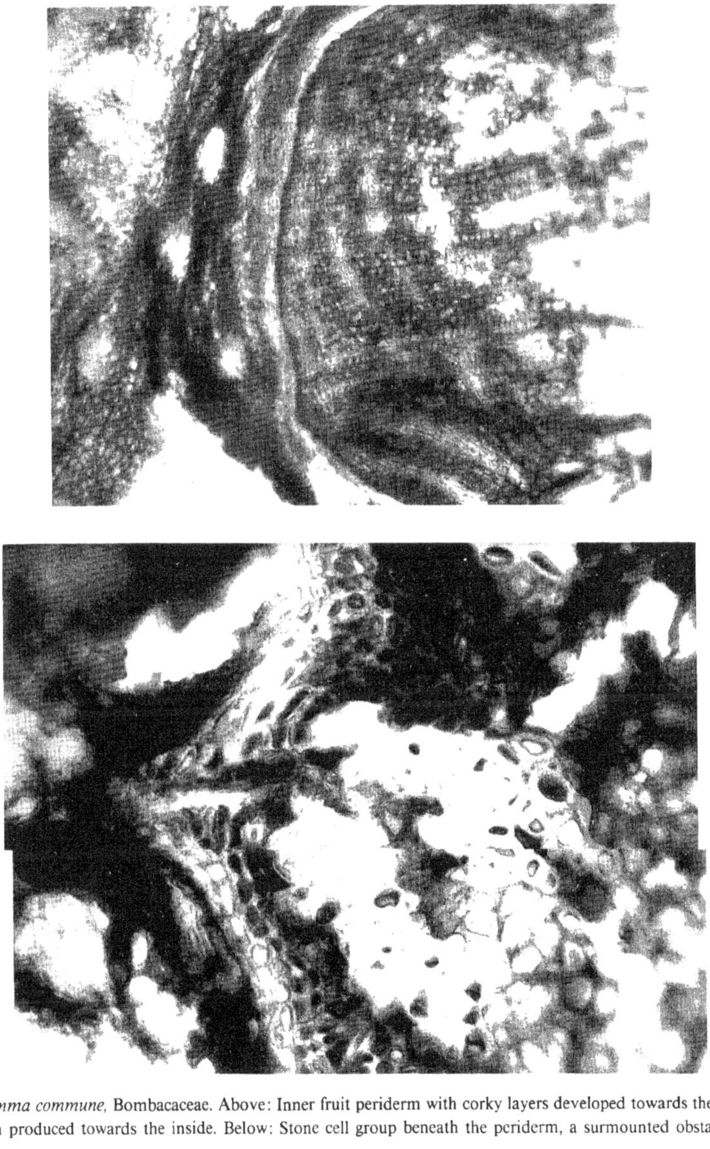

Fig. 15. Catostemma commune, Bombacaceae. Above: Inner fruit periderm with corky layers developed towards the outside and a very broad phelloderm produced towards the inside. Below: Stone cell group beneath the periderm, a surmounted obstacle.

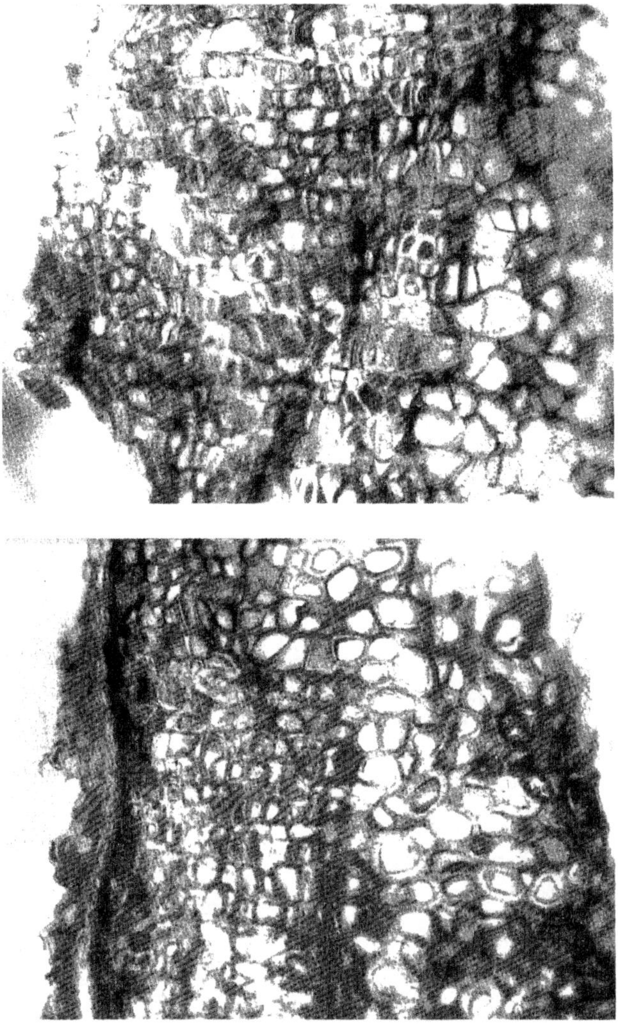

Fig. 16. Catostemma Commune. Above: Phelloderm, each row composed of more than 30 cells. Below: Phelloderm cells partly sclerified.

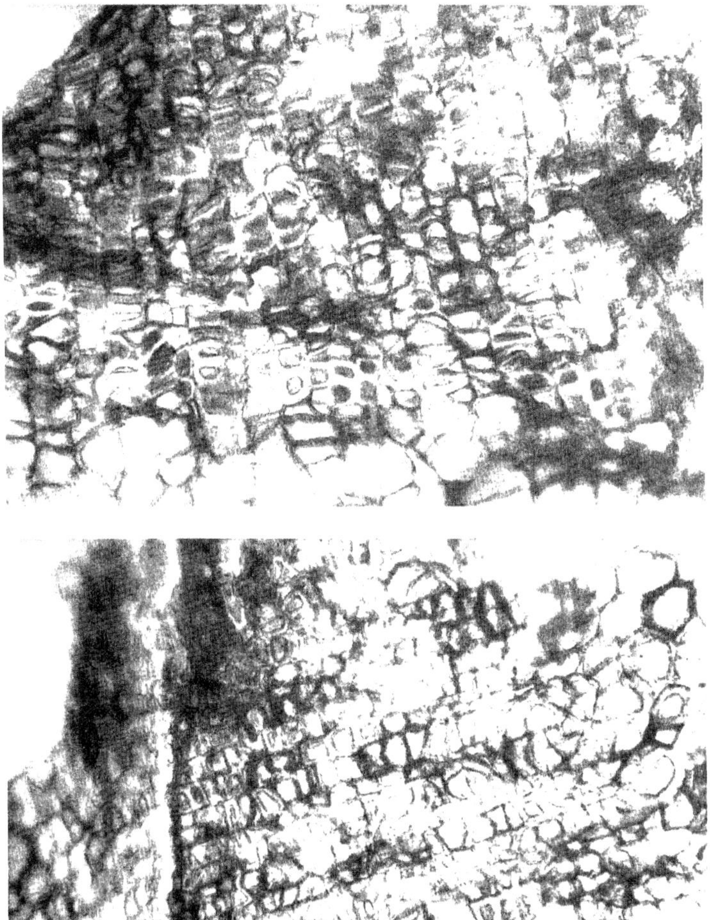

Fig. 17. Catostemma commune. Very large sclerified periderms.

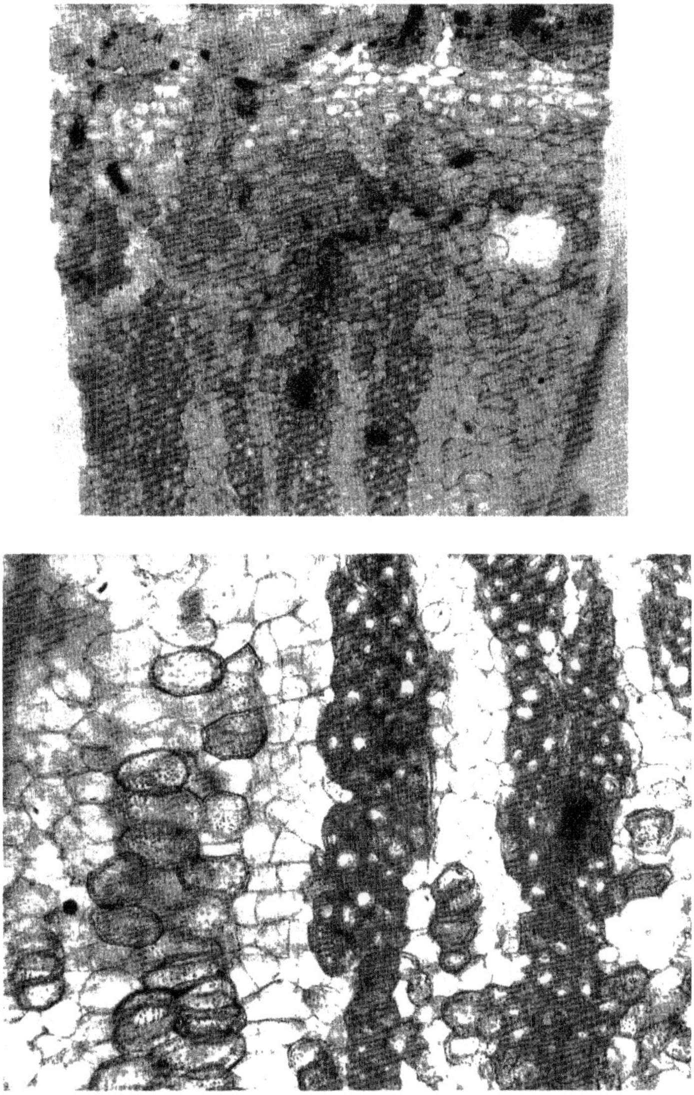

Fig. 18. Catostemma commune. Above: Sclerenchymatous "rays" of the pericarp alternating with radially extended parenchymatpus "rays". Below: The parenchyma cells extend tangentially and later divide by anticlinal walls. Some parenchyma cells developed thickened walls and sclerified.

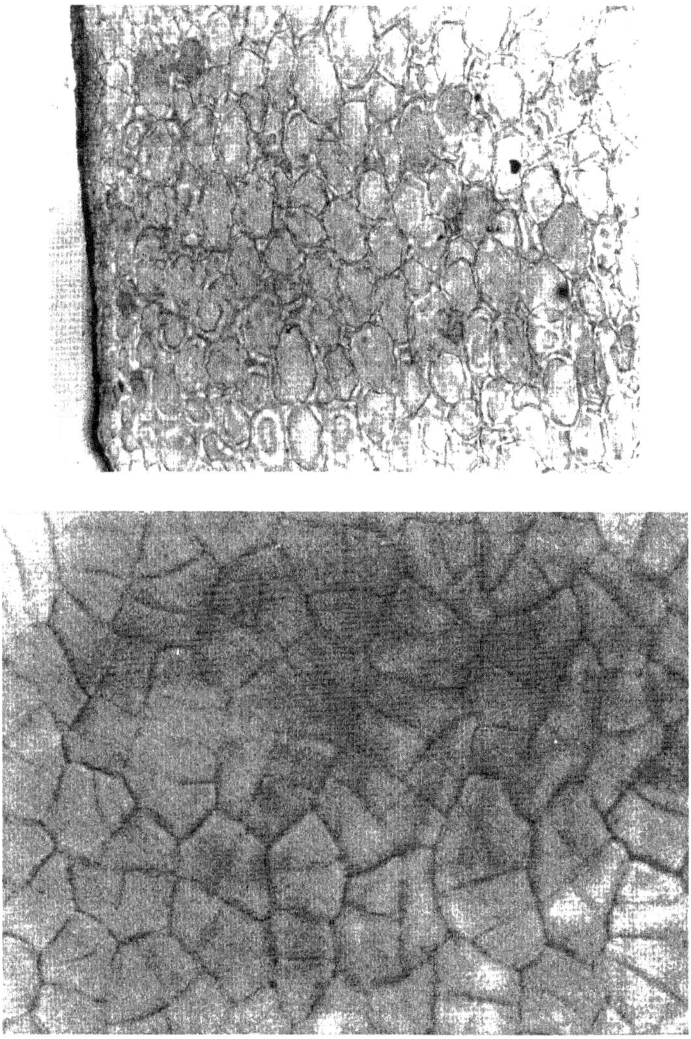

Fig. 19. Cordia exaltata, Boraginaceae. Above: Outer part of the fruit wall with collenchyma. Below: Outer fruit epidermis in surface view. Mother cells are subdivided by several (1-3) anticlinal walls.

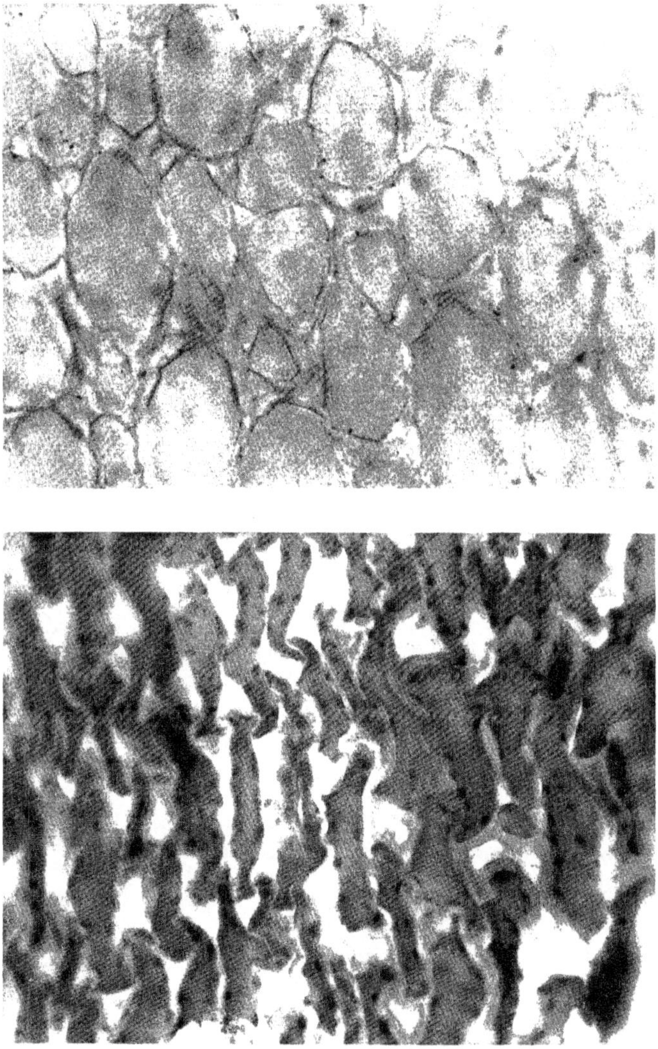

Fig. 20. Cordia exaltata. Above: Angular collenchyma. Below: Irregular cartilaginous collenchyma.

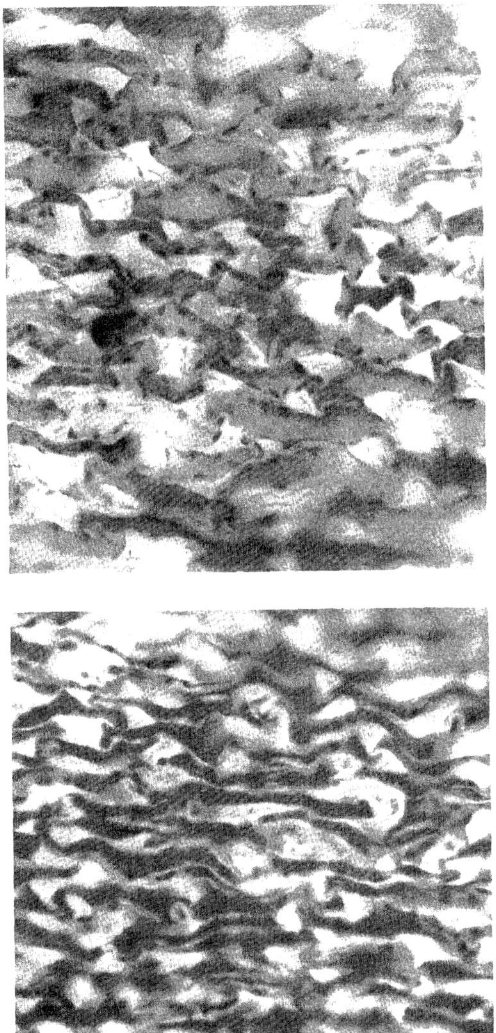

Fig. 21. Cordia exaltata. Above: Collenchyma partly compressed. Below: Almost completely compressed and tangentially extended collenchyma of the inner fruit part.

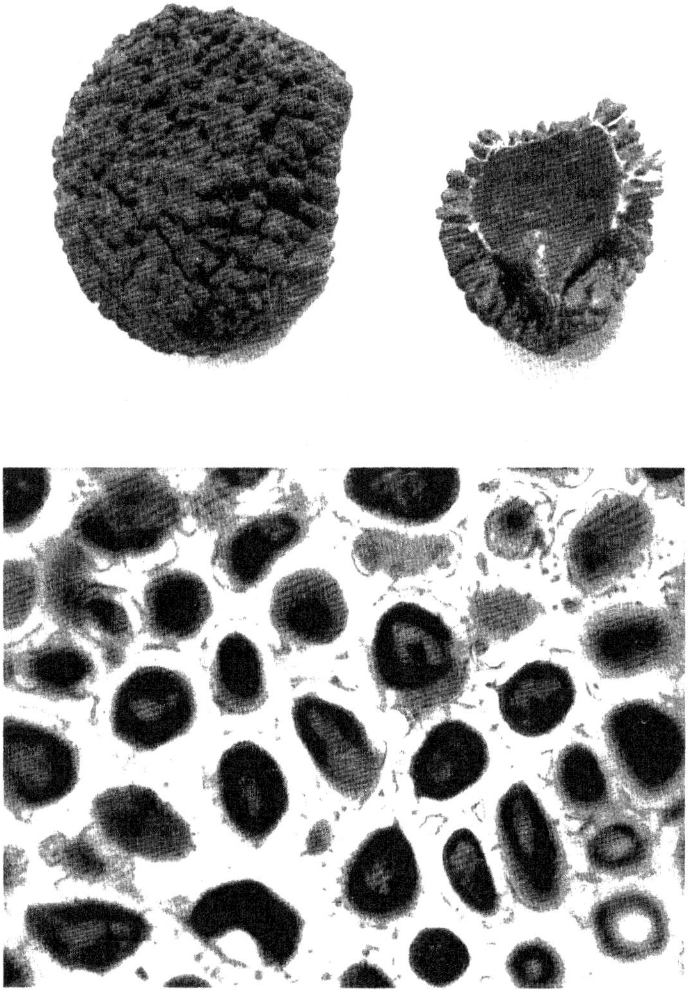

Fig. 22. Caryocar nuciferum. Above: Endocarp surface (left) and endocarp halve showing length of protuberances. Below: Thick-walled sclerenchymatous cells of the protuberances.

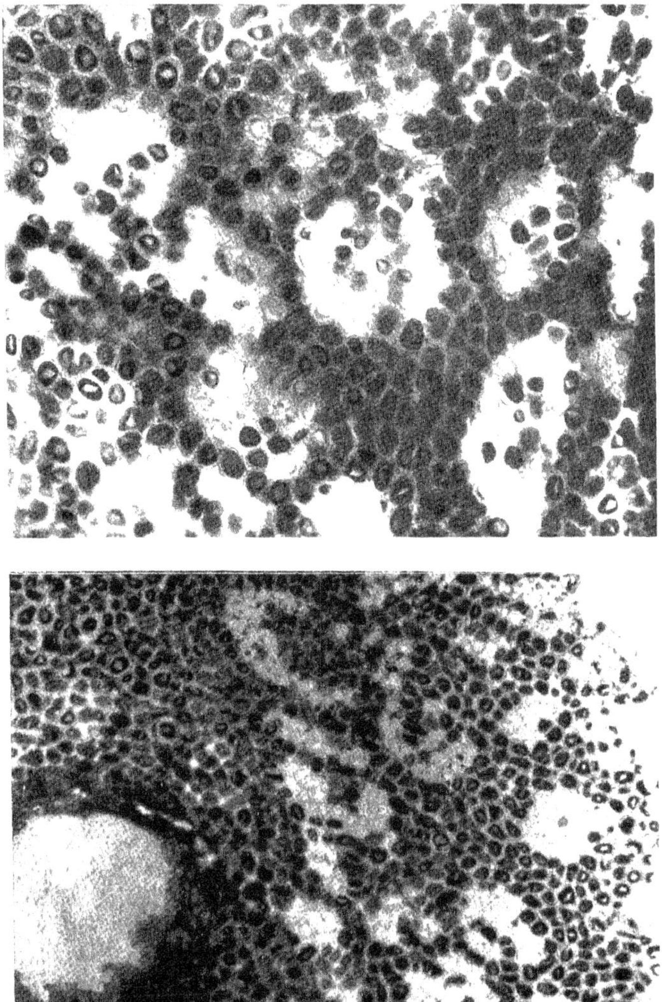

Fig. 23. Caryocar nuciferum, protuberances cut transversely.

Figs. 24-29. Cochlospermum, Cochlospermaceae. Fig. 24. Dehisced fruits and seeds. The seed is helicoidally bent and covered with long hairs along the dorsal side.

Fig. 25. Cochlospermum. Different stages of dehiscence. First a tangential splitting of the exo-mesocarp from the endocarp takes place, then loculicidal dehiscence of the united exo-mesocarp (fig. 24) and finally partial septicidal opening of the endocarp becomes obvious.

Fig. 26. Cochlospermum. After decay of the parenchymatous tissue, the septs finally consist in a strong network of veins.

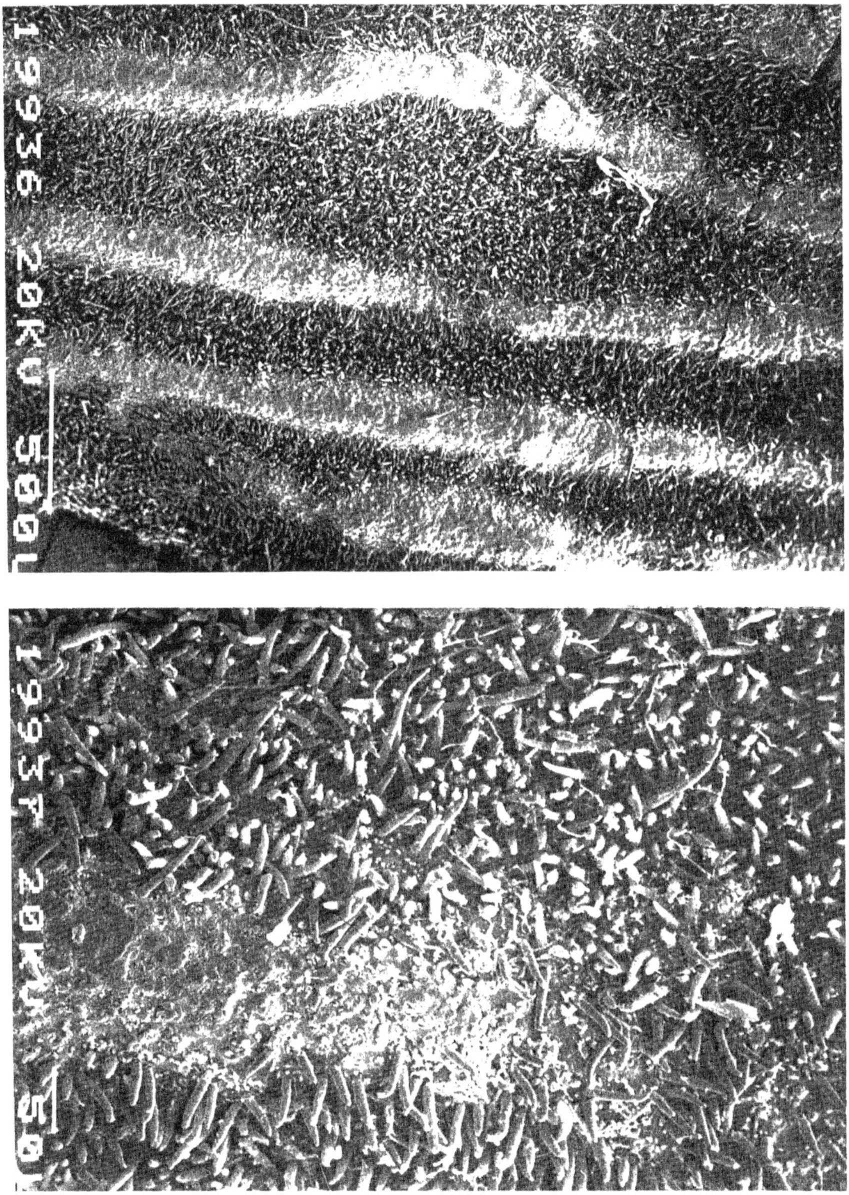

Fig. 27. Cochlospermum. Above: Pericarp outside showing the ribs, as seen in the scanning microscope. Below: Velvety hairs of the pericarp surface.

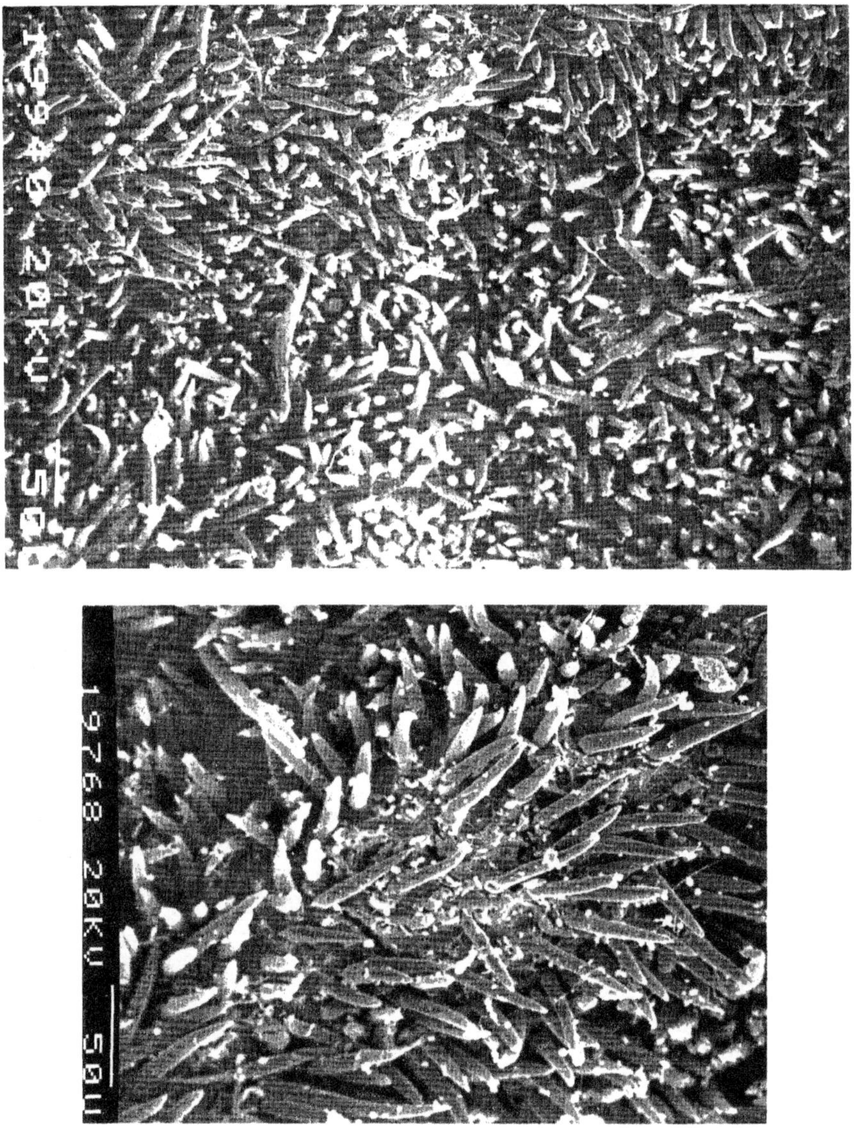

Fig. 28. Cochlospermum, hairs as seen in the S.M.

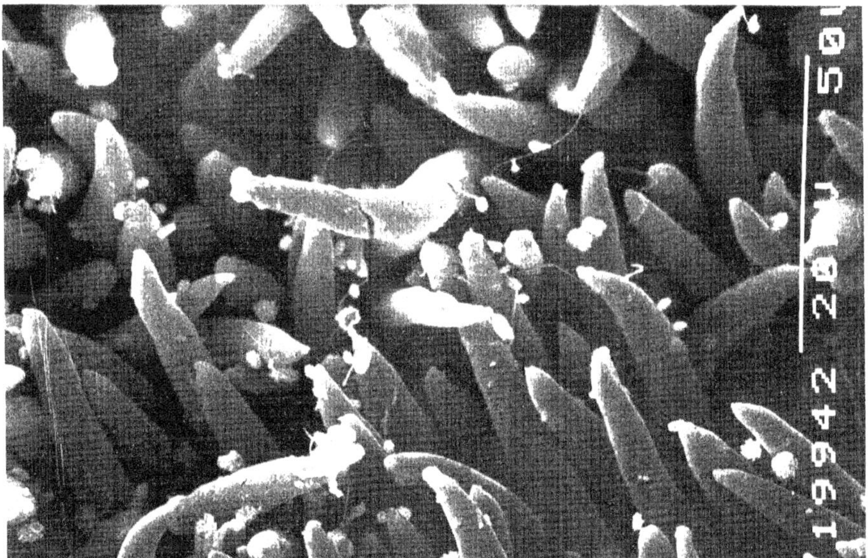

Fig. 29. Cochlospermum, hairs under higher magnification covered with fungi.

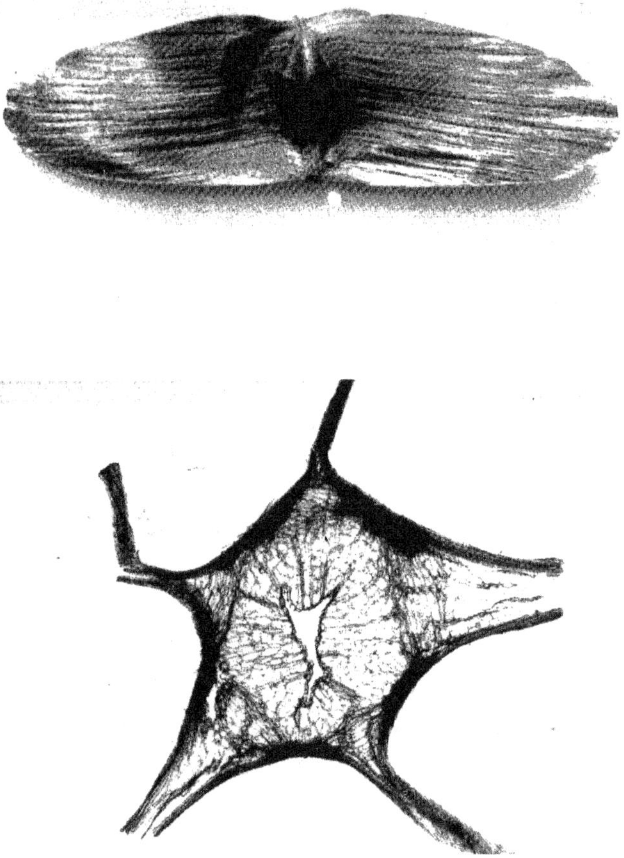

Fig. 30. Above: *Terminalia guayanensis,* Combretaceae. Samara with 2 large wings. Below: *Terminalia amazonia, Transverse section revealing the 5* wings of the 5-angular samara.

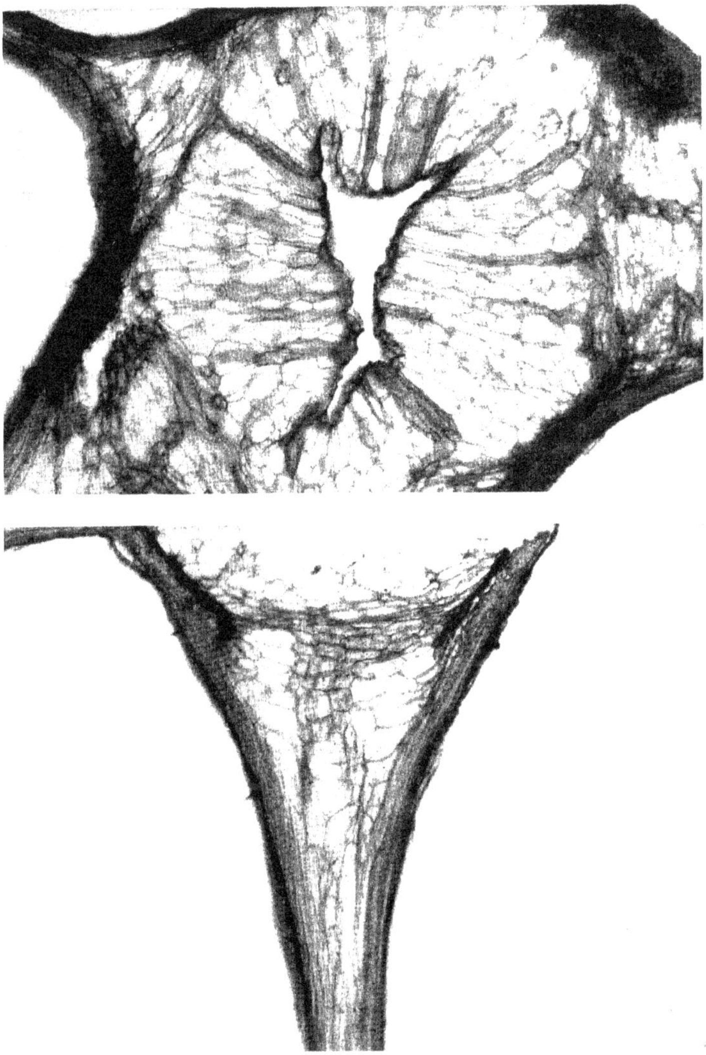

Fig. 31. Terminalia amazonia. Above: Central fruit body with radial cell rows originating from the activity of a ventral meristem. Below: Wing, mainly composed of thin-walled prosenchymatous elongated parallel to the surface of the wing.

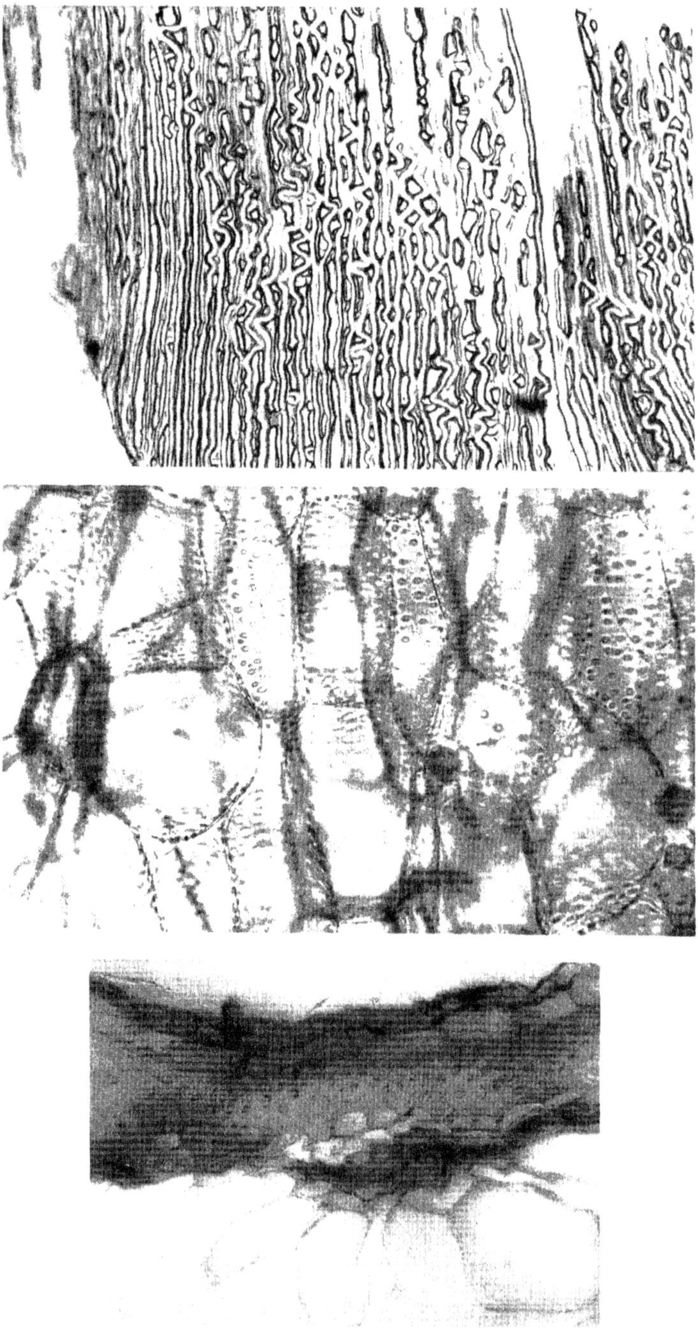

Fig. 32. Terminalia amazonia. Above: Strengthened wing periphery composed of subepidermal collenchymatous cell layers with thickened partly wavy walls. The cells are extended parallel to the surface. Middle. Cells of the fruit main body with numerous pits. The cells are dead. Below: Vascular bundle covered by a sclerenchymatous plate, a hypoderm and the outer fruit epidermis.

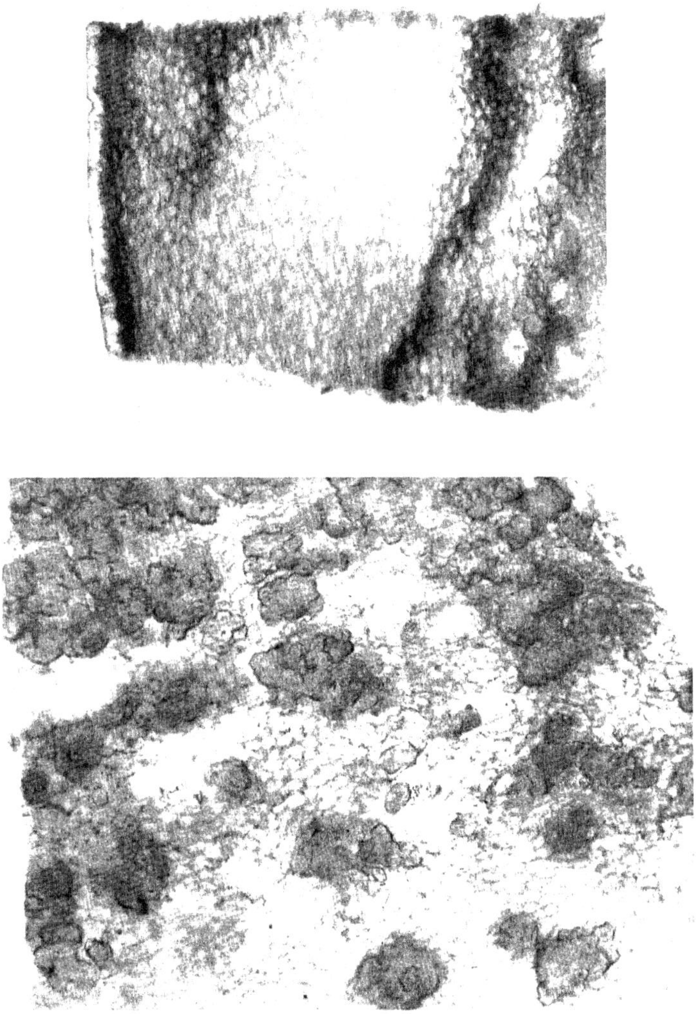

Fig. 33. Diospyros guianensis, Ebenaceae. Above: Outer fruit part composed of parenchyma. Outer epidermis thick-walled. Below: Inner fruit part with numerous stone cell groups.

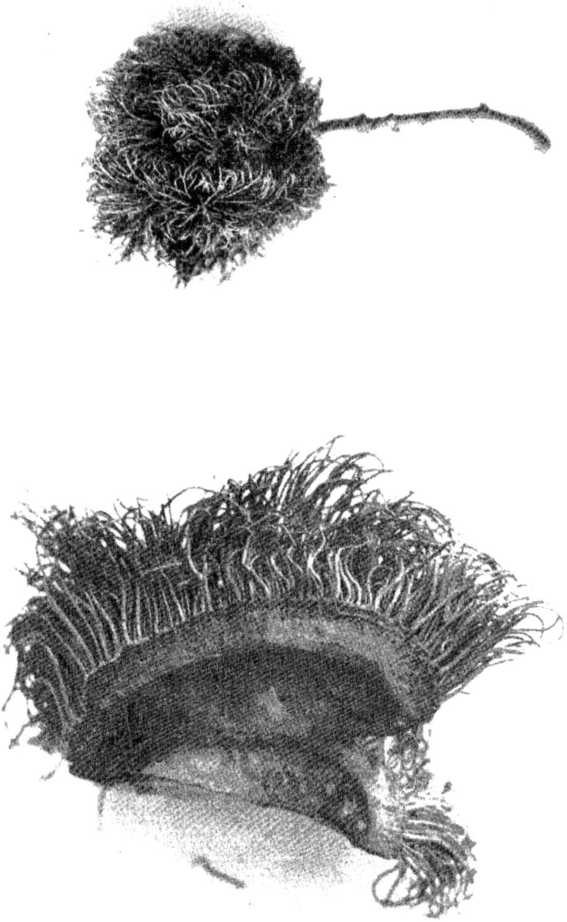

Fig. 34. Sloanea grandiflora, Elaeocarpaceae. Above: Complete fruit covered densely with bristles. Below: Opened fruit showing pericarp wall and length of the bristles.

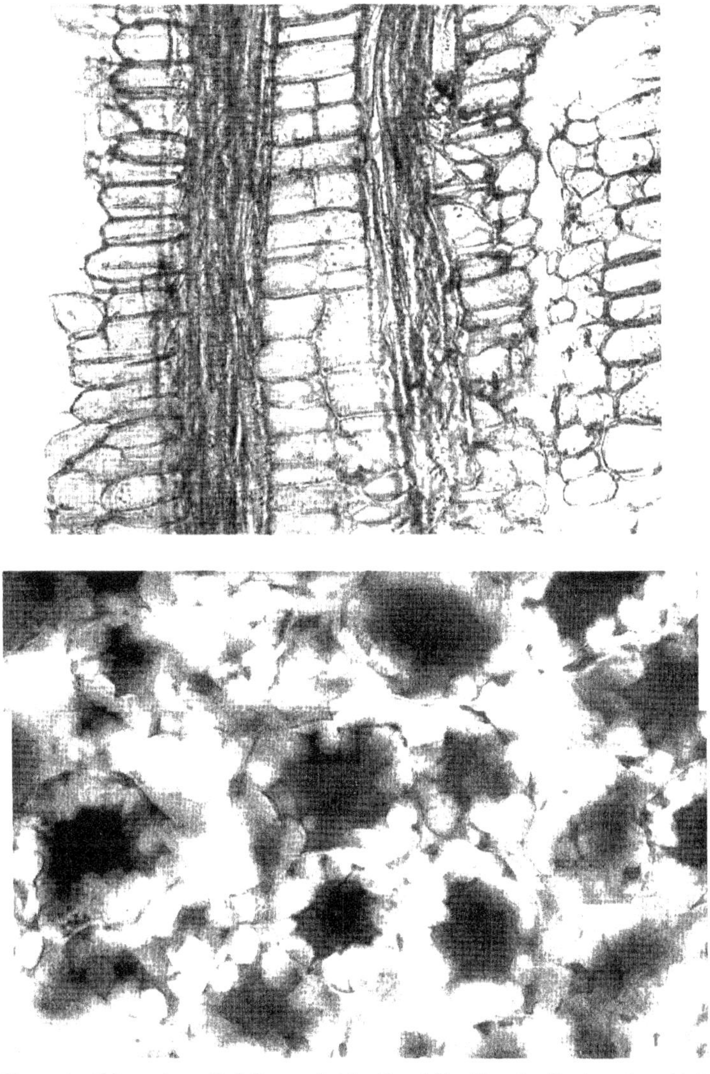

Fig. 35. Above: *Sloanea laurifolia*, pericarp. Radially extended bundles of fiber-like sclereids alternating with tangentially extended parenchyma cells. Below: *Saccoglottis cydonioides*, Humiriaceae. Roundish parenchyma cells of the fundamental tissue densely filled with large starch grains arranged in the form of a wreath.

Fig. 36. Above: *Piranhea longepedunculata*, Euphorbiaceae. Long floral peduncles and central column. 2 dehisced and separated locules each with one seed. Middle. *Amanoa guianensis*. Valve with broken off endocarp. Below: *Hura crepitans*, dehisced carpel (locule) with 2 valves.

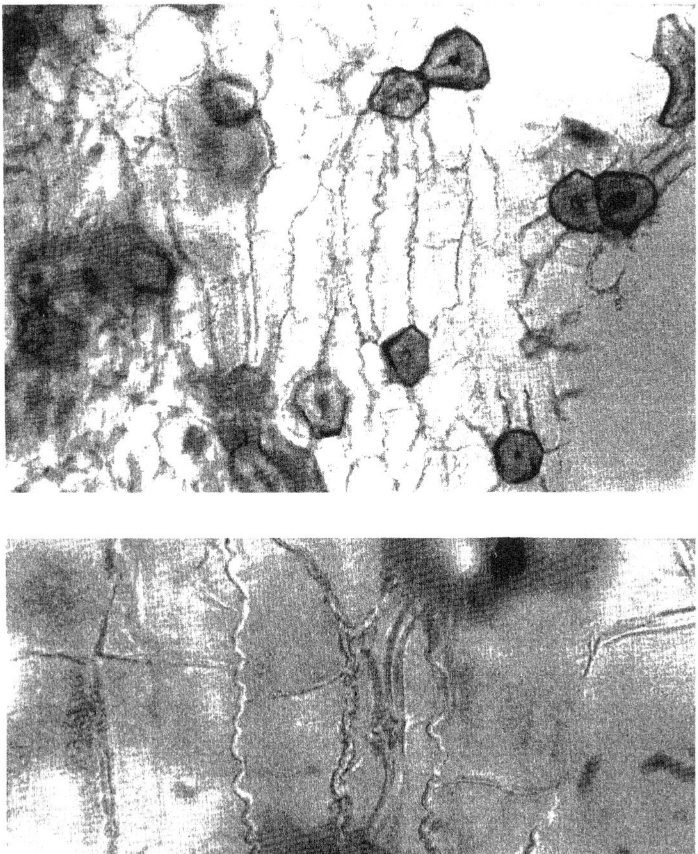

Fig. 37. Drypetes variabilis, Euphorbiaceae. Above: Radially extended parenchyma cells and interspersed stone cells of the pericarp. Below: Wavy walls of parenchyma cells.

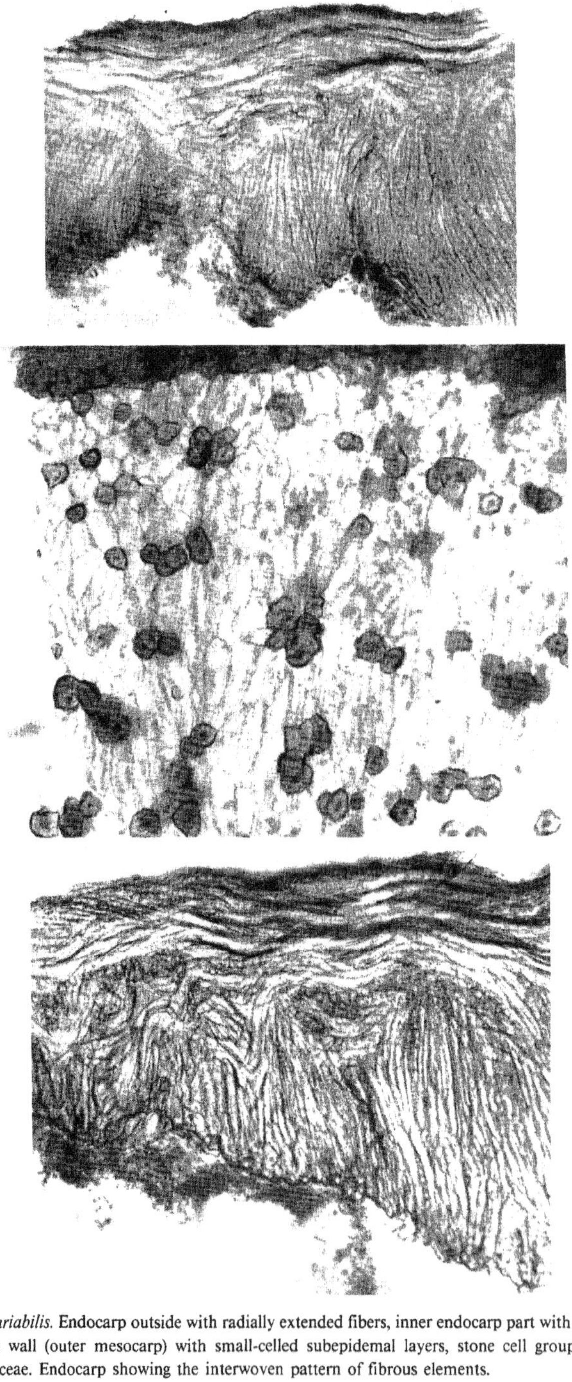

Fig. 38. Above: *Drypetes variabilis.* Endocarp outside with radially extended fibers, inner endocarp part with tangentially extended fibers. Middle. Outer part of fruit wall (outer mesocarp) with small-celled subepidemal layers, stone cell groups and parenchyma. Below: 'Kerosen negro', Euphorbiaceae. Endocarp showing the interwoven pattern of fibrous elements.

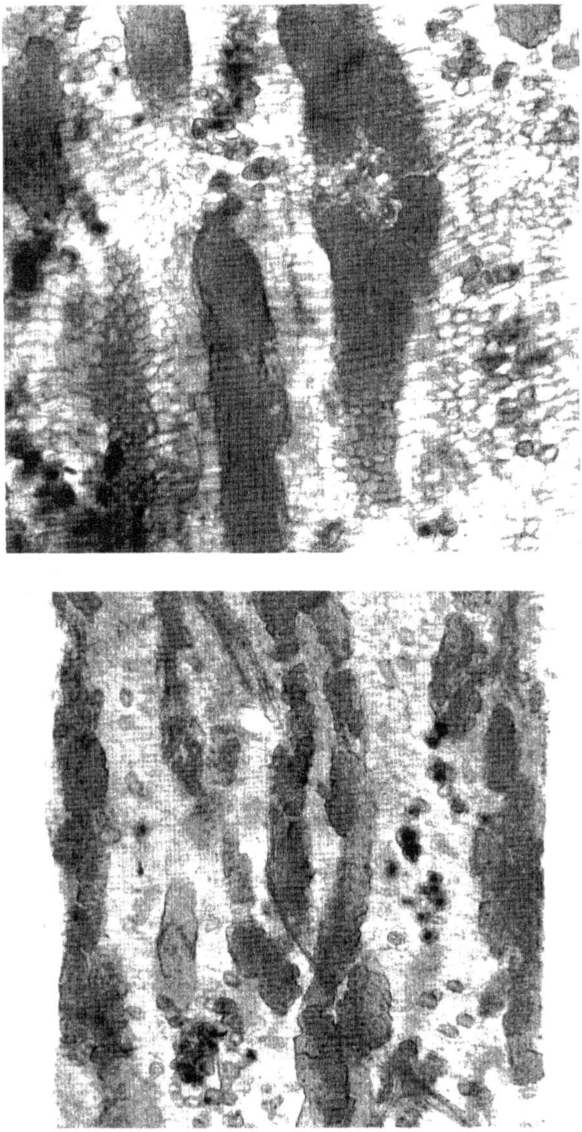

Fig. 39. Amanoa guianensis, Euphorbiaceae. Below. Radially arranged sclerenchymatous plates of the pericarp. Above: Parenchyma cells between the sclerenchymatous plates tangentially extended and subdivided by anticlinal walls to form tangential cell rows.

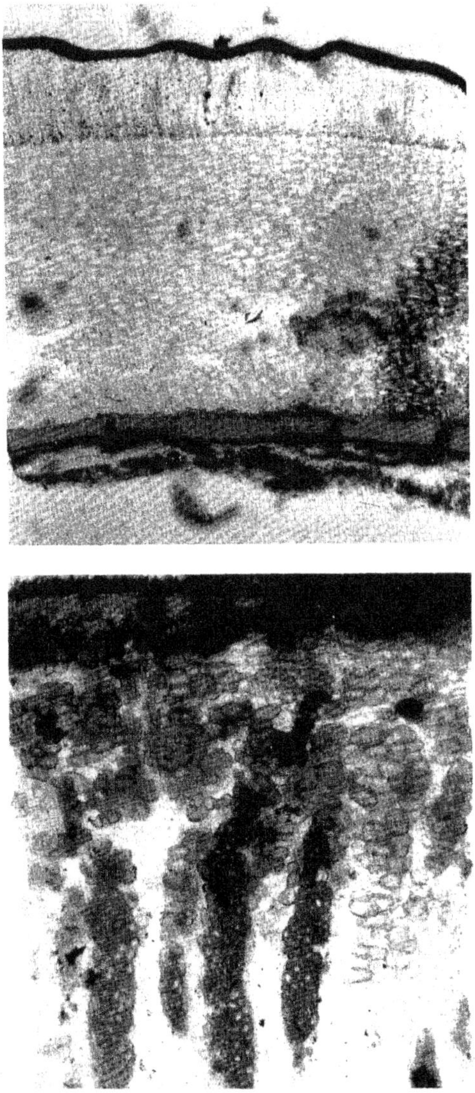

Fig. 40. Above: *Ouratea guianensis,* Ochnaceae. The palisade-shaped outer epidermis serving as a water reservoir occupies up to 1/3 of the entire pericarp transection. Beneath follows parenchyma in which vascular bundles are embedded. The endocarp is composed of tangentially arranged fiber-like sclereids. Below: *Catostemma commune,* Bombacaceae. Epidermis with hairs. Subepidermal zone abundantly including sclerenchyma. Towards the inside ground parenchyma and radially extended sclerenchymatous plates of the pericarp.

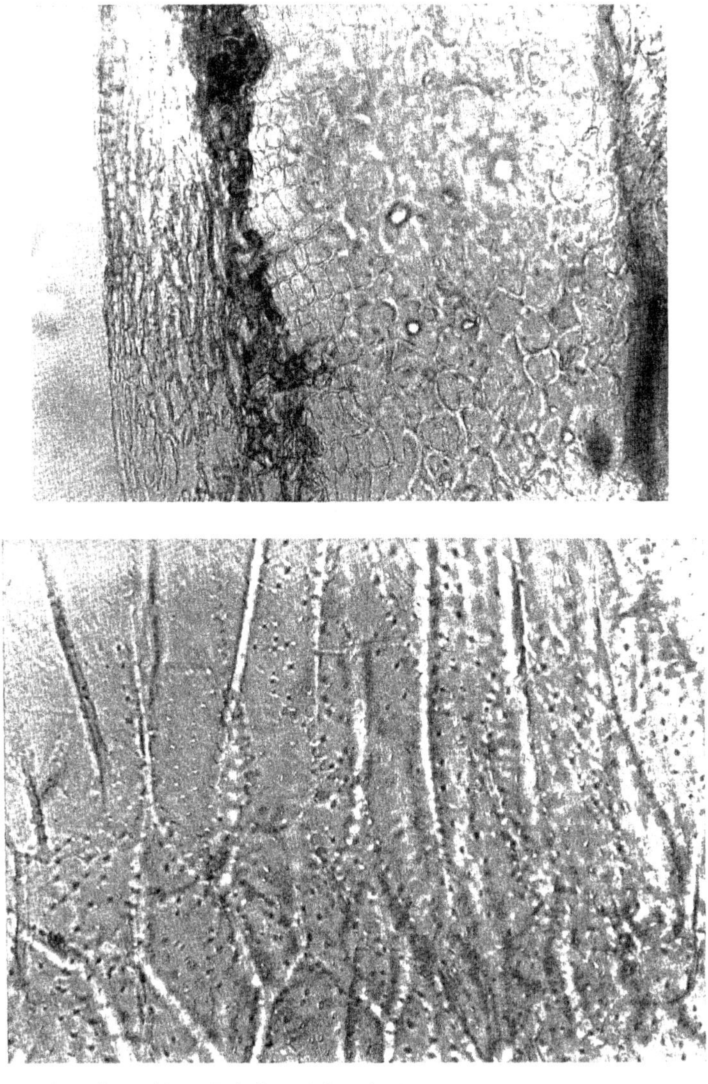

Fig. 41. Laetia procera, Flacourtiaceae. Above: Beginning cork formation beneath the subepidermal cartilaginous zone. Below: Radially extended parenchyma cells with pits serving as a water storing tissue. The cells are dead.

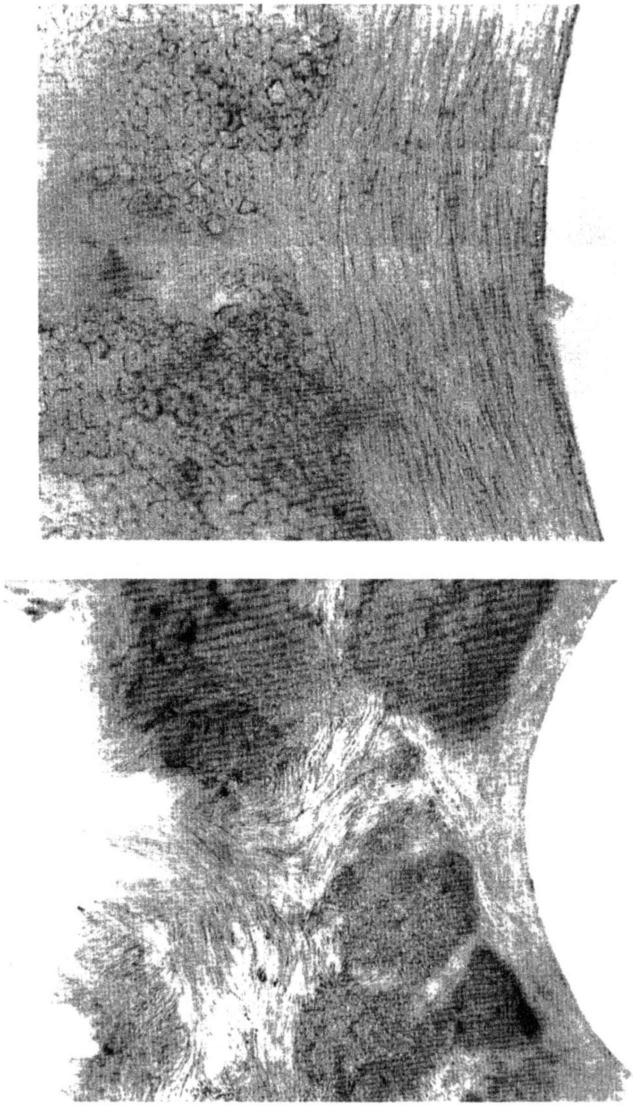

Fig. 42. Spondias mombin, Anacardiaceae. Above: Innermost endocarp part with fibers running parallel to the locular cavity. Below: Outer part of the endocarp with fibers arranged in bundles.

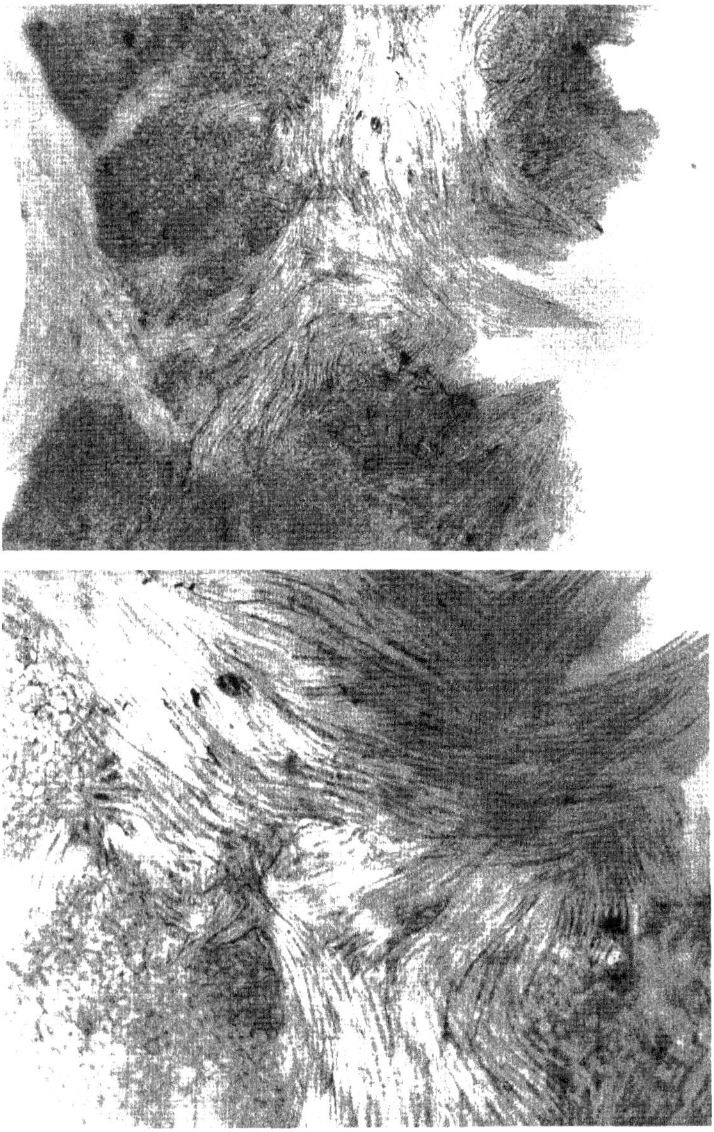

Fig. 43. Spondias mombin. Fiber bundles of the endocarp crossing each other in an interwoven manner.

Fig. 44. Clusia, Guttiferae. Above: Side view of the 11-carpellary capsule. Below: Front view of a 13-carpellary capsule. The endocarp in the center has separated from the exo-meso-carp and dehisced loculicidally, while the exo-mesocarp separated septicidally.

Fig. 45. Above: *Rheedia* sp. (Cozoiba picuda), Guttiferae. Drupe halved to show the endocarp with 3 locules each furnished with a single seed. Below: *Rheedia* sp. (Cozoiba negra). Drupe densely covered with protuberances.

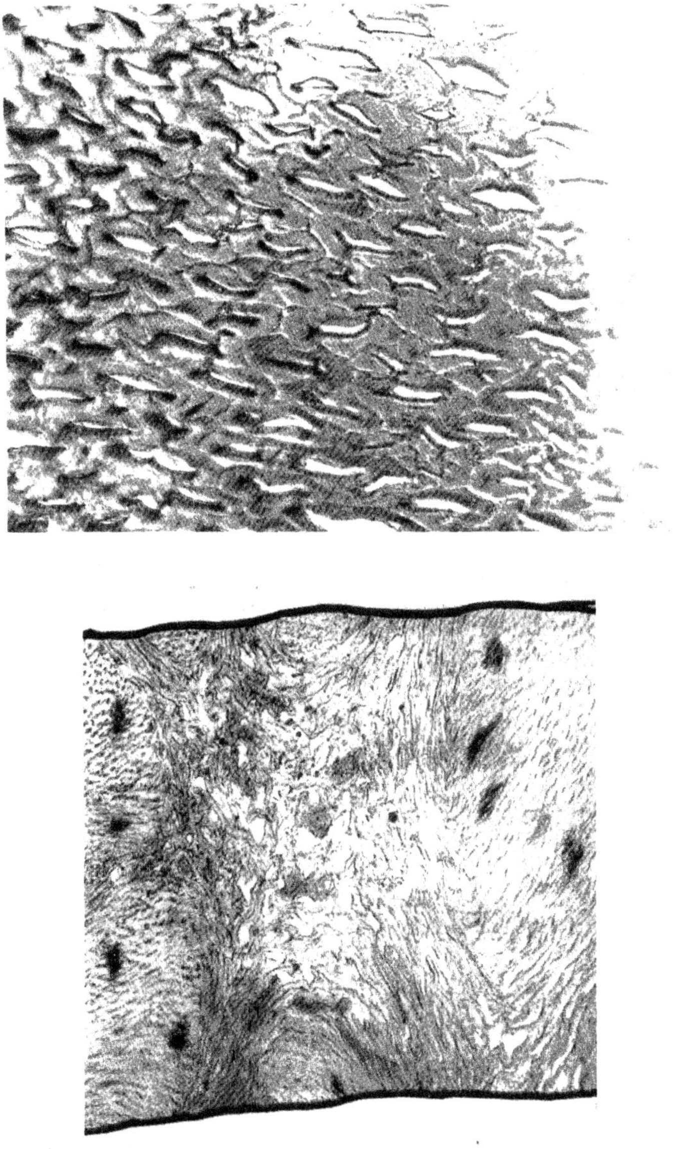

Fig. 46. Clusia, Guttiferae. Below: Entire fruit wall mainly consisting of cartilaginous tissue. Above: Thick-walled cells of the cartilaginous collenchyma with partly compressed lumen.

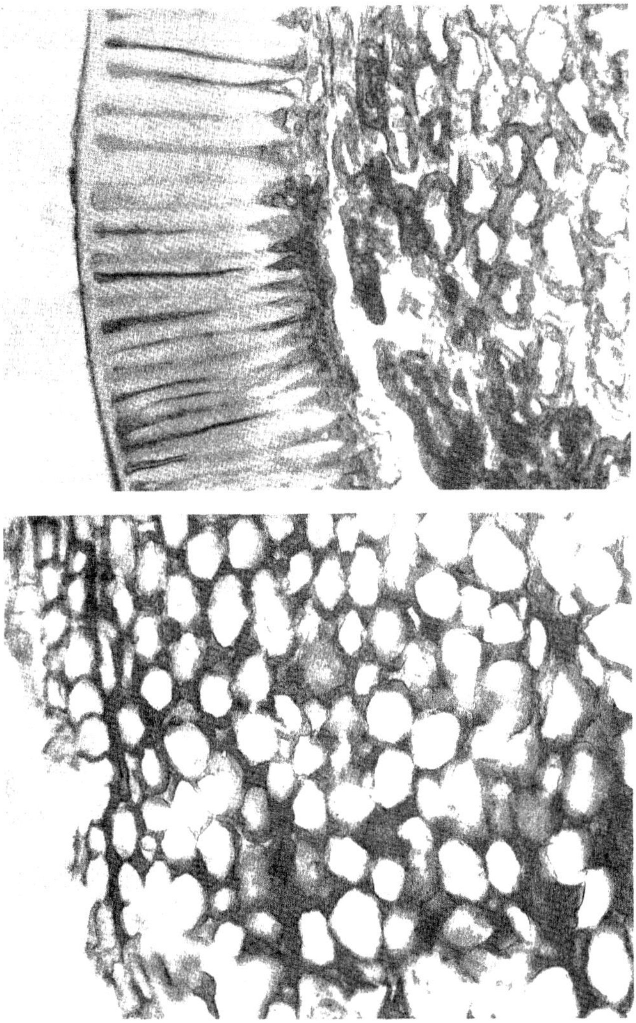

Fig. 47. Rheedia sp. (Cozoiba picuda). Above: Palisade-shaped outer epidermis of the pericarp. Below: Angular collenchyma.

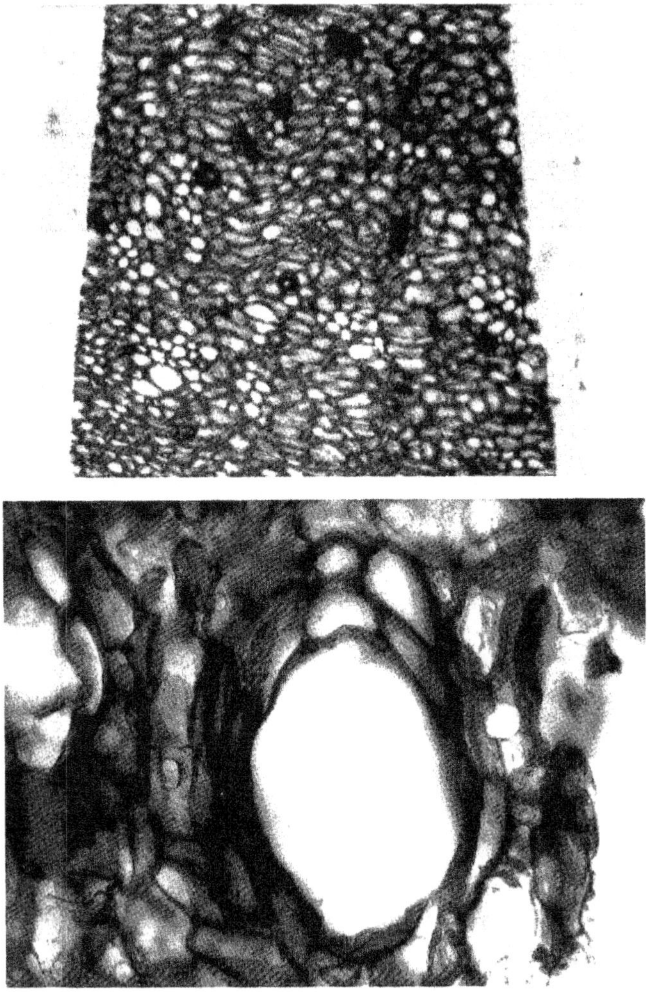

Fig. 48. Above: *Rheedia* sp. (Cozoiba picuda). Pericarp with scattered secretory tubes and parenchyma cells tangentially extended by the forces of dilatation growth. Below: *Caraipa richardiana*. Large secretory canal surrounded by epithelium.

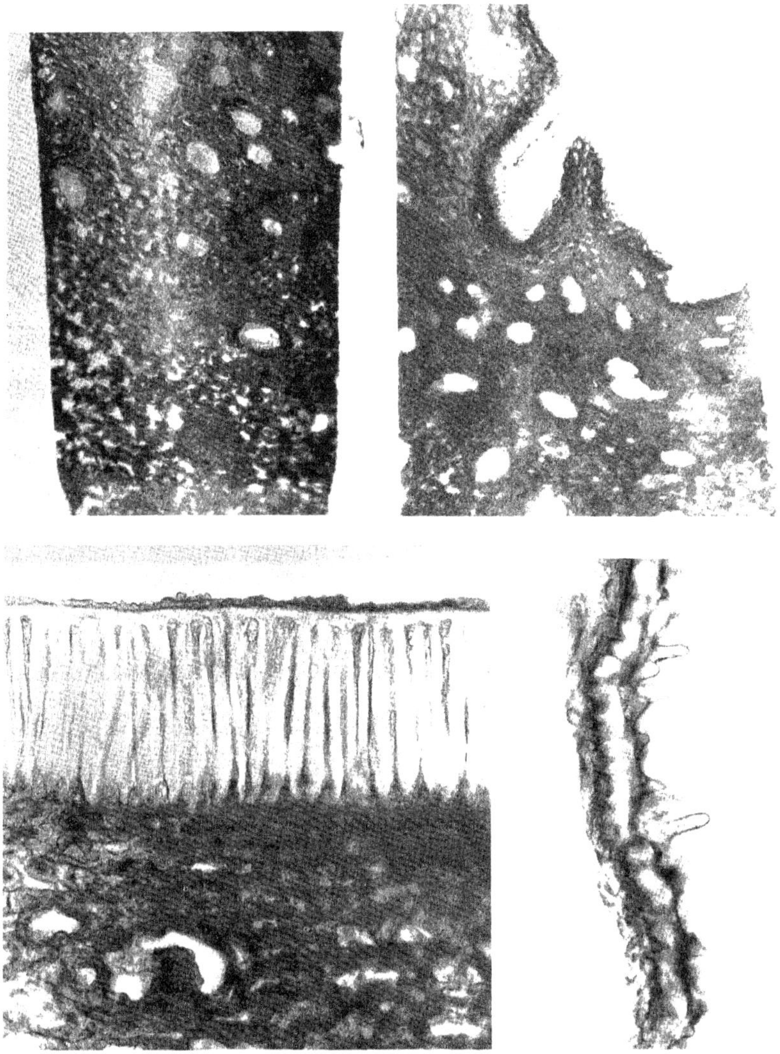

Fig. 49. Rheedia sp. (Cozoiba negra). Above, left: Pericarp composed of collenchymatous tissue in which large secretory cavities are irregularly dispersed. Right: Excrescences of the pericarp covered by an outer epidermis with horseshoe-shaped wall thickenings. Below, left: Cozoiba picuda, palisade-shaped epidermis. Right: Hairs of the outer epidermis of Cozoiba negra.

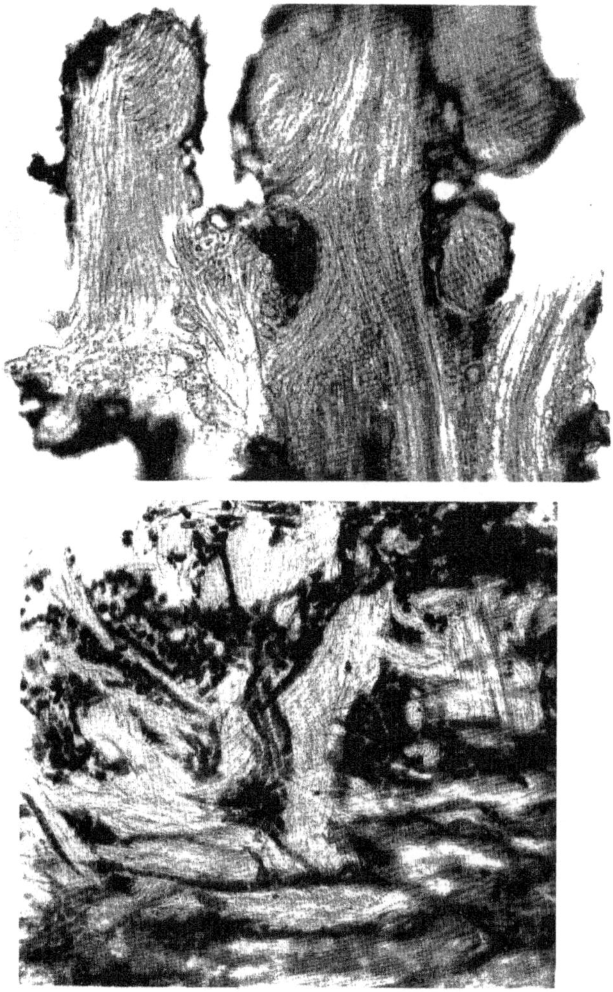

Fig. 50. Saccoglottis cydonioides, Humiriaceae. Endocarp with fiber bundles crossing one another in different directions.

Fig. 51. Above: 'Laurel marrón', Lauraceae. Opened drupe enclosed within a large cupule. Below: *Ocotea duotincta, Lauraceae. Cupules of fruits.*

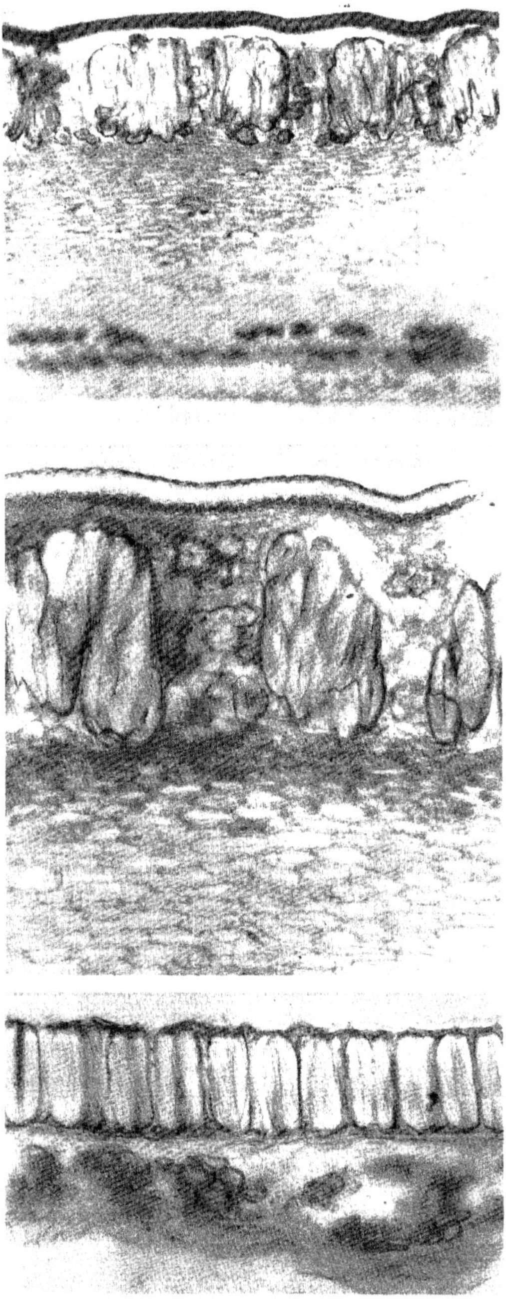

Fig. 52. Nectandra grandis, Lauraceae. Above and middle: Fruit wall composed of parenchyma in which large sclerenchymatous plates of a characteristic shape are embedded. Below. Outer fruit epidermis with heavily thickened anticlinal walls.

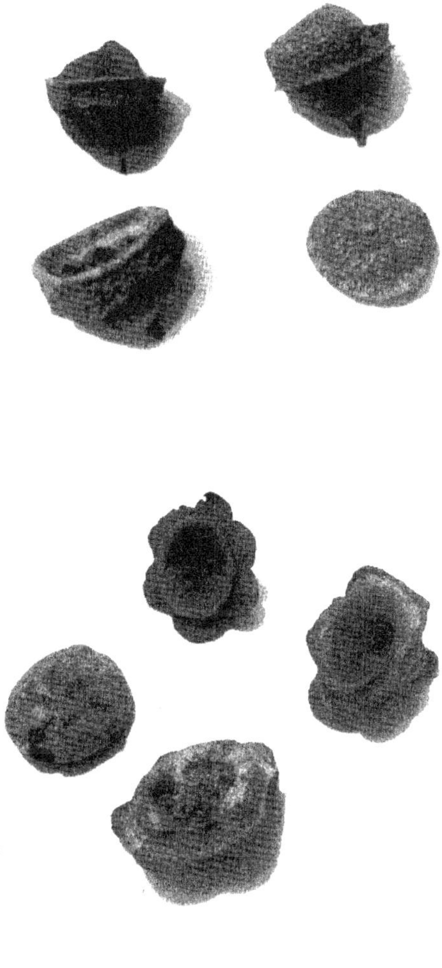

Fig. 53. Above: *Eschweilera grata,* Lecythidaceae. Pyxidia closed and opened with lid. Below: *Eschweilera* cf. *trinitensis.* Opened pyxidia surrounded by woody sepals and lid aside.

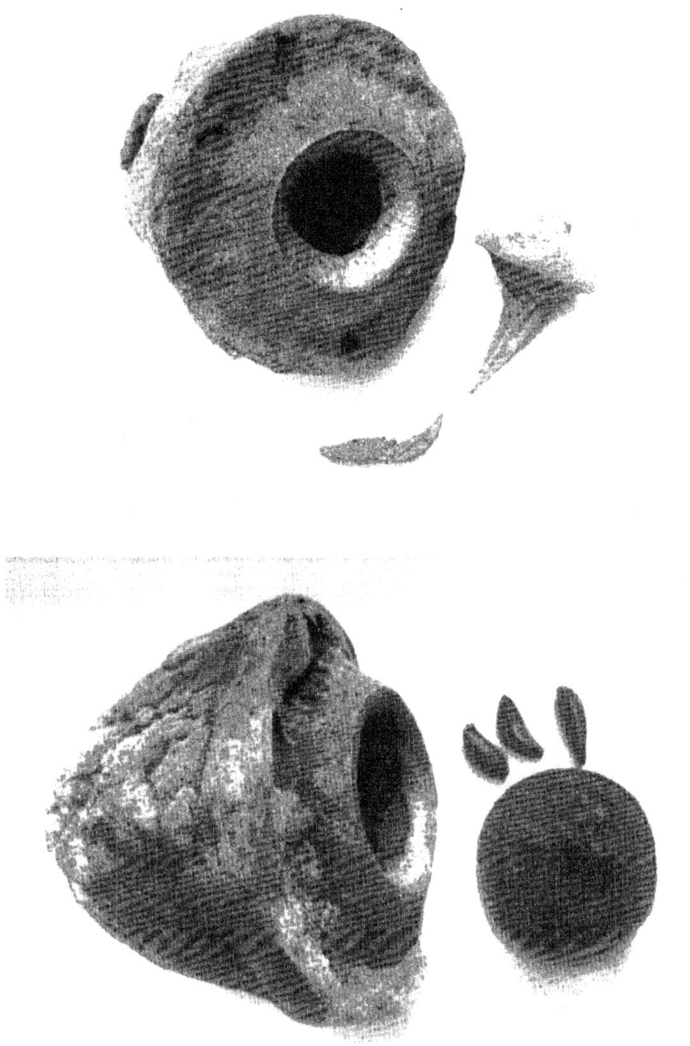

Fig. 54. Lecythis davisii, Lecythidaceae. Above: Front view of the pyxidium with central column. Below: Side view showing lid and seeds.

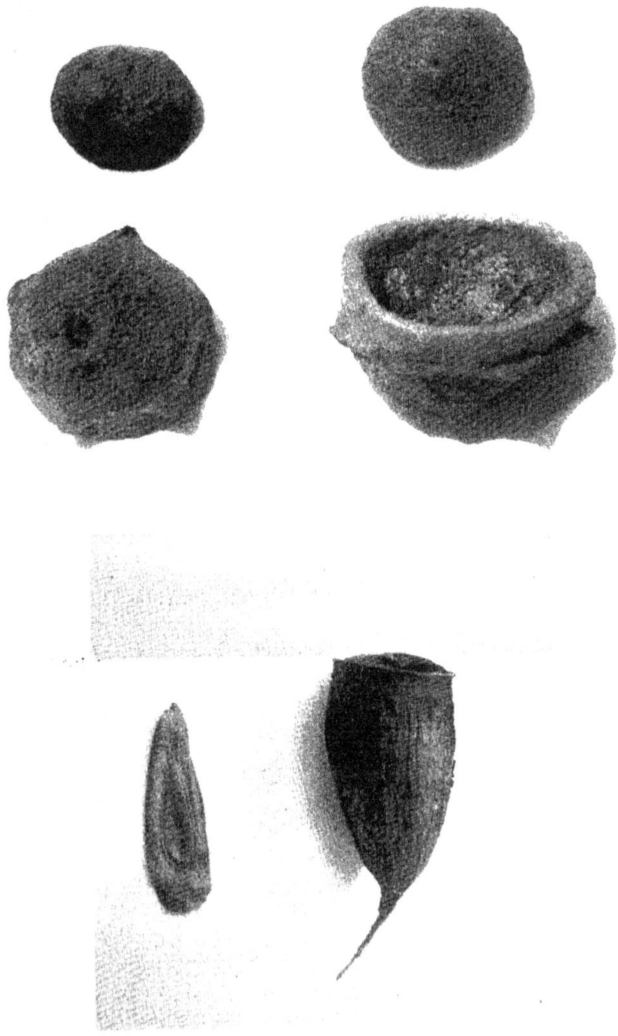

Fig. 55. Above: *Eschweilera* sp. (Majaguillo erizado), Lecythidaceae. Pyxidia seen from below and above with corresponding lids. Below: *Couratari multiflora*, Lecythidaceae. Cylindrical pipebowl-shaped pyxidium and a winged seed.

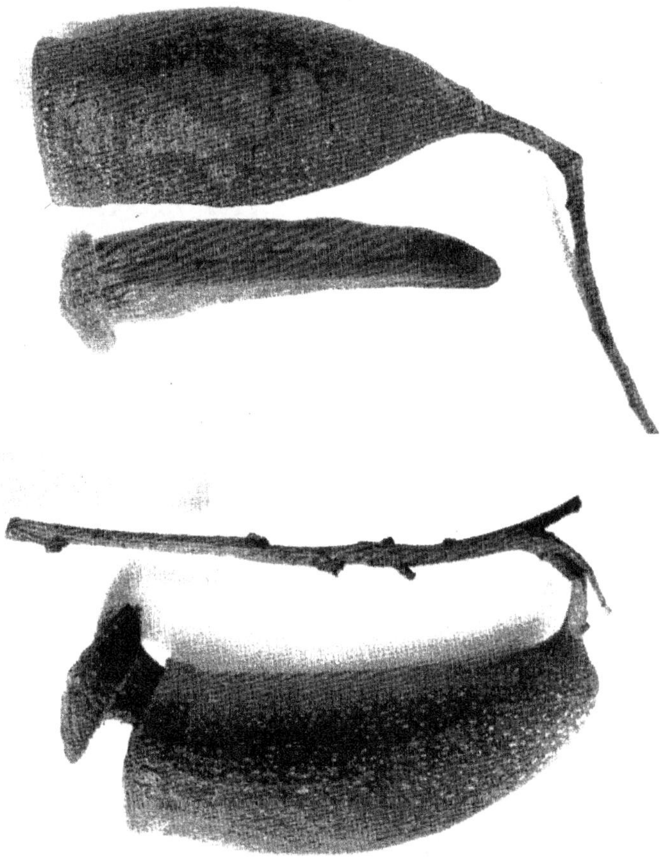

Fig. 56. Couratari pulchra, Lecythidaceae. Above: Dehisced pyxidium with central column and attached lid. Below: Partly opened pyxidium with lid showing numerous lenticels (light dots).

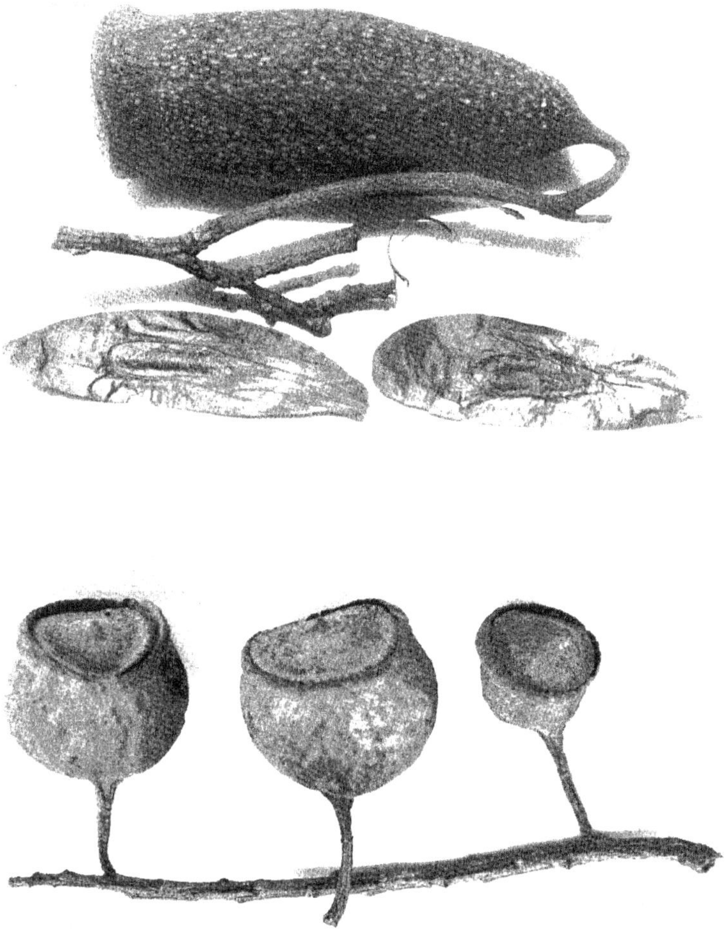

Fig. 57. Above: *Couratari pulchra,* pyxidium and 2 winged seeds. Below: Still closed pyxidia of *Gustavia augusta.*

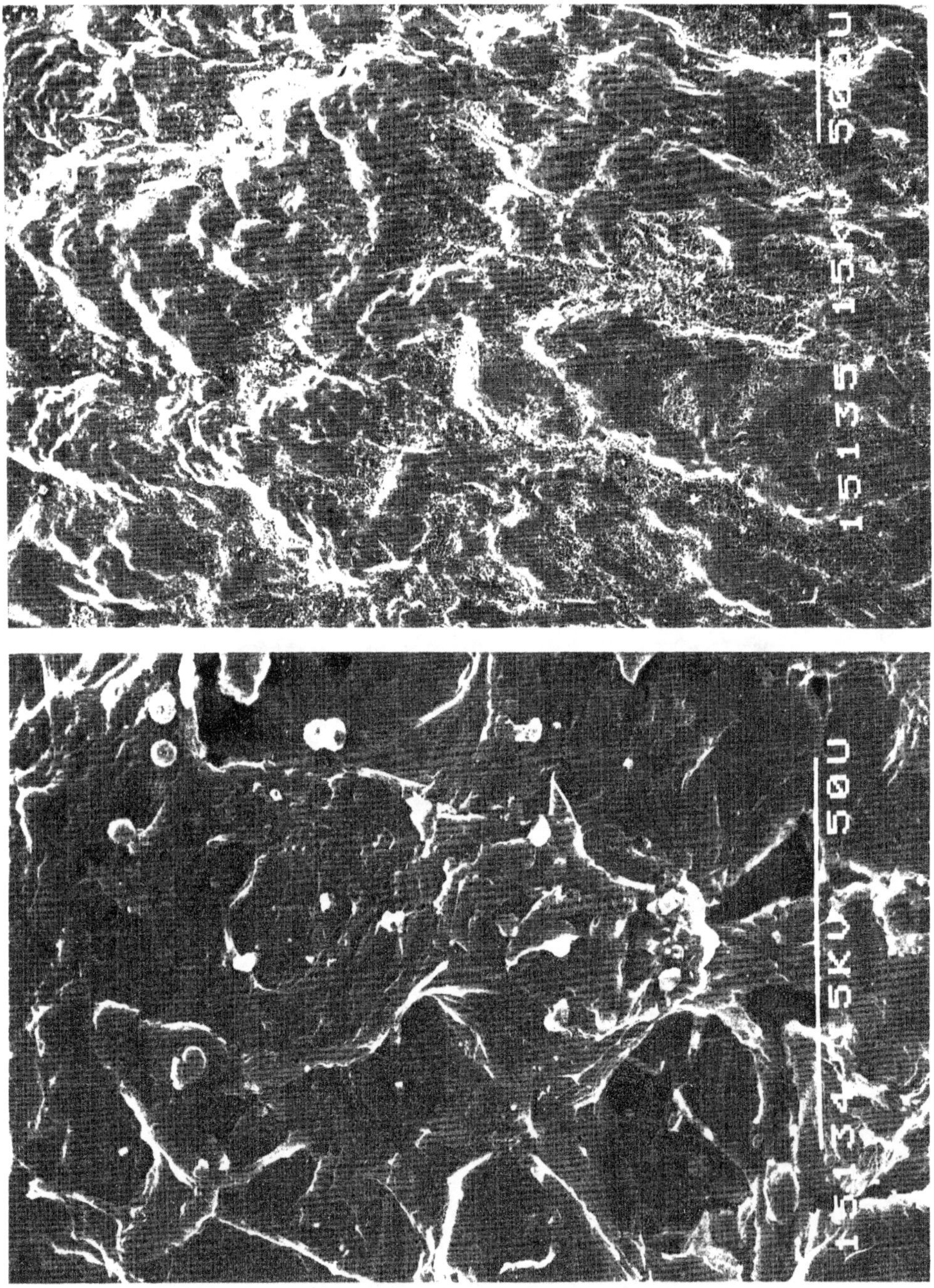

Fig. 58. Lecythis davisii. Above: Rough surface of the pyxidium. Below: Attacked by fungi.

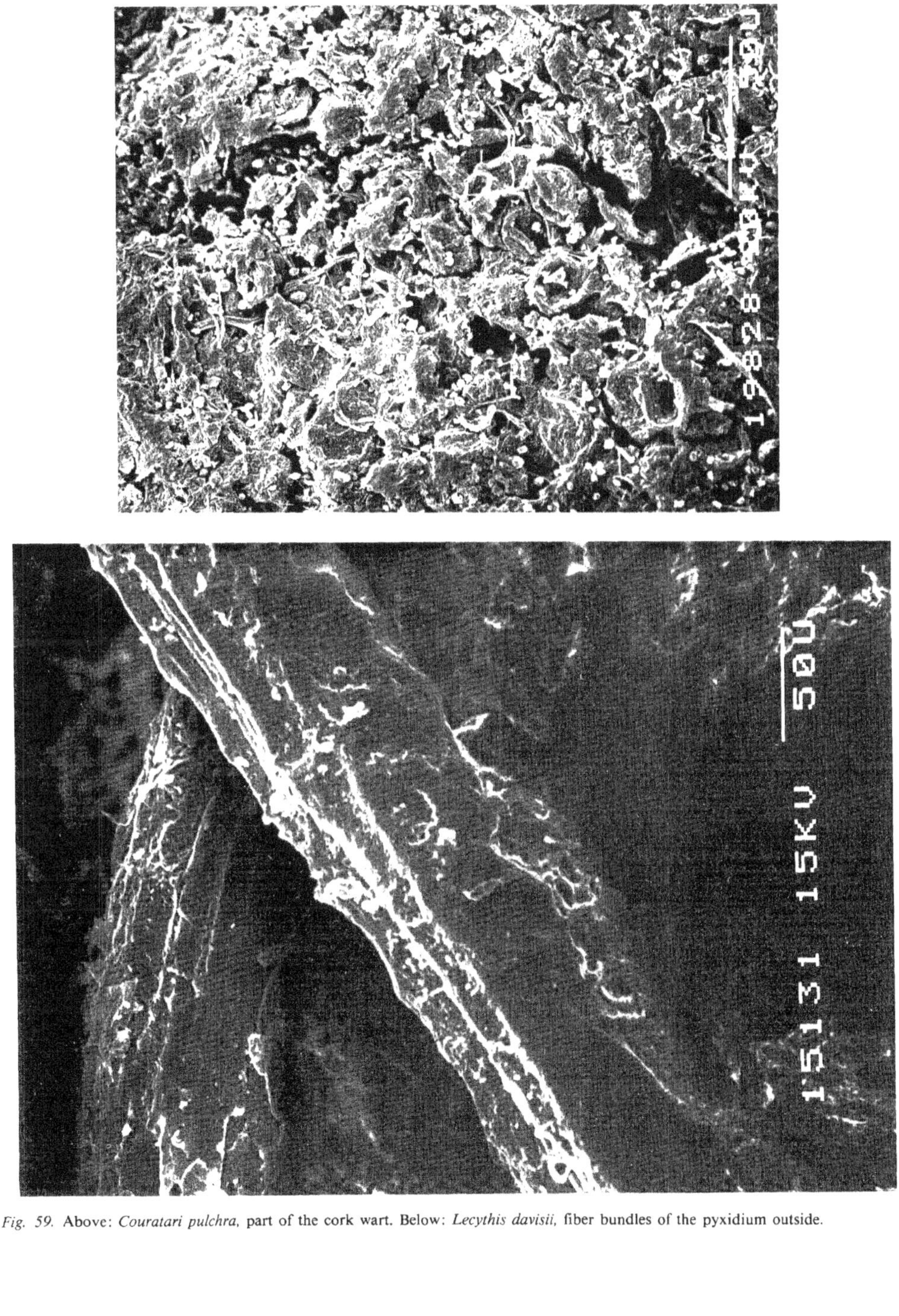

Fig. 59. Above: *Couratari pulchra*, part of the cork wart. Below: *Lecythis davisii*, fiber bundles of the pyxidium outside.

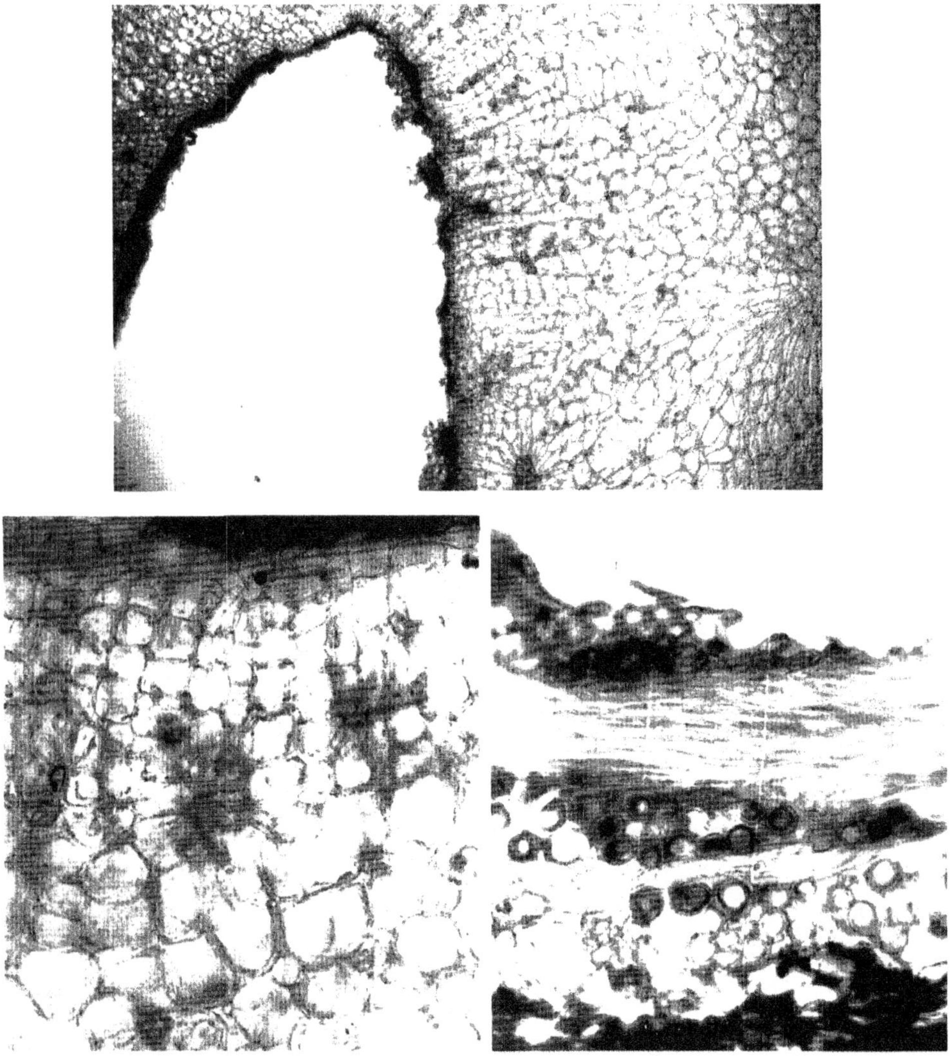

Fig. 60. Above: *Gustavia augusta,* Lecythidaceae. Meristem of the future dehiscence line. Below, left: Radial cell rows of the meristem of *Gustavia augusta.* Right: *Couratari multiflora.* Pericarp wall with alternating layers of fiber bundles and layers of parenchymatous cells (roundish).

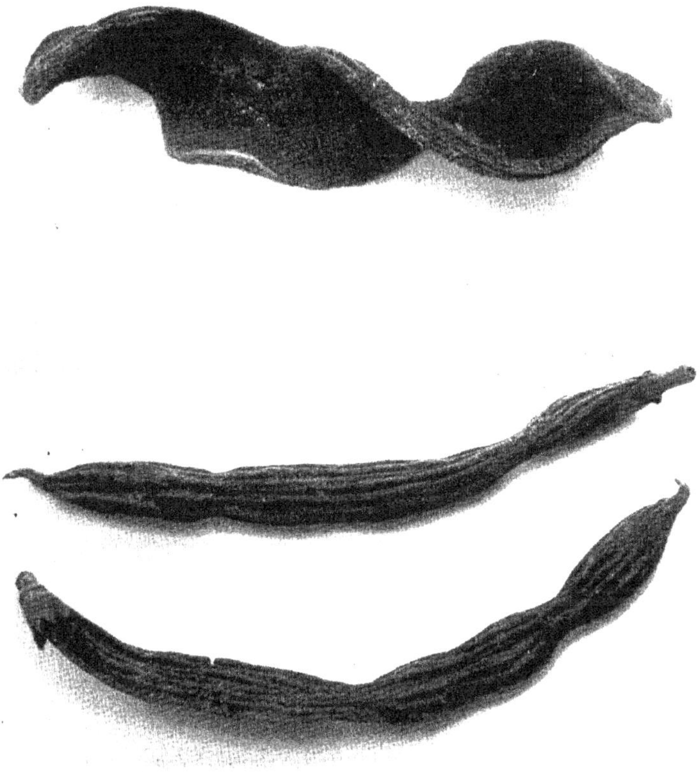

Fig. 61. Above: *Inga fastuosa,* Mimosaceae. Helicoidally bent pod with a velvety surface. Below: *Inga scabriuscula.* Elongated pod with longitudinal ribs. Calyx persistent.

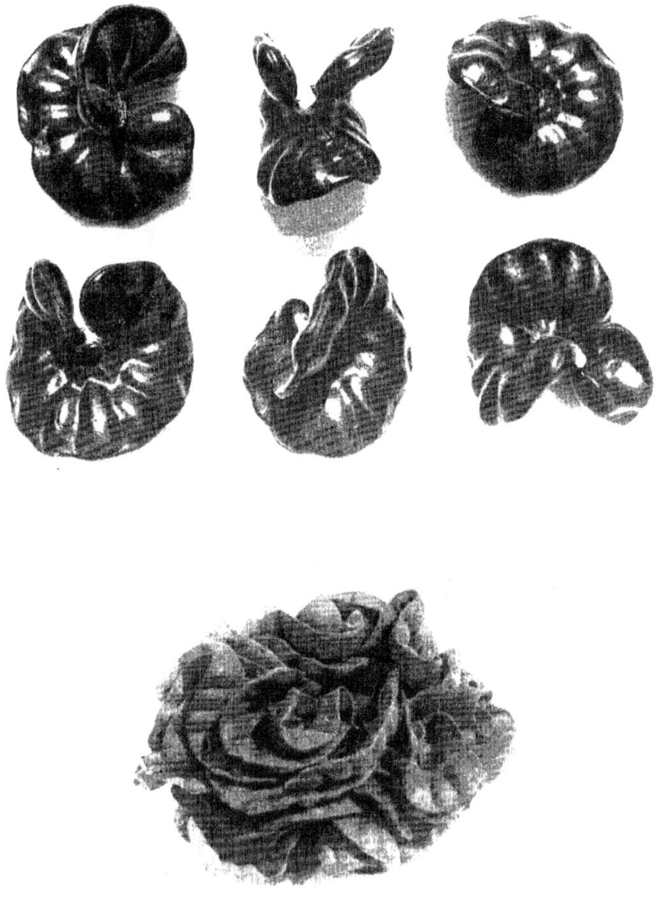

Fig. 62. Above: *Enterolobium cyclocarpum,* Mimosaceae. Helicoidally bent pods with shiny surface. Below: *stryphnodendron polystachyum, numerous flattened and bent pods are assembled in a pseudo-infrutescence.*

Fig. 63. Above: *Enterolobium* sp. (Caro blanco), Mimosaceae. Somewhat bent woody indehiscent pods. Below: *Parkia pendula.* Dehisced pods with 2 valves and imprints of the seeds.

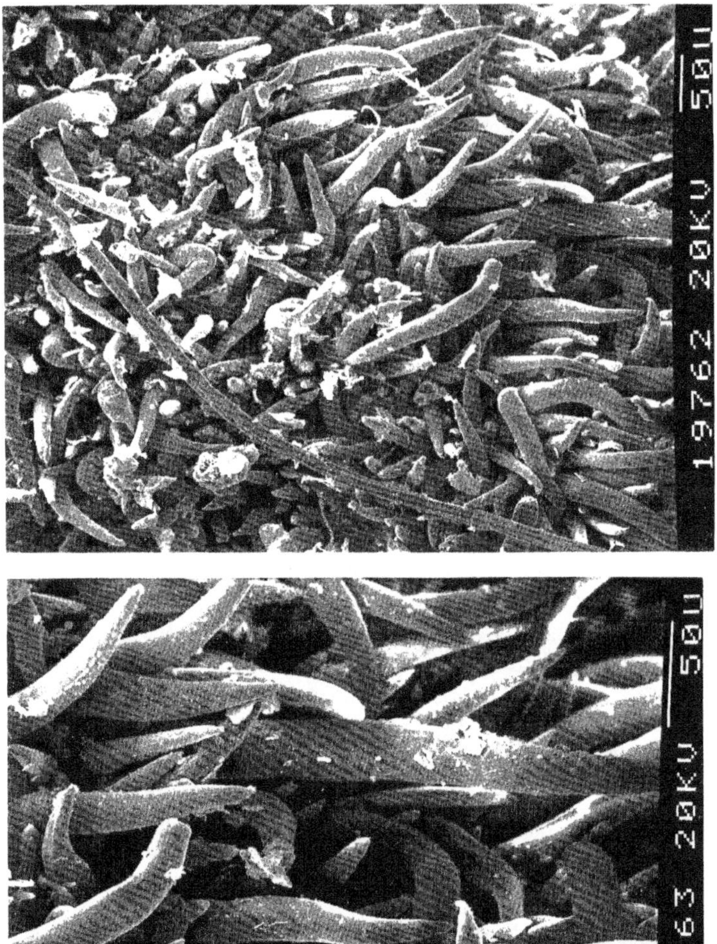

Fig. 64. Inga scabriuscula, Mimosaceae. Hairs of the velvety surface.

Fig. 65. Swartzia schomburgkii, Caesalpiniaceae. Dehisced fruits with seeds. The seeds remain attached to the fruit by a long often curled funicle. Seeds with a fleshy whitish aril. Above: single-seeded fruits. Below: Fruits with 2 seeds.

Fig. 66. Above: *Swartzia schomburgkii.* Cauliflorous stem with inflorescences. Below: *Macrolobium acaciaefolium.* Ovoid pods with delicate venation.

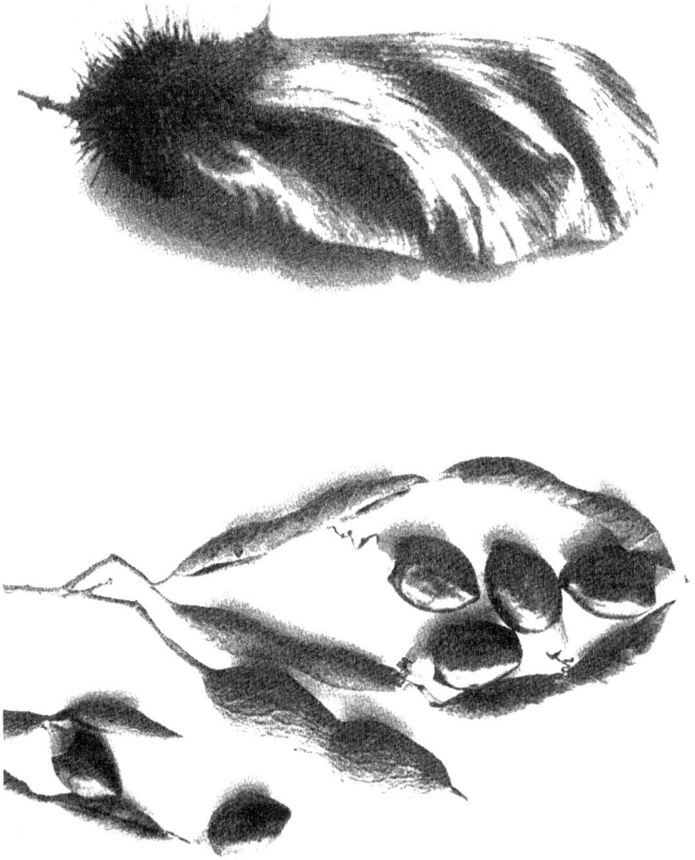

Fig. 67. Above: *Centrolobium paraense,* Papilionaceae. Samara with a large wing and spines on the base. Below: *Swartzia schomburgkii.* 2-seeded fruits closed and dehisced. Note the delicate somewhat prominent venation on the fruit surface. Dehiscence takes place in a strange manner: The dehisced valves remain united on their tips. The funicles of the 2 seeds are alternately attached to the carpels on opposite sides. Each valve thus bears one seed, but the entire structure becomes elongated to the double by the union of the valve tips. In this way the seeds become better exposed to birds and bats.

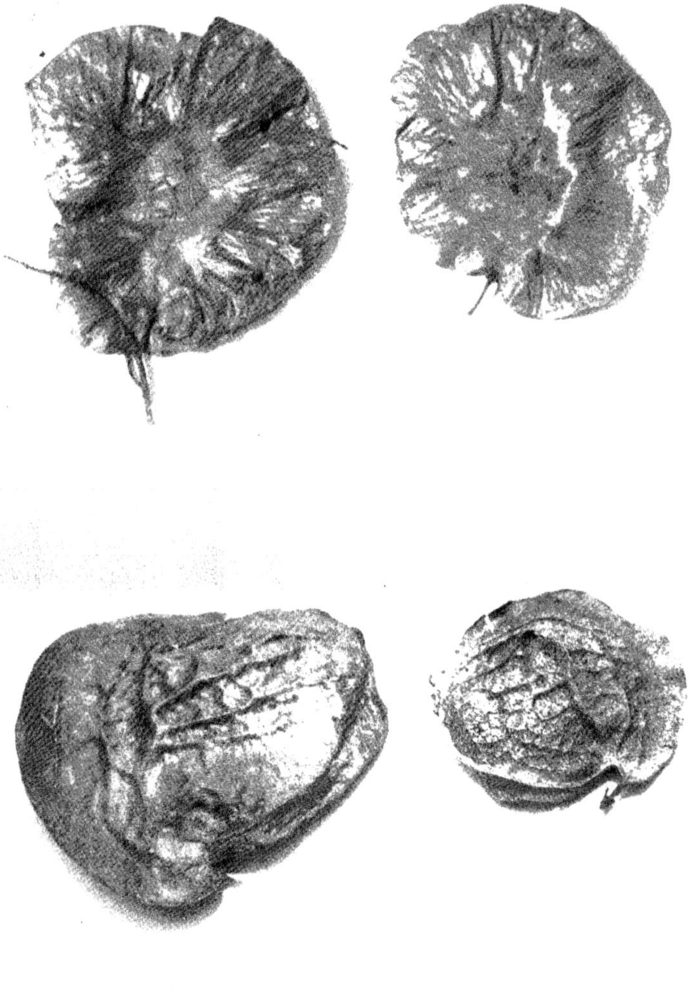

Fig. 68. Above: *Pterocarpus rohrii,* Papilionaceae. Samara with a large papery wing surrounding the fruit body entirely. Below: *Pterocarpus officinalis.* Samara with a wing on the dorsal suture and a prominent reticulate venation.

Fig. 69. Above: *Carapa guianensis,* Meliaceae. Woody globular capsule with 4 3-angular seeds. Below: *Pterocarpus* sp. Pod halfway surrounded by a reduced wing.

Fig. 70. Above: Trichilia smithii, Meliaceae. Dehisced 3-locular capsule with marks of the seeds. The seeds diminish successively in size towards the fruit tip imprinting their outlines on the endocarp inside. Fruit surface warty. Below: *Cedrela odorata.* Dehisced capsules with 5 valves and a large central column. Seed winged.

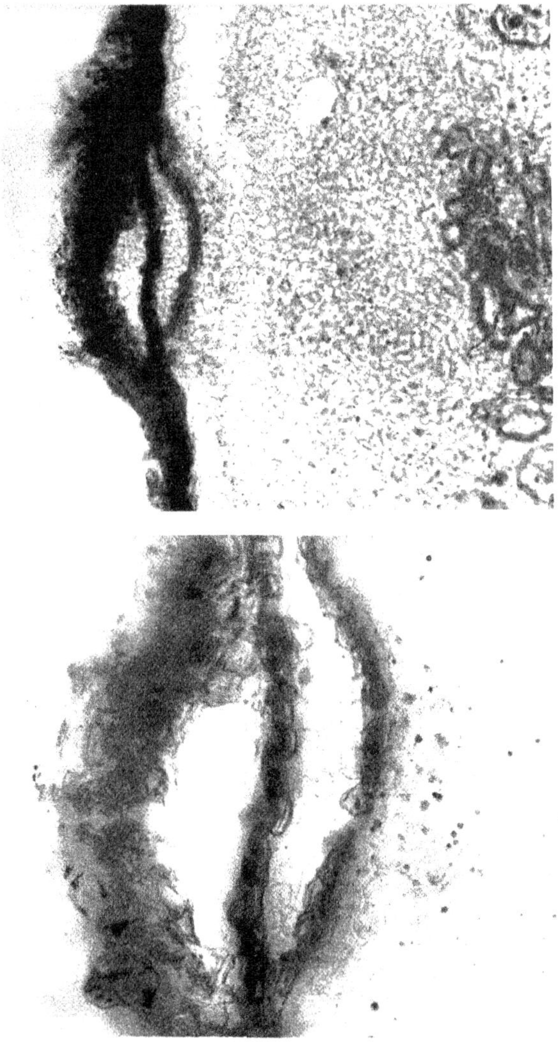

Fig. 71. Carapa sp. ('Carapa lisa'), Meliaceae. Stratified lenticels of the fruit surface. Layers of normal cork cells alternate with roundish filling cells.

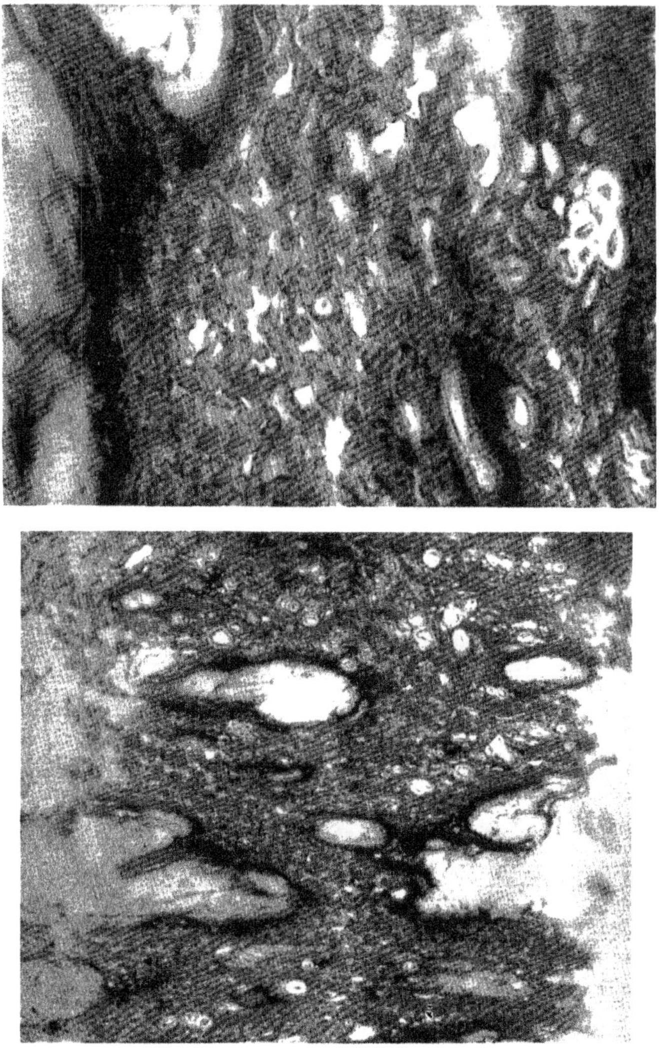

Fig. 72. Carapa sp. ('Carapa lisa'). Above: Compressed collenchymatous tissue and dispersed stone cells of the pericarp. Below: Large secretory cavities and partial tissue degeneration.

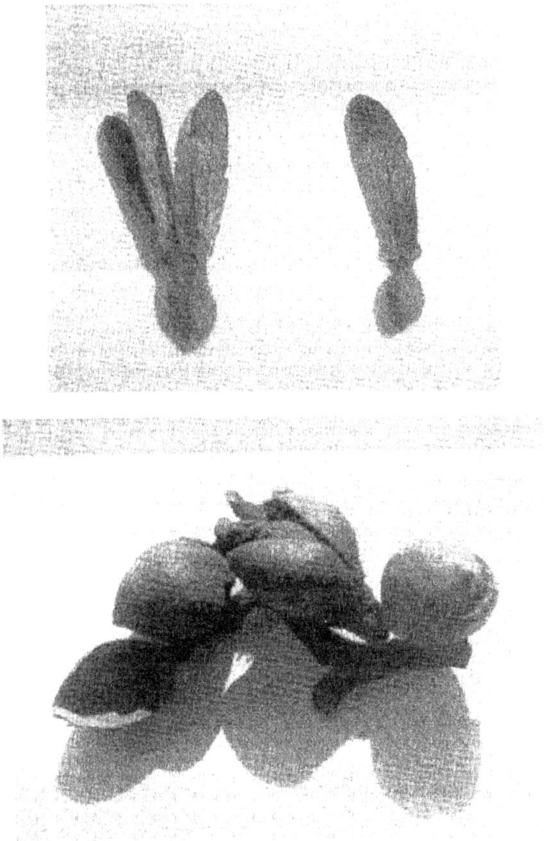

Fig. 73. Above: *Triplaris surinamensis,* Polygonaceae Samara with 3 large wings proceeding from the persistent calyx. Below: *Virola sebifera,* Myristicaceae. Closed and dehisced capsules.

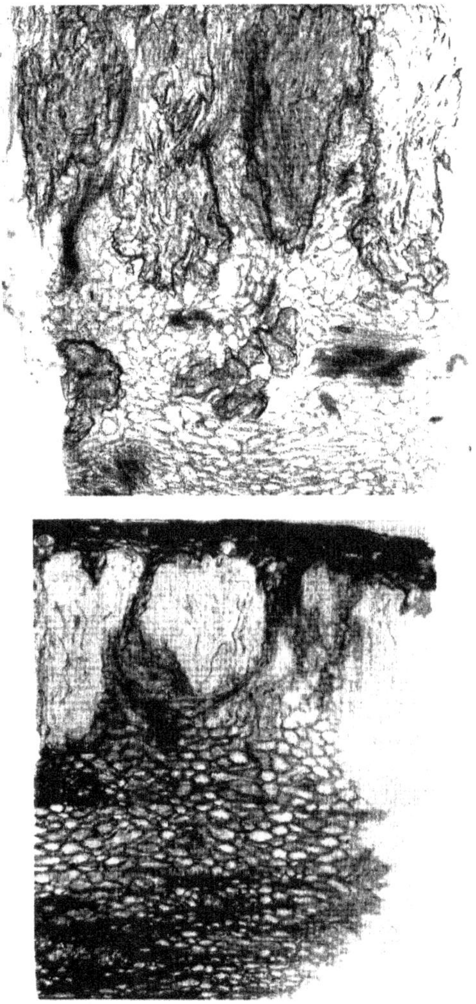

Fig. 74. Virola sebifera. Above: Inner part of the pericarp with large stone cell groups and vascular bundles. Below: Outer part of the pericarp with epidermis, small-celled subepidemal layers, stone cell plates and partly compressed parenchyma.

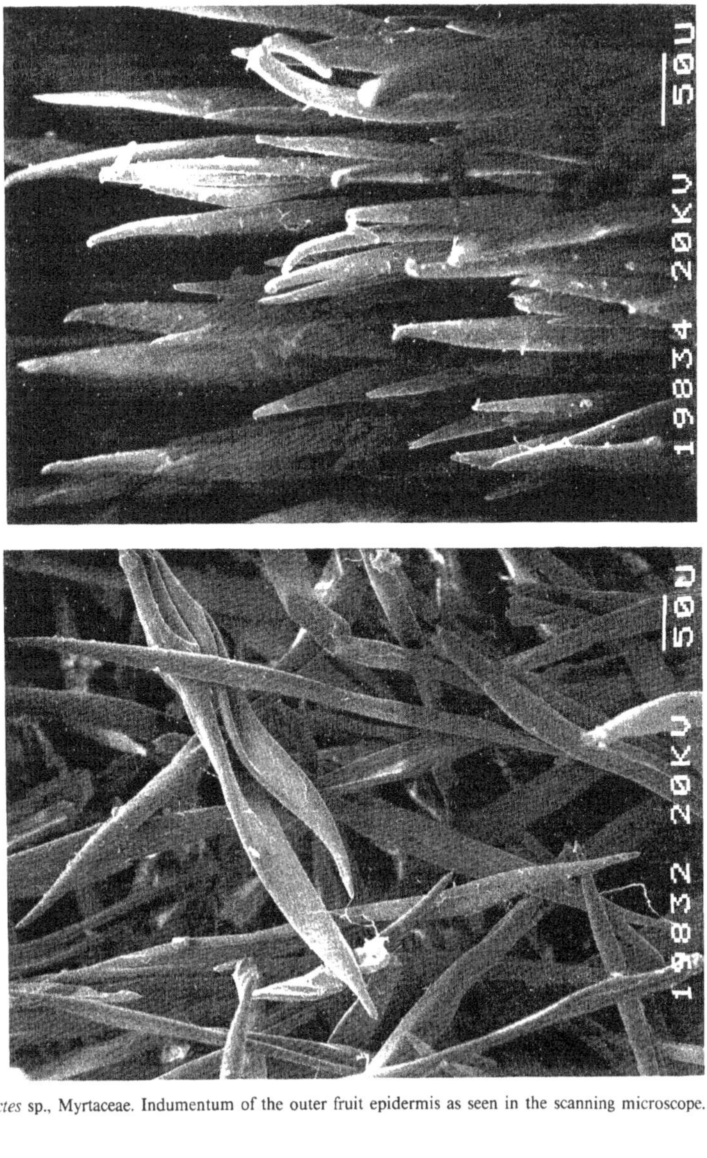

Fig. 75. Calycorectes sp., Myrtaceae. Indumentum of the outer fruit epidermis as seen in the scanning microscope.

Fig. 76. Calycorectes. Above: Bunch of hair in the outer fruit epidermis. Below: Twisted hairs, as seen in the SM.

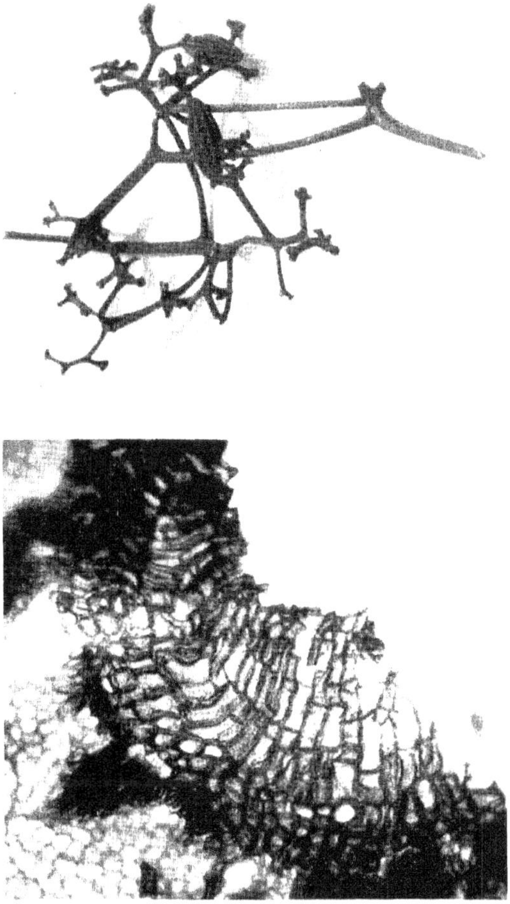

Fig. 77. Torrubia cuspidata, Nyctaginaceae. Above: Ramified inflorescence axis with 2 anthocarps. Below: Fruit base.

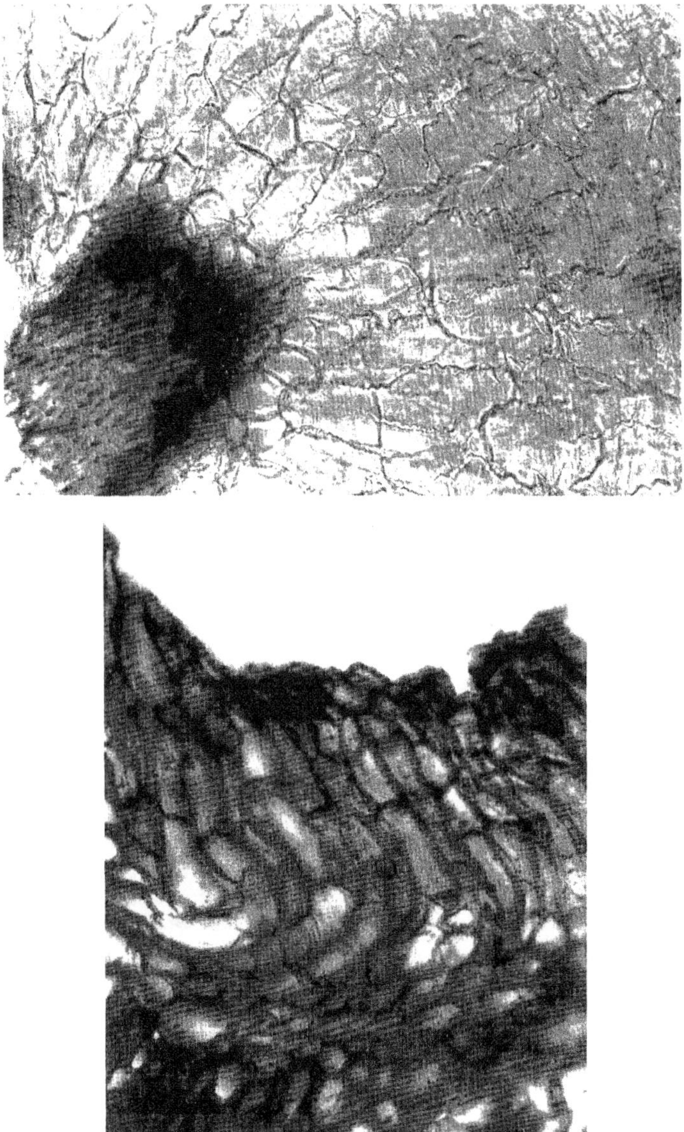

Fig. 78. Torrubia cuspidata. Above: Radially extended parenchyma cells with thin wavy walls of the anthocarp. Below: Remnants of the dorsal rib meristem of the anthocarp (note cell rows).

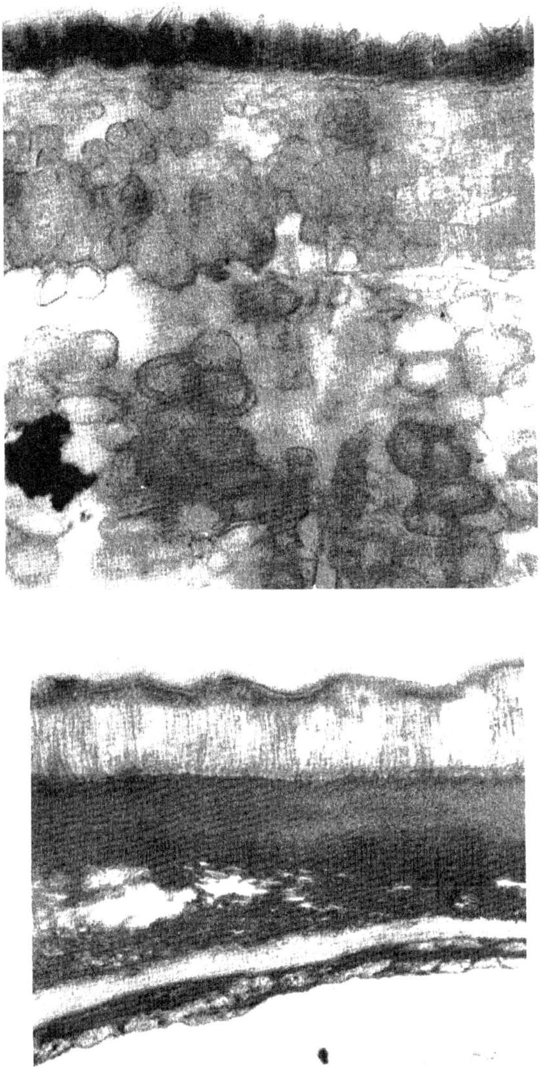

Fig. 79. Above: *Catostemma commune*, Bombacaceae. Epidermis with stellate hairs. Beneath sclerified and lignified parenchyma cells with pits. Below: *Ouratea guianensis*, Ochnaceae. T. s. pericarp. Outer epidermis, fundamental parenchyma and seed coat.

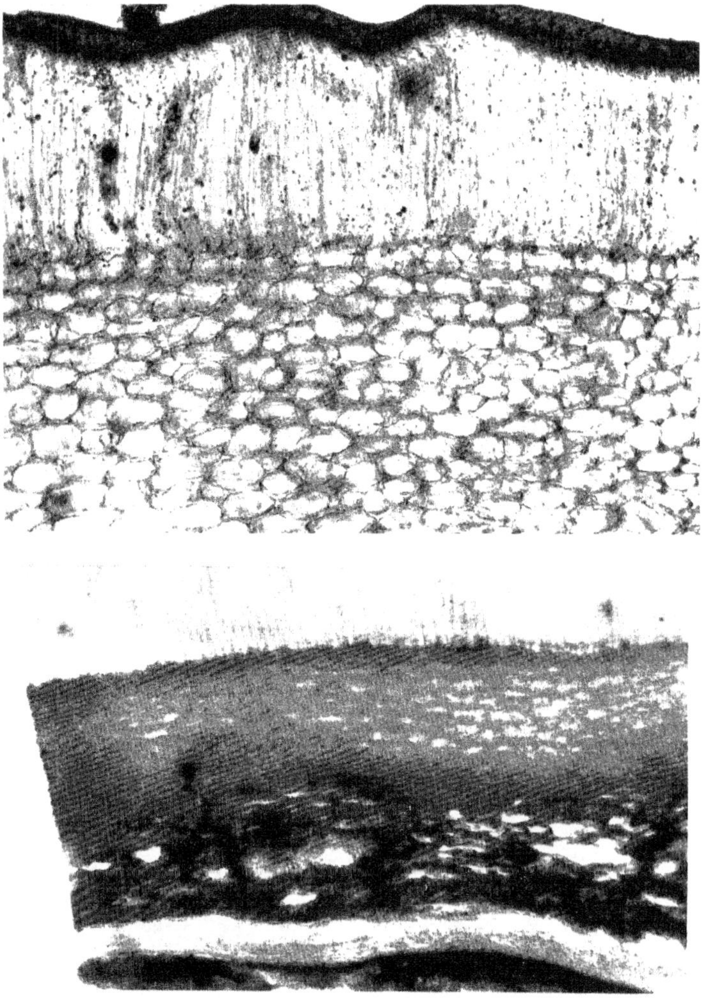

Fig. 80. Ouratea guianensis. Above: Palisade-shaped outer epidermis and roundish parenchyma cells beneath. Below: Outer and inner parenchyma of the pericarp the latter tangentially elongated and with vascular bundles. Fibrous endocarp.

Fig. 81. Above: *Manicaria (saccifera?),* Palmae. Bilocular 'double fruit'. Below: *Mauritia (flexuosa?)* in front and side view demonstrating the spirostichies of scales in 2 contrarotating helical lines.

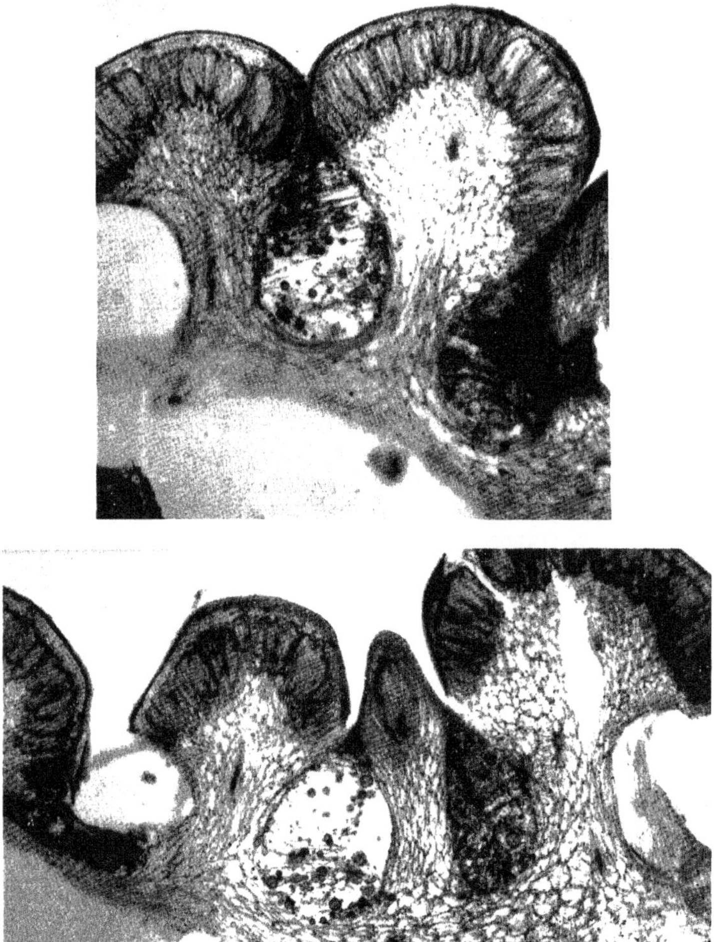

Fig. 82. Touroulia guianensis, Quiinaceae. Ribs of the fruit with 2 different tissue types, an inner parenchymatous part and a sclerenchymatous peripheral part. In between the ribs lie large oil pockets densely filled with oil drops.

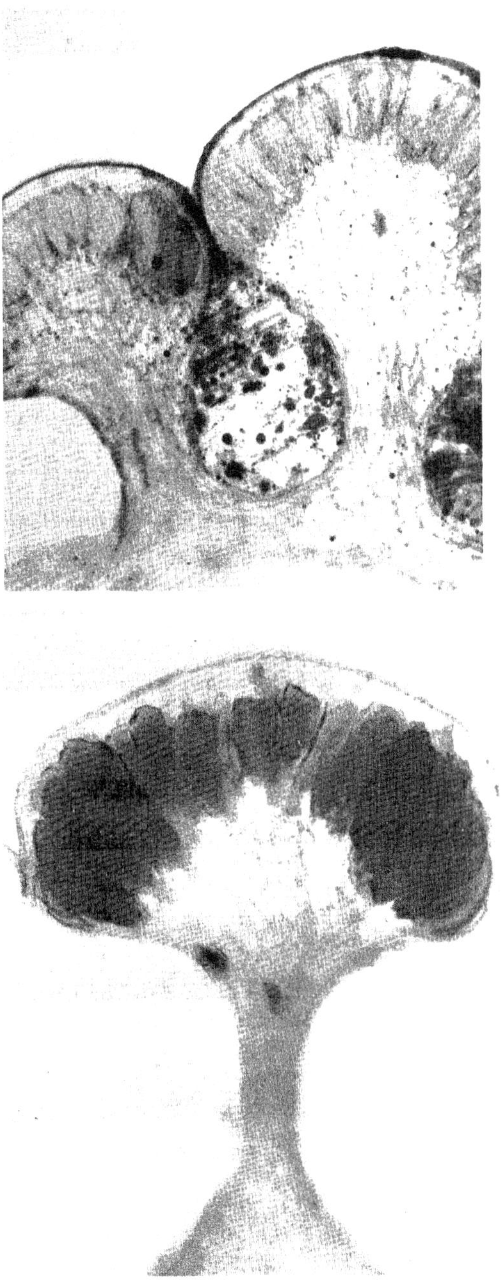

Fig. 83. Touroulia guianensis. Above: Fruit periphery with ribs and oil pockets. Below: The sclerenchymatous plates of the rib periphery in dark.

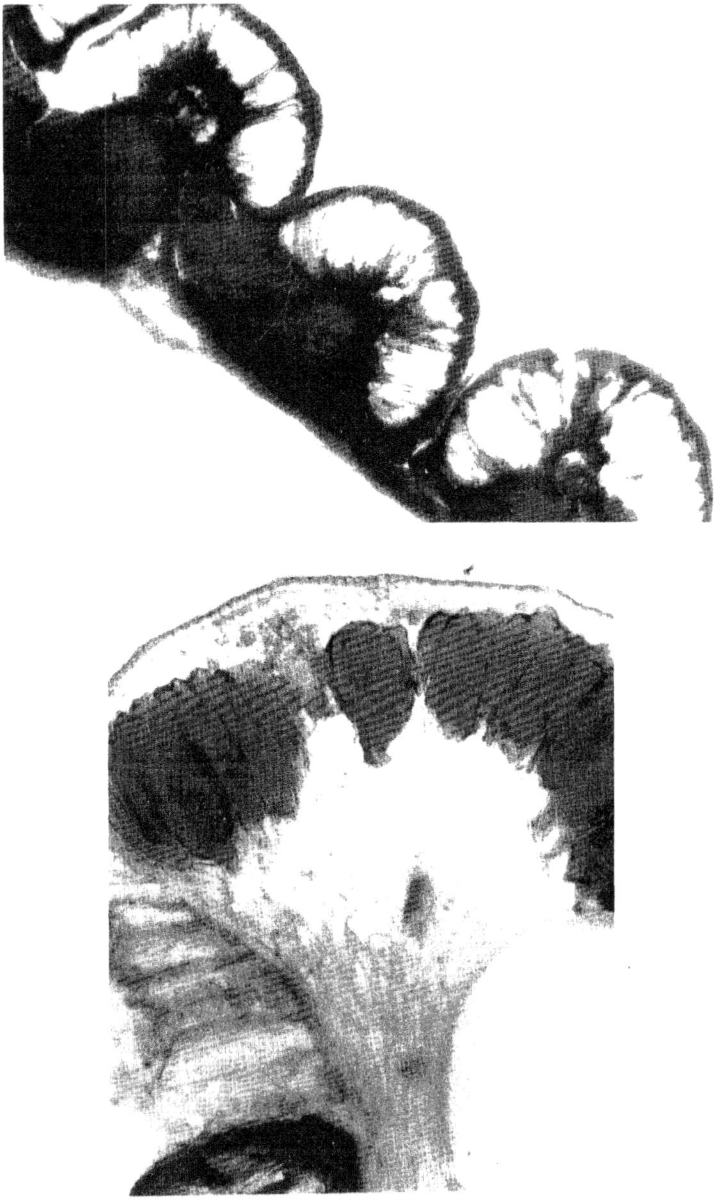

Fig. 84. Touroulia guianensis. Above: ribs of the fruit periphery cut at the fruit base. Below: Vascular bundle in the center of the rib.

Fig. 85. Fruits of the Rosaceae. Above: *Licania densiflora.* Pear-shaped fruit. Middle. *Licania alba.* Below: *Hirtella davisii.*

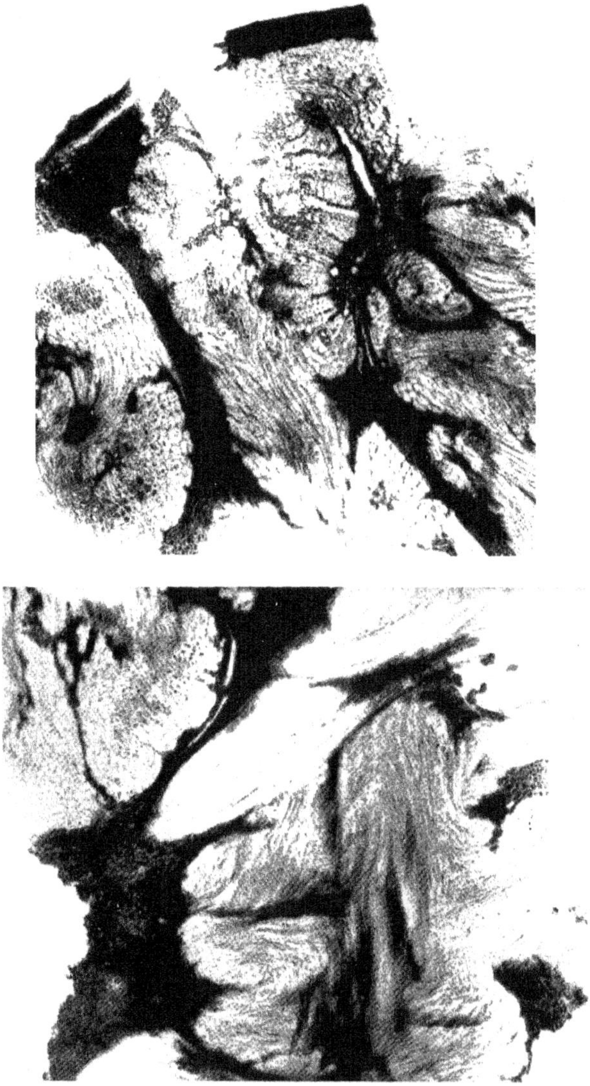

Fig. 86. Licania alba, Rosaceae. Fibrous endocarp. Fiber bundles crossing one another. Vascular bundles in the endocarp.

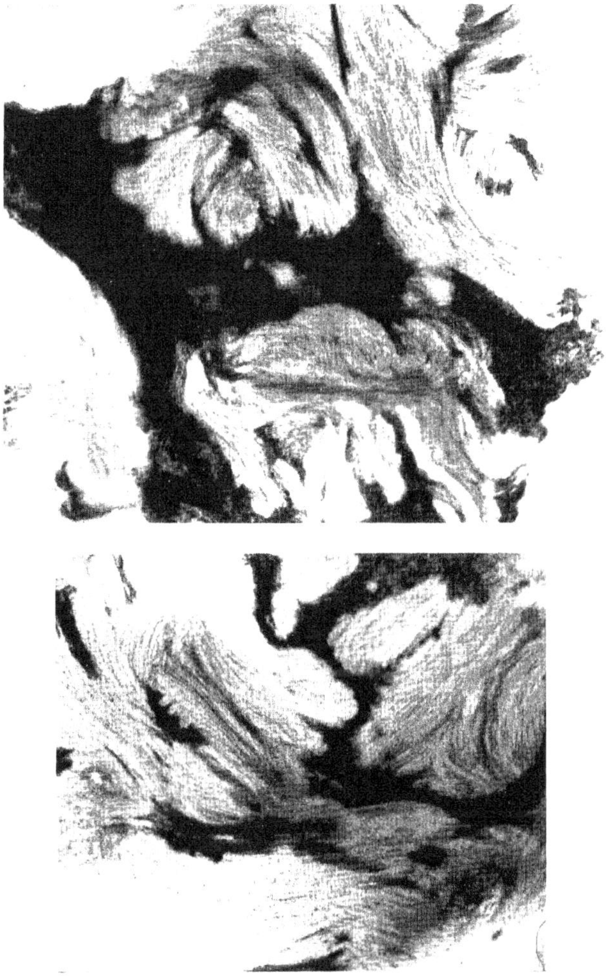

Fig. 87. Licania alba. Endocarp with crossing fiber bundles.

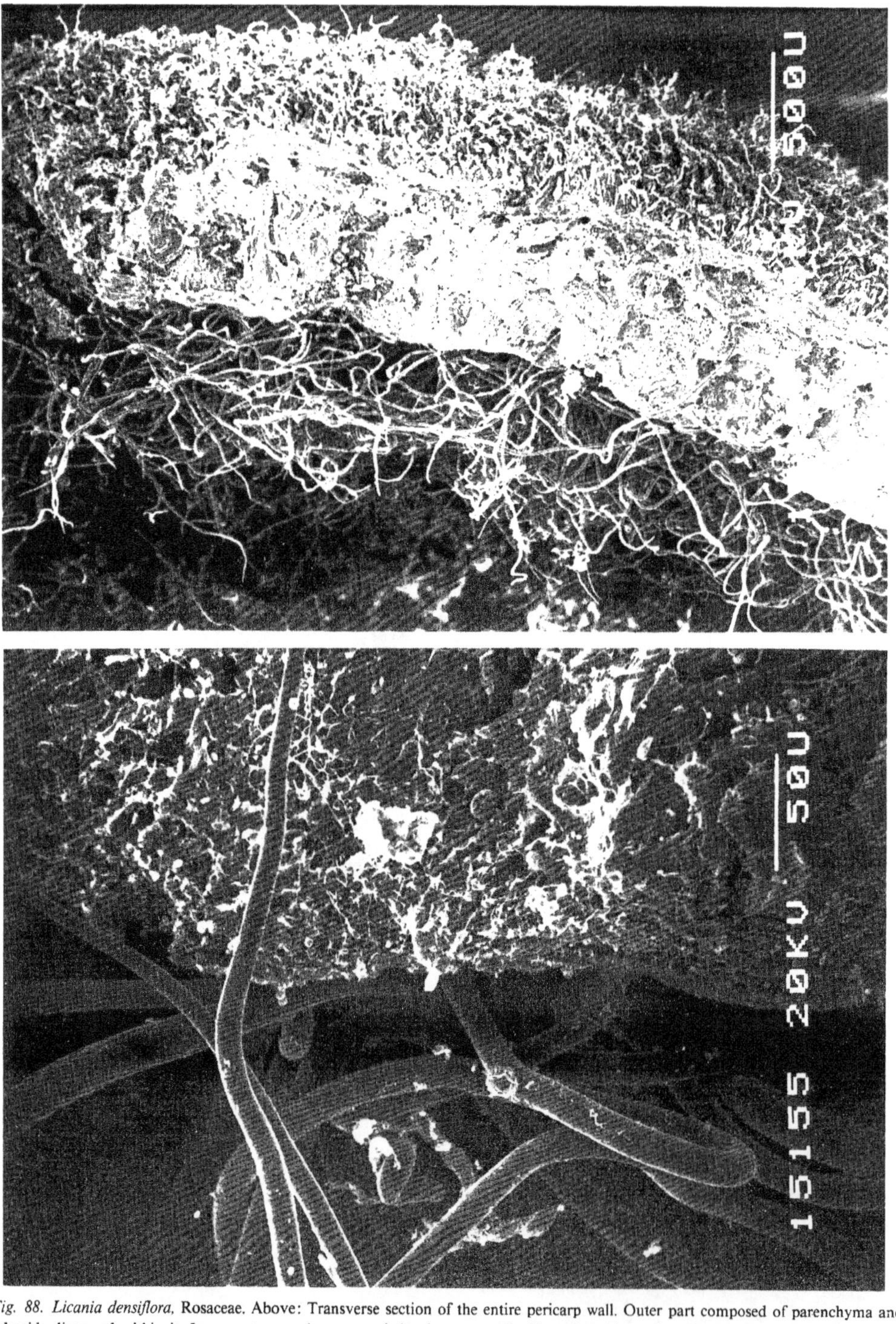

Fig. 88. Licania densiflora, Rosaceae. Above: Transverse section of the entire pericarp wall. Outer part composed of parenchyma and sclereids dispersed within it. Inner part or endocarp consisting in stone cells. The sclerenchyma bundles crossing one another. Inner epidermis with very long hairs. Below: Innermost pericarp part with hairs. Scanning microscope.

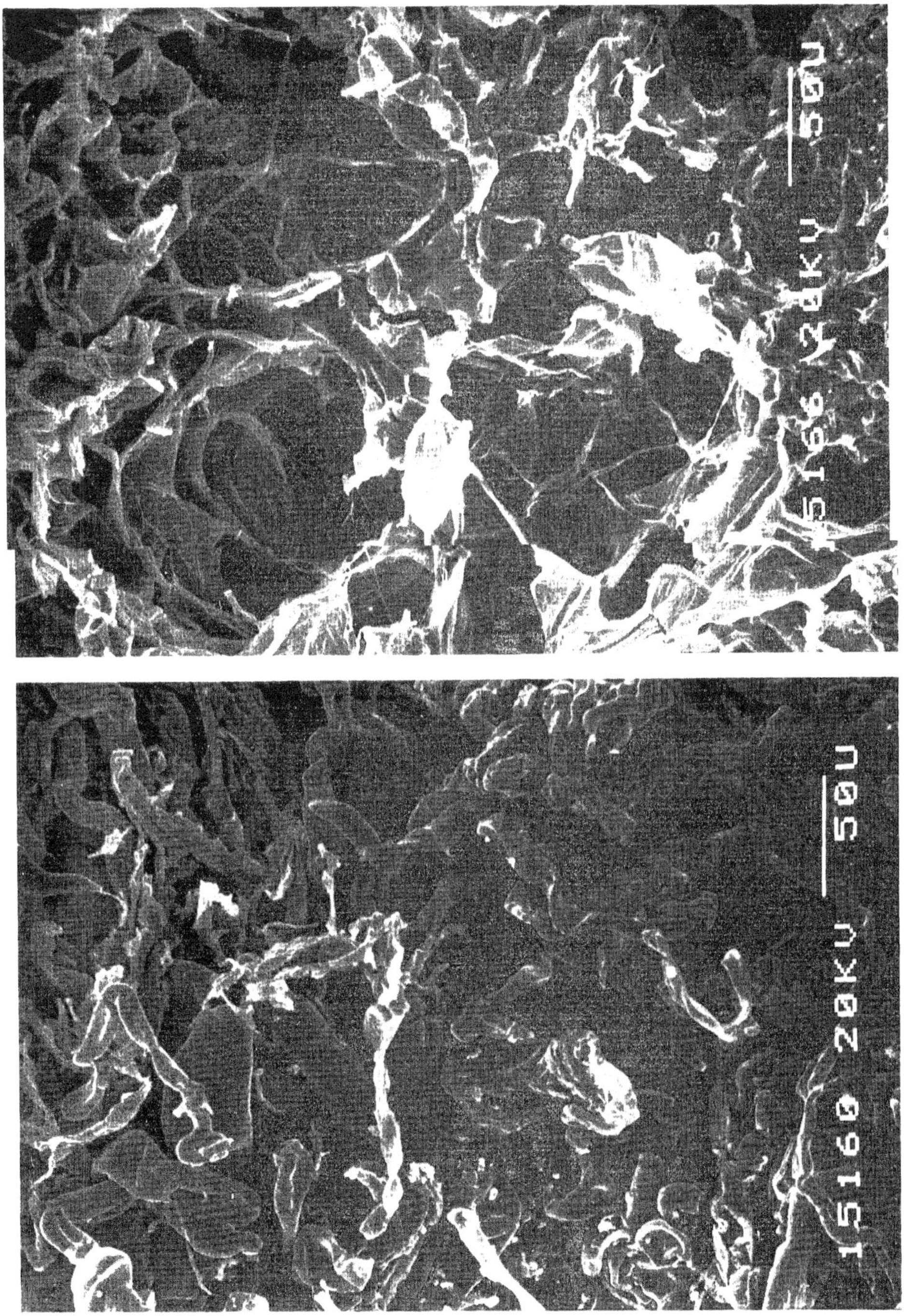

Fig. 89. Above: *Licania alba*, Rosaceae. Outer part of the pericarp. Below: *Licania densiflora*, pericarp outside in t.s. with hairs. Scanning microscope.

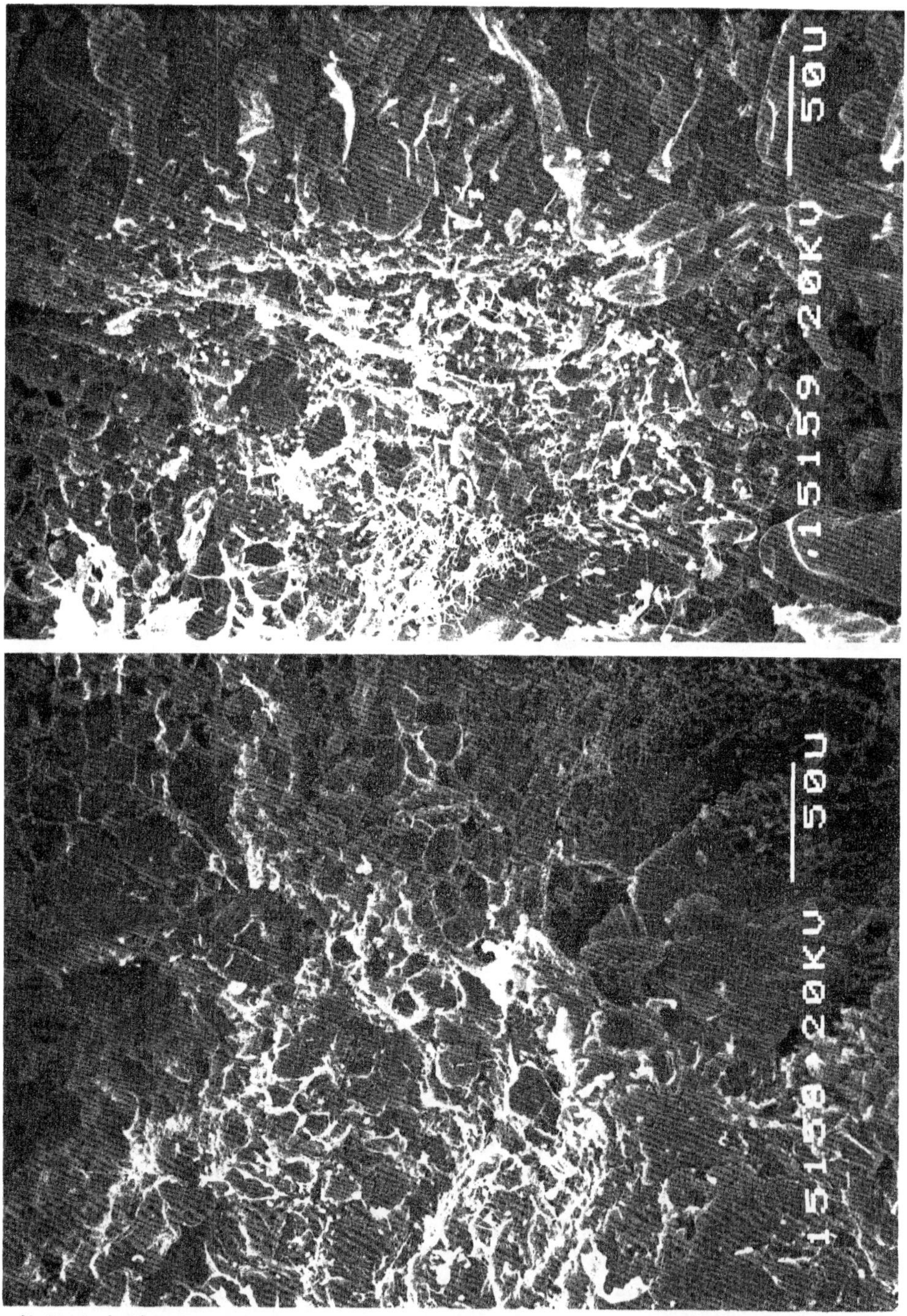

Fig. 90. Licania densiflora. T.s. pericarp. Above: Outer part with hairs. Below: Middle region. Scanning microscope.

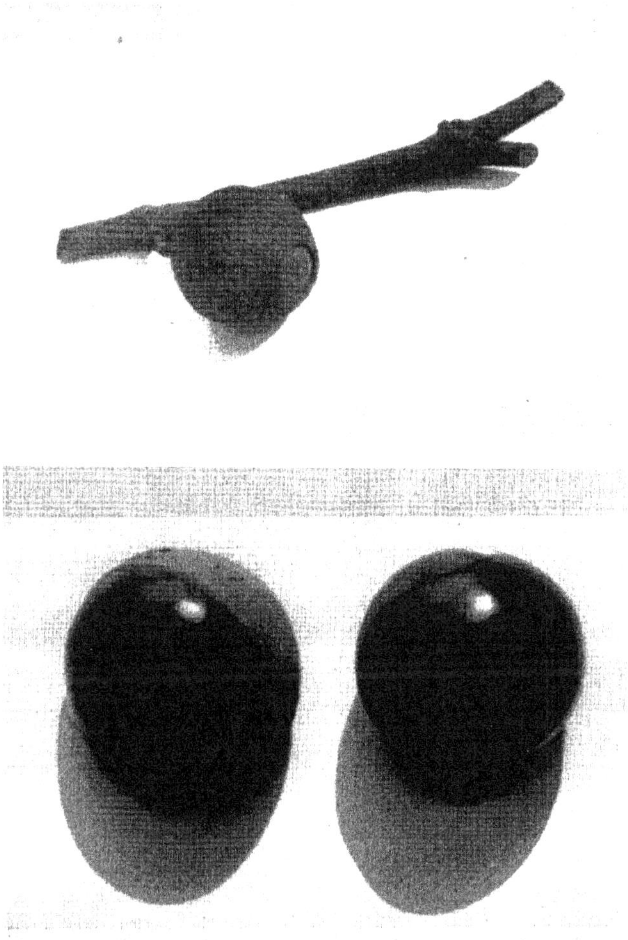

Fig. 91. Above: *Mouriria huberi*, Melastomaceae. Globular berry surrounded by its calyx cup ("cupule"). Below: "Mantequillo", Sapotaçeae. Seeds resembling a horse-chestnut with prominent hilar region.

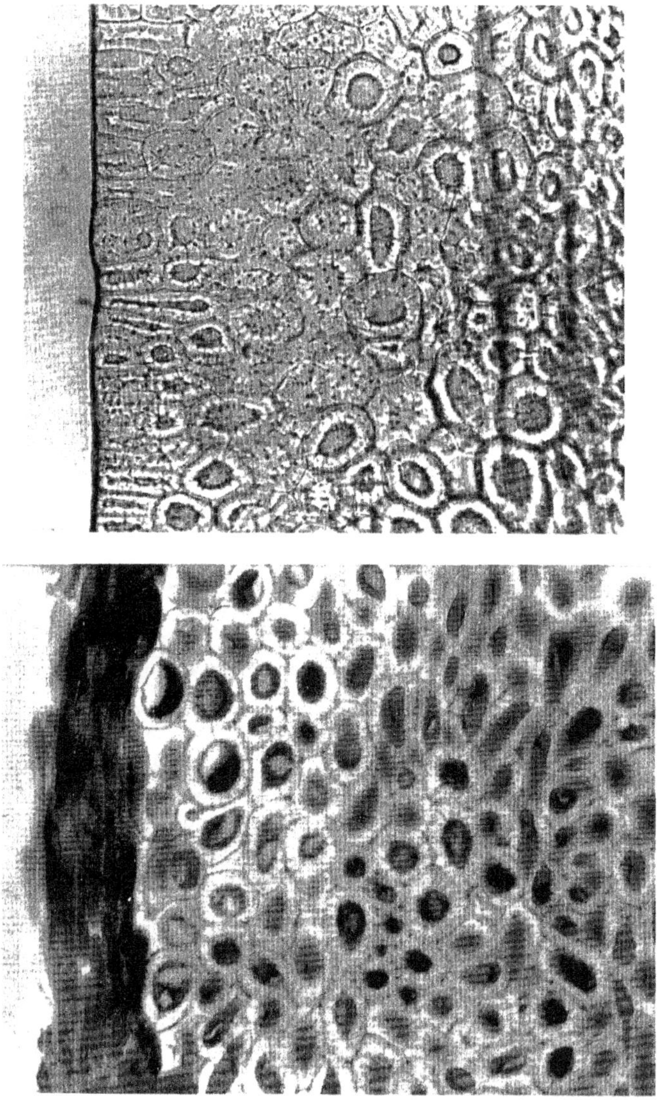

Fig. 92. "Mantequillo". Above: Outer part of the seed coat with cuticle (in black). Below: Inner part of the seed coat.

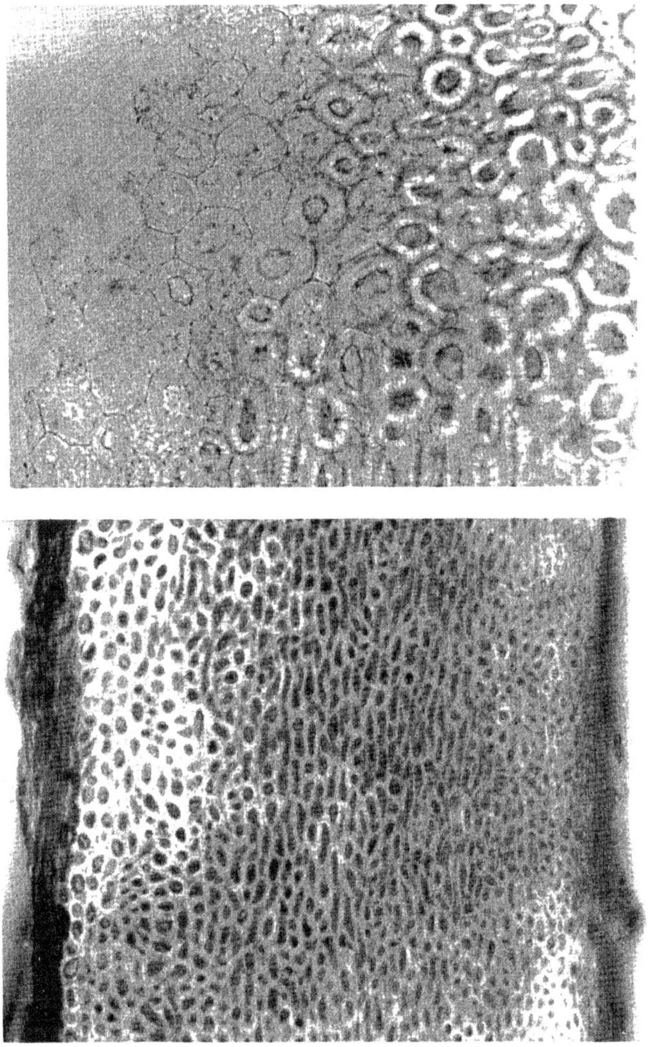

Fig. 93. "Mantequillo". Below. Entire t. s. of the very thick seed coat. Above. Middle region of the seed coat with roundish thick-walled cells.

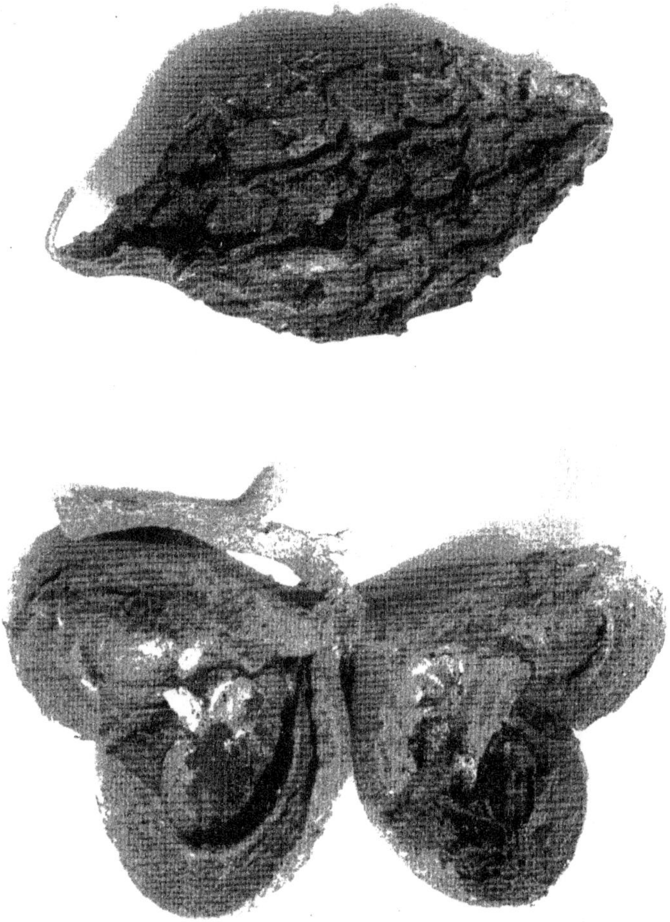

Fig. 94. Below: *Sterculia rugosa,* 2 dehisced mericarps. Above: *Herrania lemniscata,* Sterculiaceae. Capsule with reticulate surfac structure.

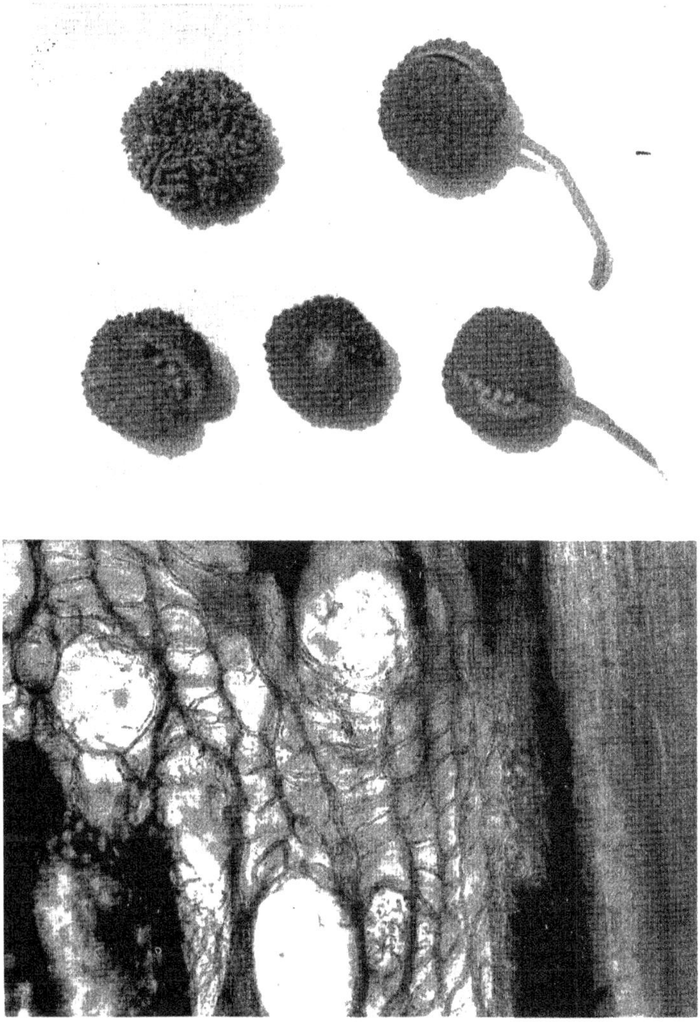

Fig. 95. Above: *Guazuma ulmifolia,* Sterculiaceae. Globular capsules with surface warts partly dehisced and exposing small seeds. Below: *Sterculia pruriens.* Tangentially elongated parenchyma cells of the mesocarp divided by anticlinal walls and forming tangential cell rows. Scattered in between are large slime cells.

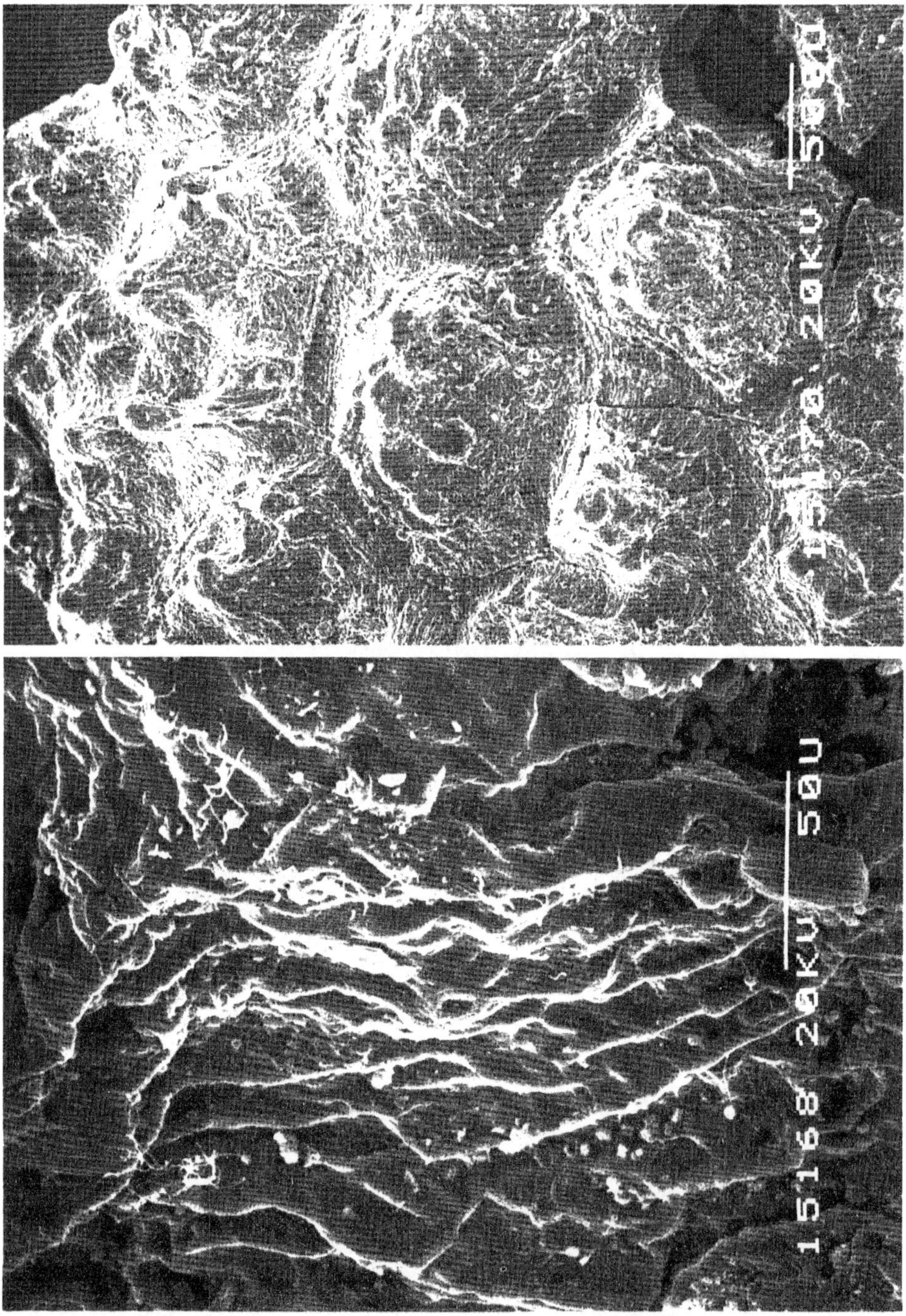

Fig. 96. Guazuma ulmifolia, Sterculiaceae. Above: Surface protuberances. Below: Surface of protuberance. Scanning microscope.

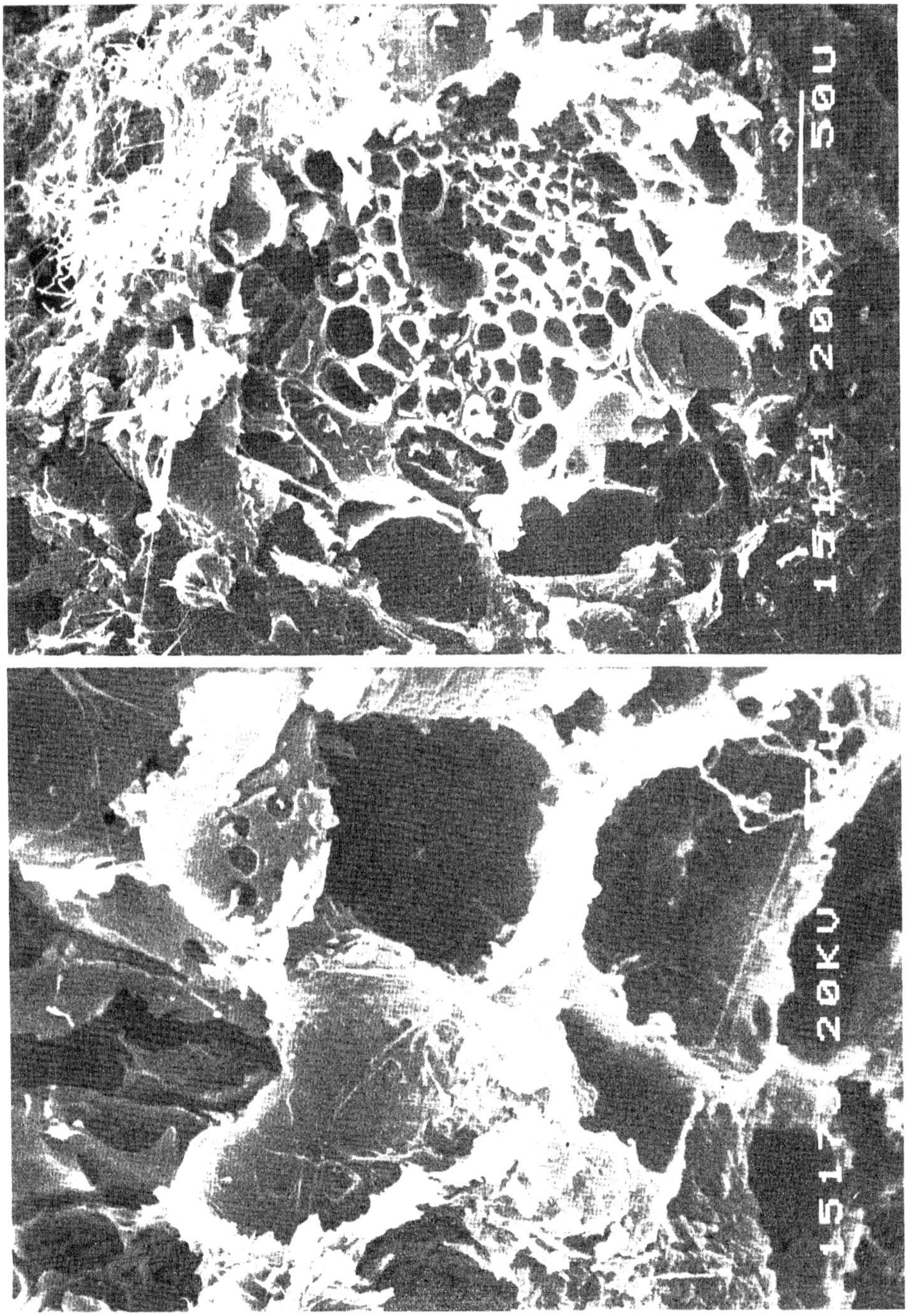

Fig. 97. Guazuma ulmifolia. Above: T. s. of a protuberance with vascular bundle. Below: Parenchyma cells under higher magnification. Scanning microscope.

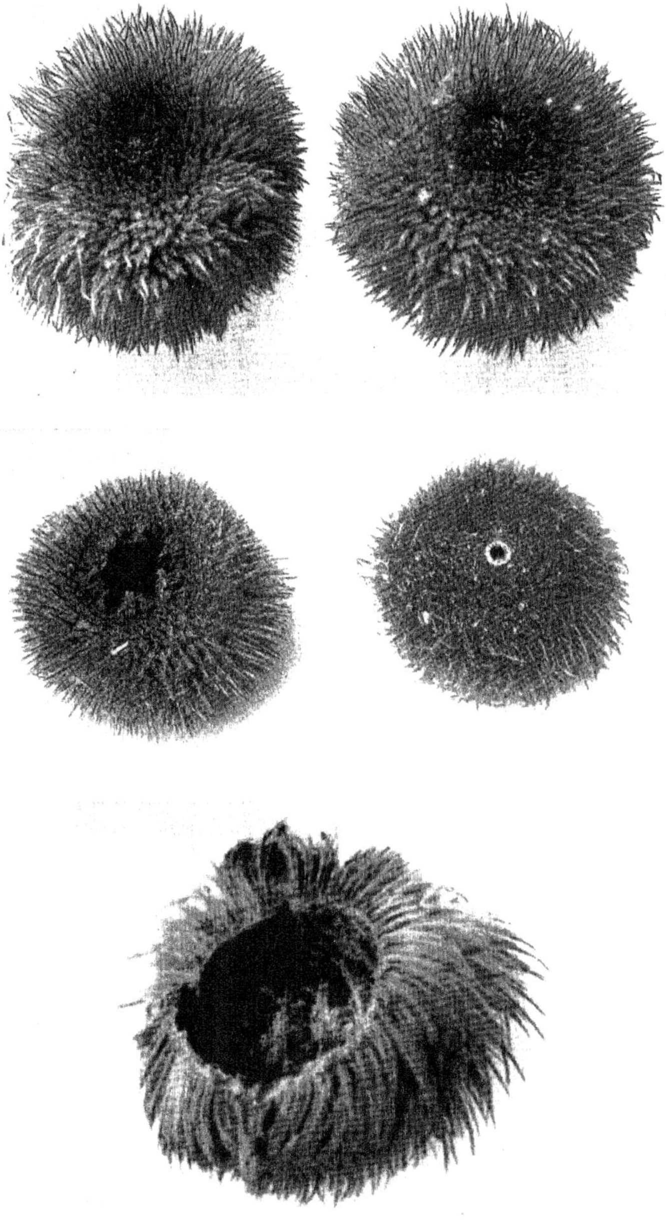

Fig. 98. Above: *Apeiba* sp. (Cabeza de negro), Tiliaceae. The fruit densely covered with bristles resembles the "Head of a negro". Middle. View from above with apical opening and from below. *Below: Apeiba tibourbou.* Opened fruit with long ash-blond bristles resembling a wig.

Fig. 99. Luehea candida, Tiliaceae. The woody capsules decorated with short spines open with 5 valves loculicidally.

Fig. 100. Apeiba echinata, Tiliaceae. The spiny fruit resembles a sea-urchin. Above: View from above with apical opening. Below: View from below.

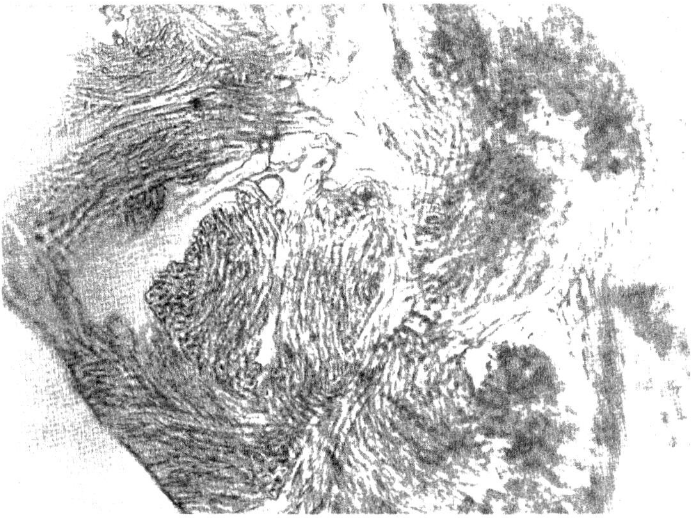

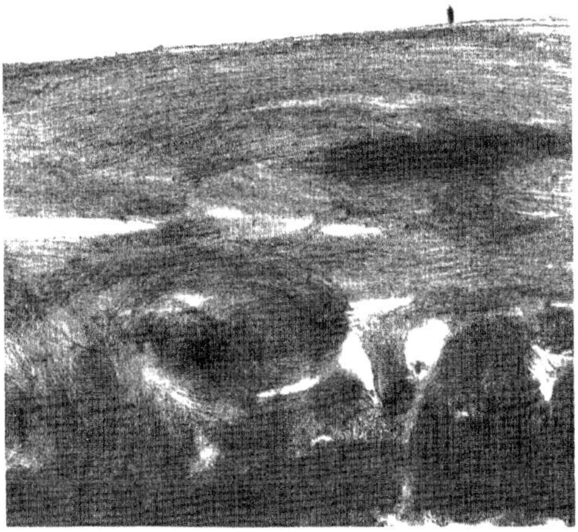

Fig. 101. Above: *Apeiba* sp. (Cabeza de negro). Fiber groups of the pericarp crossing one another in different directions. Below: *Apeiba echinata.* Fruit endocarp with fiber bundles running in different directions.

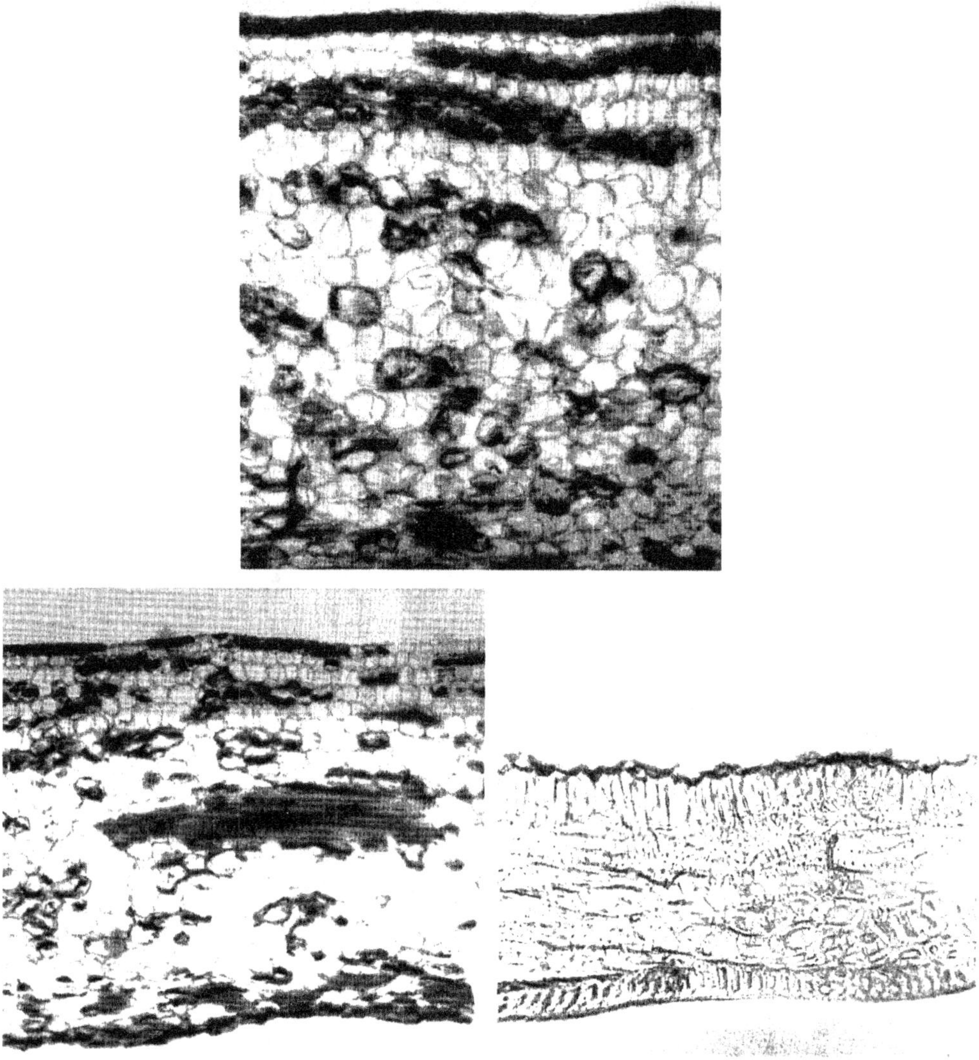

Fig. 102. Paypayrola longifolia, Violaceae. Above and middle: Parenchymatous part of the fruit wall with tangential cell rows arising through dilatation growth. Below: Fibrous endocarp with an outer and inner layer of longitudinally elongated sclereids, while the sclereids of the multilayered central part are tangentially extended.

Fig. 103. Above: *Vochysia tetraphylla,* Vochysiaceae. Still closed and dehisced fruit. Below: *Qualea dinizii, Vochysiaceae.* Dehisced capsules.

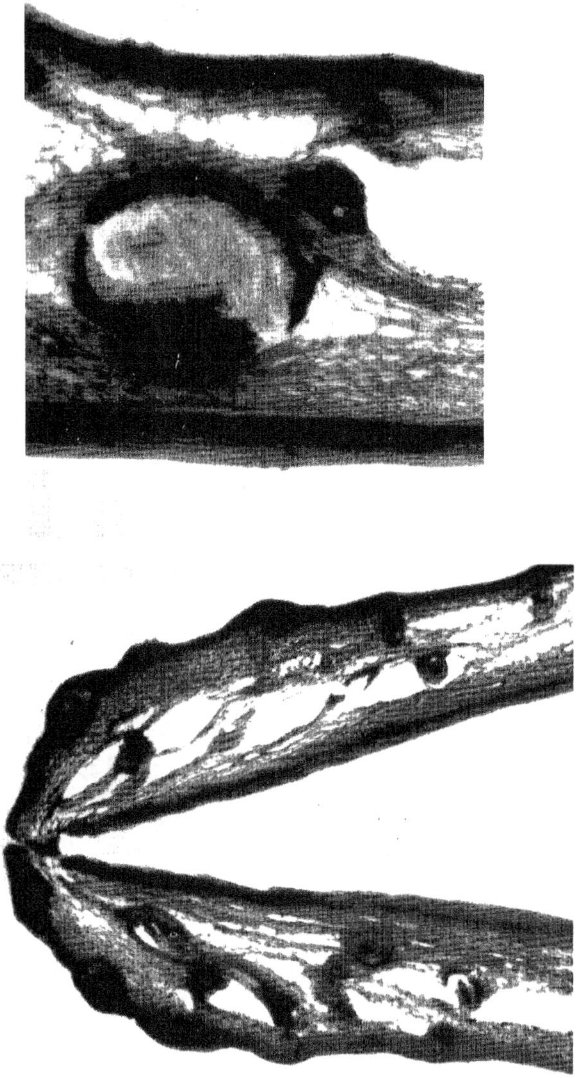

Fig. 104. Vochysia lehmannii. Below: Entire view of locule. Above: Fruit wall with slime cavity.

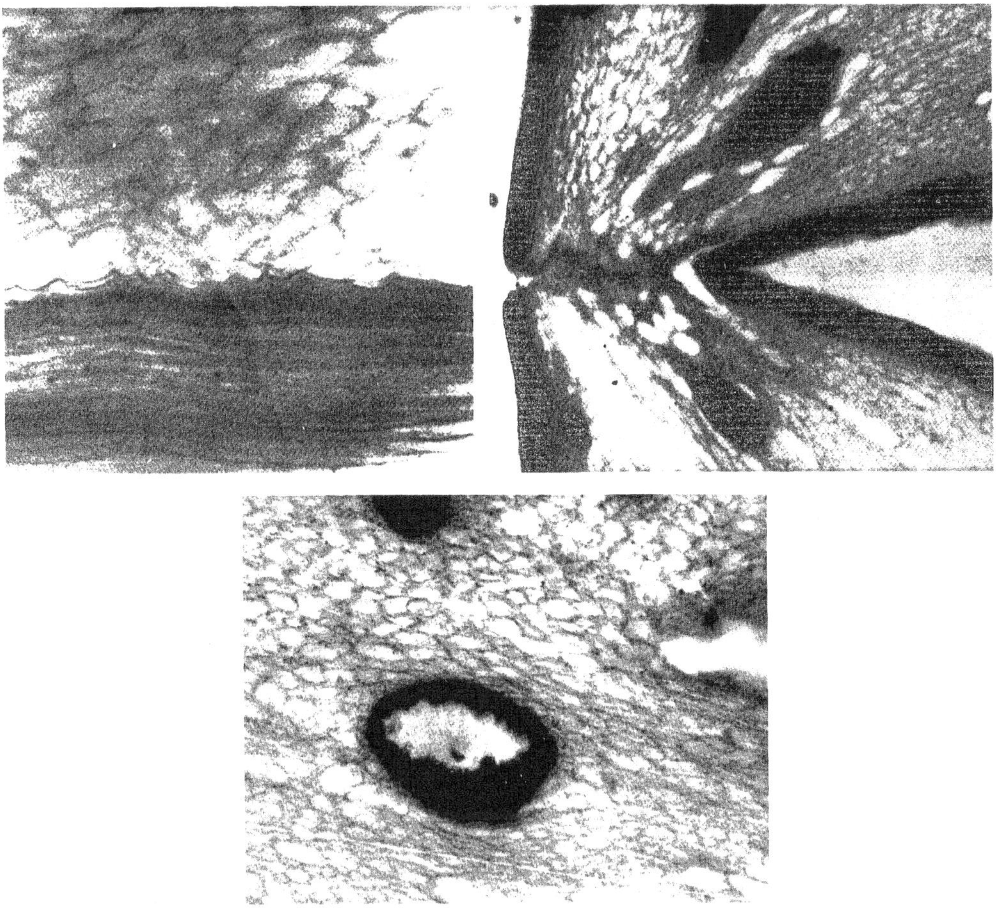

Fig. 105. Vochysia lehmannii. Above: Above left: Fibrous endocarp composed of prosenchymatous cells. Right: Dorsal suture of the locule. Below: Concentric vascular bundle (hadrocentric).

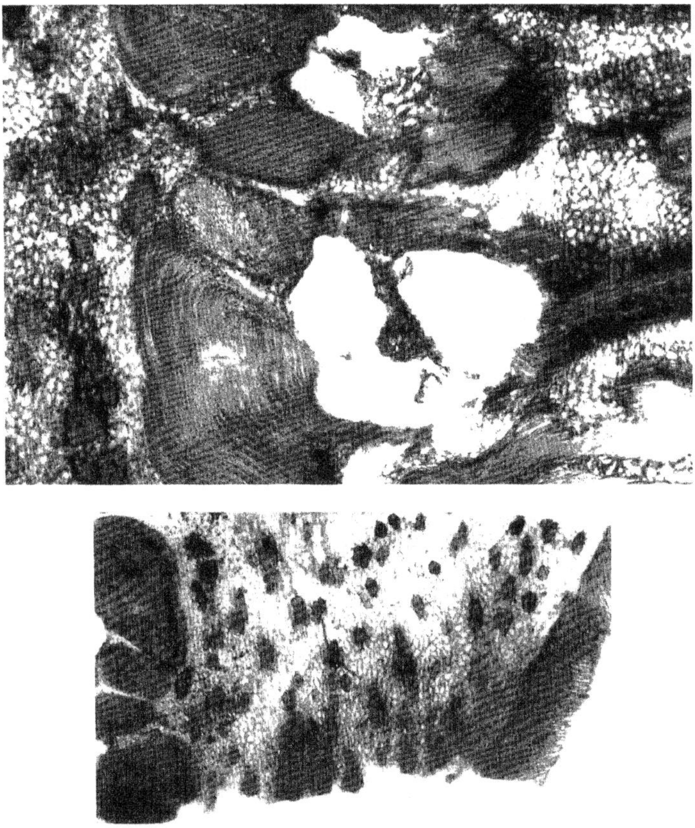

Fig. 106. Qualea dinizii, Vochysiaceae. Above: Xylem enclosing slime cavity. Below: Entire view of pericarp wall with fibrous endocarp (towards the right), parenchyma with scattered groups of stone cells and vascular bundles (towards the left).

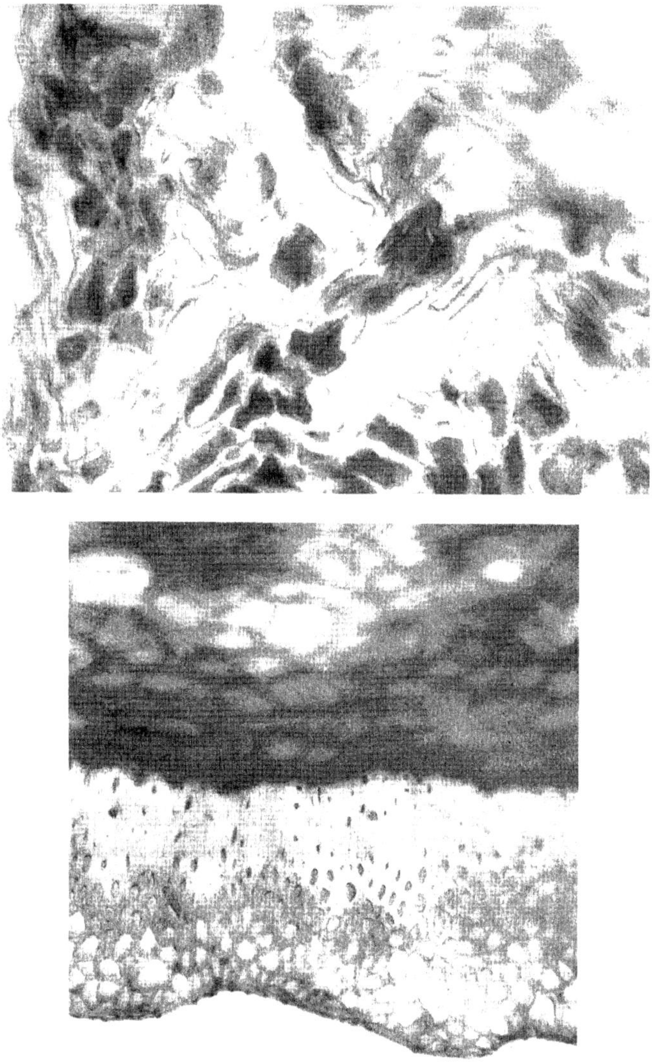

Fig. 107. Qualea dinizii, Vochysiaceae. Above: Cartilaginous collenchyma of the outer pericarp part. Below: Endocarp in transverse section.

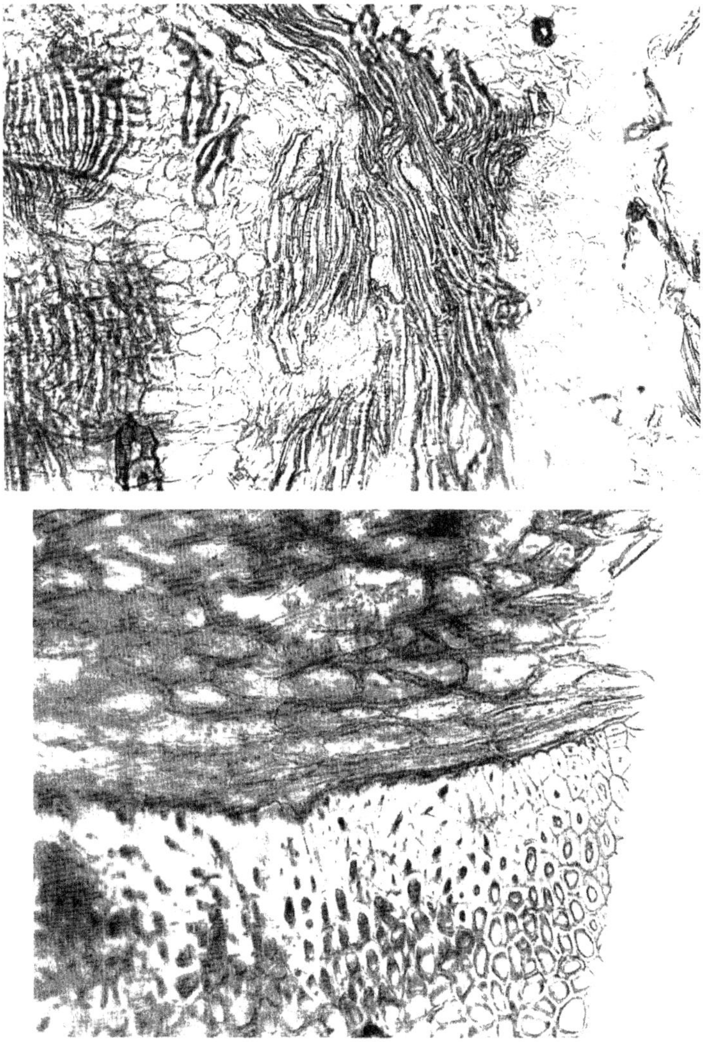

Fig. 108. Qualea dinizii. Above: Fiber sclereids of the outer mesocarp part with irregular outlines. Below: Parenchyma of the mesocarp and endocarp.

INDEX OF VERNACULAR PLANT AND ANIMAL NAMES

INDEX OF SCIENTIFIC PLANT AND ANIMAL NAMES

GENERAL INDEX

GPSR Compliance

The European Union's (EU) General Product Safety Regulation (GPSR) is a set of rules that requires consumer products to be safe and our obligations to ensure this.

If you have any concerns about our products, you can contact us on ProductSafety@springernature.com

In case Publisher is established outside the EU, the EU authorized representative is:

Springer Nature Customer Service Center GmbH
Europaplatz 3
69115 Heidelberg, Germany

Batch number: 10809439

Printed by Printforce, the Netherlands